P9-CQN-154

Pharmaceutical Innovation

The Chemical Heritage Foundation Series in Innovation and Entrepreneurship

Everybody Wins! A Life in Free Enterprise
by Gordon Cain

Pharmaceutical Innovation: Revolutionizing Human Health
edited by Ralph Landau, Basil Achilladelis, and Alexander Scriabine

PHARMACEUTICAL INNOVATION

Revolutionizing Human Health

EDITED BY

RALPH LANDAU,

BASIL ACHILLADELIS,

AND

ALEXANDER SCRIABINE

CHEMICAL HERITAGE PRESS

PHILADELPHIA

Printed in the United States of America.

The Chemical Heritage Foundation Series in Innovation and Entrepreneurship records, analyzes, and makes known the human story of chemical achievement.

For information about CHF publications write
Chemical Heritage Foundation
315 Chestnut Street
Philadelphia, PA 19106-2702, USA
Fax: (215) 925-1954

Library of Congress Cataloging-in-Publication Data

Pharmaceutical innovation : revolutionizing human health / edited by Ralph Landau, Basil Achilladelis, and Alexander Scriabine.
p. cm.
Includes bibliographical references and index.
ISBN 0-941901-21-1
1. Pharmaceutical industry—Technological innovations. 2. Pharmaceutical biotechnology. 3. Drug development. 4. Pharmaceutical industry—Economic aspects. I. Landau, Ralph. II. Achilladelis, Basil, 1937– . III. Scriabine, Alexander.

RS122.P42 1999
615′.19—dc21 99-052121

∞The paper used in this publication meets the minimum requirements of the American National Standard for Information Sciences—Permanence of Paper for Printed Library Materials, ANSI Z39.48-1984.

Contents

Preface

By Ralph Landau

This book originated from a study I began in 1993 that dealt with the chemical industry. It culminated in a volume titled *Chemicals and Long-Term Economic Growth*, published in 1998 by John Wiley & Sons in conjunction with the Chemical Heritage Foundation and edited by Ashish Arora, Nathan Rosenberg, and me.

That book, as its title suggests, focused primarily on the economic lessons of the first science-based industry and covered the 150 years since William Henry Perkin's discovery of the first synthetic dyestuff in 1856 in Great Britain. The book also deals in substantial detail with the technical and commercial factors that led to the worldwide success of this industry, now the largest manufacturing industry in the United States and the second largest in Europe.

The chemical industry is number 28 in the U.S. Department of Commerce System of Industrial Classification (SIC). It includes a large number of diverse products—about seventy thousand of them—in such fields as fibers, plastics, cosmetics, paints, and many others, all of which contribute to a higher standard of living. Included in SIC 28 is the pharmaceutical industry (SIC 283). Although our book included all components of SIC 28 in its purview, it became clear as we pursued our studies that the pharmaceutical sector of the chemical industry really deserved a separate extensive study.

The reasons for this are sixfold. First, although pharmaceuticals arose from the dyestuffs industry in the nineteenth century, it soon developed its own distinguishing characteristics. Second, the chemical industry is deemed to have reached maturity, which is certainly not the case for the still rapidly growing pharmaceutical sector. Third, the pharmaceutical industry is clearly the most innovative sector of the chemical industry, and its expenditures on research and development are now a substantial part of the entire expenditure by SIC 28, as we cite in the introduction to this book. Fourth, unlike most of the chemical industry, the pharmaceutical industry has close relations with consumers and the entire health care system, which has become a major expenditure of every industrial country (in the United States it makes up about 14 percent of the gross domestic product). Fifth, the industry is now leading with whole new fields of discovery, such as biotechnology and possibly even genetic manipulation. Finally, it has a total market capitalization value that exceeds most of the chemical industry, and its leading members in the United States, such as Merck and Pfizer, have recently had market capitalizations of $153.4 billion and $114.5 billion, respectively, although these have fallen more recently. Compare this with a market capitalization of $69.4 billion for DuPont, the largest chemical company, whose annual sales in 1997 at $45 billion were much larger than the sales of the aforementioned pharmaceutical companies ($23.6 billion for Merck and $12.5 billion for Pfizer). These statistics indicate that the pharmaceutical industry is seen by investors as one still full of promise.

Thus, it seemed propitious to undertake this volume to recount the remarkable record of innovation and accomplishment in this industry, which, while still innovating, stands at the threshold of even newer technologies that will further affect the health and welfare of humanity.

We have cast this volume less as an economic study and more as a technical and commercial history of how this industry has reached its present prized status among science-based industries. We have chosen to deal with the more restricted areas, as the overall story is still very little known to the general public. Nevertheless, some of the more general analyses in our chemical book may still be applicable to the pharmaceutical industry.

Our purpose, therefore, is to bring to an audience of informed laypersons how we have arrived at our present state in the healing of humanity's ills through the commercialization of science and technology, which is what innovation is, and which has been, over the long sweep of history, the fundamental source of wealth creation and rising standards of living. Nevertheless, we believe that many people within the industry may learn useful background information for their work, and policy makers may better understand the remarkable record that this industry has created in the century since its true scientific origins. Financial industry people and investors can also profit from our study, and students of medicine and many doctors and other health professionals may find it useful.

The book is written with only minimal use of scientific terminology. The few scientific terms used in this book are listed in the glossary. It is a story of the economic importance of an industry as well as its many contributions to human health.

Acknowledgments

The editors would like to acknowledge the invaluable help that the Pine Tree Trust, and to a lesser extent the Sloan Foundation, offered in the financing of this project. We are especially grateful for the invaluable assistance throughout this volume of Lewis Gasorek. We also thank Judi Wind for helping to assemble some of the industry data, and Angela Rey, Lillian Berko, Gisele Glynn, William Schaeffler, Christine Scriabine, and Barbara Botti for invaluable assistance in preparing the manuscript. Pfizer, Inc., Merck & Co., and Syntex provided timely consultations and data that contributed to our understanding of this complex industry. Finally, we thank Martha Gottron for her invaluable editorial assistance in whipping the manuscripts of the various authors into the shape you see in the following pages.

The authors of the chapter on taxonomy (Chapter 4) would like to acknowledge the significant contributions of Pfizer, IMS America (Cognizant), and all individuals who helped review the commercial and scientific inputs to this process.

The research described in Chapter 5 was funded by four pharmaceutical companies and by the Sloan Foundation through MIT's Program on the Pharmaceutical Industry. Their support is gratefully acknowledged. We would also like to express our appreciation to all of those firms that contributed data to the study and to Allan Afuah, Gary Brackenridge, and Nori Nadzri for their outstanding work as our research assistants.

Lastly, the editors are also indebted to Patricia Wieland, Frances Coulborn Kohler, and Shelley Wilks Geehr of the Chemical Heritage Foundation for their work on the final stages of preparing the book for publication.

Introduction

BY RALPH LANDAU

FOR A FULLER UNDERSTANDING OF THE pharmaceutical industry, a further look at its relation to the chemical industry as a whole is useful. Although part of the chemical industry, pharmaceuticals constitute a dynamic industry in their own right. This introduction offers some background for the material contained in this book.

The Pharmaceutical Industry

Comparing the accomplishments of the pharmaceutical industry to those of the chemical industry as a whole makes its individuality clear. In 1997 the sales volume of the American pharmaceutical industry was $122 billion, while the chemical industry overall took in $392 billion. The ratio of domestic sales to sales abroad for the pharmaceutical industry has run about two to one over the last year or so. The pharmaceutical industry contributes nearly 1.2 percent to the gross domestic product, while the chemical industry as a whole contributes 1.9 percent. The pharmaceutical industry in 1998 invested about $6 billion in capital expenditures and $21 billion in research and development. The ratio in the chemical industry is the reverse, which demonstrates how significant an investment the pharmaceutical industry makes in research. A look at the two industries' relative profits, however, presents a different picture. The pharmaceutical industry contributes over 40 percent of the profits of the entire chemical industry, and its market capitalization is very high (exceeding $1 trillion in 1998). It contributes nearly 20 percent of its sales value to R&D, while the chemical industry proper contributes only 3 to 4 percent of its sales. Regulations (the necessity of establishing the safety and efficacy of new chemical entities, or NCEs) and the high rate of failure necessitate such a heavy expenditure. The pharmaceutical industry employed about 338,000 persons in 1997—a substantial percentage of the total chemical industry employment of 1,034,000.

Pharmaceuticals are also very much a U.S. success story, contributing in the last twenty-five years approximately three times as many major global drugs as any one of its nearest competitor countries, such as Japan, the United

Table 1. Top Thirteen Pharmaceutical Companies Worldwide in 1998 (in millions of U.S. dollars)

Company	Pharmaceutical Sales	Profit	Total R&D
Glaxo Wellcome (U.K.)	13,274	4,461	1,954
Merck (U.S.)	12,840	7,637	1,821
Pfizer (U.S.)	12,230	3,575	2,060
Bristol-Myers Squibb (U.S.)	11,851	3,291	1,577
Novartis (Switzerland)	10,558	3,145	1,900
Roche (Switzerland)	10,467	N/A	1,997
American Home Products (U.S.)	8,902	2,488	1,320
Johnson and Johnson (U.S.)	8,569	3,016	1,353
Hoechst Marion Roussel (Germany)	8,243	888	1,500
SmithKline Beecham (U.K.)	7,710	2,220	1,395
Schering-Plough (U.S.)	6,695	2,200	1,000
Rhône-Poulenc (France)	6,173	1,023	N/A
Bayer Group (Germany)	5,100	N/A	900

Kingdom, Germany, Switzerland, and France.[1] The figures for the top thirteen pharmaceutical companies in the world in 1998 (as based on company reports and our estimates) appear in Table 1. The large number of U.S. companies represented there, compared to those in any single other country, supports this contention.

As the cost of developing an NCE has escalated to over $300 million in recent years, restructuring to create larger companies has become common. Thus Novartis is the result of a merger of two Swiss companies, Ciba-Geigy and Sandoz, and Glaxo Wellcome, a merger of Glaxo and Burroughs Wellcome, very nearly merged with a third U.K. firm, SmithKline Beecham, which itself had been a merger of a U.S. and a U.K. firm. The creation of Hoechst Marion Roussel involved the acquisition of Dow Marion Roussel from the Dow Chemical Company and consolidation with the French subsidiary Hoechst Roussel. Since then, Hoechst itself is in the process of merging with Rhône Poulenc and shedding most of its chemical business. Bristol-Myers Squibb was obviously an earlier consolidation, and Merck has acquired the marketing company Medco. Table 2 gives a summary of some of these activities in the last decade.

There are other alliances and many historical trends of longer duration that, although touched on in this book, would merit fuller treatment than is proper in it. Most U.S. pharmaceutical companies basically began not as chemical companies, but as manufacturing and marketing apothecaries. But in Europe the true developers of early medicines were the chemical companies, and these companies are still important factors in the pharmaceutical industry. The relation

[1] More data can be obtained from *Opportunities and Challenges for Pharmaceutical Innovation—Industry Profile* (Washington, D.C.: U.S. Pharmaceutical Research and Manufacturers of America, 1996), as well as from the annual statistical volumes of the Chemical Manufacturers Association.

between pharmaceuticals and chemicals is particularly close at Bayer and Hoechst. Imperial Chemical Industries (ICI), like Bayer and Hoechst, succeeded in pharmaceuticals (so much so that it spun off this activity into Zeneca in 1992 and became again primarily a specialty chemical company). Attempts to combine these industries have fared unevenly in the United States. DuPont found it difficult to enter the pharmaceutical industry and so merged its activities in that area with Merck. DuPont then bought out Merck's share in 1998 to regain 100 percent of the company, now DuPont Pharmaceuticals, and announced that it would focus on life sciences; its market capitalization rose on the prospect. Dow's adventures in pharmaceuticals ended when it sold Dow Marion Roussel to Hoechst. Monsanto has fashioned itself into a life sciences company, but its acquisition of the pharmaceutical company G. D. Searle has not yet contributed significantly to its future vision, which focuses on biotechnology and plant sciences. Monsanto was nearly acquired by American Home Products, but after this deal collapsed, Monsanto's future is less clear. In short, managements have increasingly found it a challenge to handle widely diversified businesses and are narrowing their focus. Pharmaceutical management is highly specific, and the industry's managers need a long history of skills acquired from working in it.

The pharmaceutical industry is also heavily international– disease knows few boundaries—and managing global activities under different political and cultural regimes is very demanding. Every major nation has a health care system of some kind. Not all capitalist countries are equally friendly to the industry: In some, pharmaceutical companies are subject not only to regulations but also to actual controls. As the costs of health care mount because of increased access and an aging population, the differences between cultures and policies become more salient. Thus, in Japan, doctors, not pharmacies, dispense medicines, which explains why the Japanese ingest more medicines per capita than inhabitants of other countries do. In Germany, on the other hand, doctors are restrained by an annual cap on what they can prescribe for medicines; otherwise they must pay out of their own pockets. Other differences affect corporate behavior: Many countries have inadequate protection of intellectual property, primarily patents, and thus major drug companies will not invest in their economies. Knowledge of these differences and experience in handling them are vital for the most successful management of the pharmaceutical industry.

The Analytical Framework

Most of this book is contained in its first two chapters, the first by Basil Achilladelis and the second by Alexander Scriabine, which look at the historical evolution of the industry from two different perspectives.

Achilladelis is a natural products chemist and a historian of science and technology who studied technological innovation in the chemical and pharmaceutical industries for more than thirty years. He has examined the pharmaceutical industry from a critical perspective and carefully analyzed its impact on the

Table 2. Mergers and Acquisitions in the Pharmaceutical Industry

Year	Merger/Acquisition
1999	Hoechst AG *with* Rhône-Poulenc Rorer
1998	Sanofi SI *with* Synthelabo
	Zeneca *with* Astra
1997	Hoffman-La Roche *with* Boehringer Mannheim
	Nycomed *with* Amersham
1996	CibaGeigy *with* Sandoz
	Elan *with* Athena Neurosciences
1995	Knoll *with* Boots
	Glaxo *with* Burroughs Wellcome
	Gynopharma *with* Ortho-McNeil
	Hoechst-Roussel *with* Marion Merrell Dow
	Pharmacia *with* Upjohn
	Rhône-Poulenc Rorer *with* Fisons
	Schwarz Pharma *with* Reed & Carnick
1994	Hoffman-La Roche *with* Syntex
	Pharmacia *with* Erbamont
	Sanofi *with* Sterling (prescription drug operation)
	SmithKline Beecham *with* Sterling (over-the-counter pharmaceutical unit)
1991	SmithKline *with* Beecham
1990	Boots *with* Flint
	Pharmacia *with* Kabi
	Rhône-Poulenc *with* Rorer
1989	American Home *with* A.H. Robins
	Bristol-Myers *with* Squibb
	Dow *with* Marion
1988	Kodak *with* Sterling
1986	Schering-Plough *with* Key
1985	Monsanto *with* Searle
	Rorer *with* USV/Armour

Data from *Windhover's Health Care Strategist* (Norwalk, Conn.: Windhover Information, Inc., 1999).

economy. Scriabine is a pharmacologist and physician who has spent most of his career in the pharmaceutical industry, participated in the development of drugs, and directed the research of other scientists. In his chapter he evaluates drugs primarily on the basis of their scientific novelty and their value for the practice of medicine. The two approaches to drug evaluation lead at times to different conclusions as to the overall value of a drug. The economic value of some drugs was created by imaginative marketing, while other drugs, invaluable in the practice of medicine, were not properly promoted and did not sell well. The editors hope that the differences in the approaches to drug evaluation will complement each other and enhance the value of this book.

In chapter 1 Achilladelis examines the dynamics of technological innovation in the pharmaceutical industry in the light of current theory and historical evidence. He recognizes five periods in the development of the pharmaceutical industry that correspond to five generations of drugs. During the first (1820–1880), academic researchers, physicians, and pharmacists investigated the active principles of medicinal plants or used simple organic chemicals for the treatment of diseases. The pharmaceutical industry hardly existed: Drugs were prepared by pharmacists or physicians. The pharmaceutical industry began to emerge during the second period (1880–1930), establishing itself not only in Western Europe but also in the United States. Chemical companies established research laboratories and hired scientists to develop new drugs, isolating new chemicals from medicinal plants or synthesizing them and testing for biological effects. The first effective vaccines and synthetic drugs were marketed, and pharmacology emerged as a branch of medicine. During the third period (1930–1960), the pharmaceutical industry matured, and most companies became dependent on their own research for new products. World War II spurred Western governments to support the development of antibiotics, antimalarials, and other drugs needed in the war effort. After the war the market for pharmaceuticals expanded rapidly, with the U.S. pharmaceutical industry taking the lead. Hormones, vitamins, antibiotics, and anti-inflammatory drugs were developed and marketed. Pharmaceutical companies prospered and made substantial investments in research and development, and the industry became an important contributor to the economies of the United States and Western Europe.

The thalidomide tragedy (malformations in infants of mothers treated with thalidomide during pregnancy) at the beginning of the fourth period (1960–1980) led to stricter government regulations for the approval of new drugs and thus to substantial increases in the cost of developing them. It became unprofitable to develop drugs that lacked substantial advantage over those already available. This period nonetheless saw the introduction of a growing industry of antihypertensives, diuretics, tranquilizers, antipsychotics, antidepressants, contraceptives, new antibiotics, and cancer chemotherapeutics. The U.S. pharmaceutical industry also expanded into new markets abroad, primarily in Western Europe, but also in South America and Asia; the industry's growth rate nonetheless slowed. Pharmaceutical firms diversified "horizontally" into other related—and even unrelated—industries; as a result, expenditures on research and development leveled off and the rate of innovation slowed.

The pharmaceutical industry, under pressure from takeover threats, declining profitability, and the growth of the biotechnology industry, compounded by the restrictions introduced by health maintenance organizations (HMOs), is currently in a fifth "wave," a period of restructuring. Consequently, pharmaceutical companies have relied more than ever on "blockbuster" drugs—those with annual sales in excess of $500 million, but the profitability of the industry as a whole nonetheless declined in the 1990s: Between 1992 and 1994 the stock value of twelve large U.S. companies declined by $40 billion, primarily because of competition from generic products and pressure from HMOs and the government for lower prices. The industry responded by restructuring—"downsizing" and forming joint ventures with other pharmaceutical companies, generic manufacturers, and even HMOs. In the bull market of the later

1990s, however, the market capitalization of many pharmaceutical companies has risen precipitously. In 1999, as a correction, the industry suffered another setback in its major capitalization, although not in its profitability. There were many reasons, including serious overvaluation in boom markets. Another reason was fear of new government price controls imposed by government financing of drug purchases.

New drugs of the fifth generation include novel antihypertensives (e.g., captopril), antidepressants (e.g., fluoxetine), antimigraine drugs (e.g., sumatriptan), cancer chemotherapeutics (e.g., paclitaxel), anti-Alzheimer's drugs (e.g., tacrine), and drugs for treatment of osteoporosis (e.g., alendronate) and benign prostatic hyperplasia (e.g., finasteride).

Biotechnology companies have begun to introduce new drugs. Their first products were based on the use of new technology in the manufacturing of previously known proteins: The introduction of foreign genes into bacterial or cell cultures has facilitated the large-scale manufacturing of products like alteplase, insulin, erythropoietin, and growth hormone. This new industry put more pressure on large pharmaceutical companies to restructure their research organizations and introduce new technologies in their own laboratories or to make alliances with or outright purchases of successful biotechnology companies, or both.

In chapter 2 Scriabine categorizes the many types of drugs in the medical profession's arsenal, notes their mechanism of action, rates their effectiveness, and carefully reviews their history. The first section covers drugs that attack or block infectious agents. Vaccines, important in preventing disease, are first. Antibacterial drugs come next, beginning with sulfonamides. Antibiotics follow, the most important among them being penicillin, a result of the major contribution made to the war effort during World War II by the U.S. pharmaceutical industry. Tetracyclines, discovered during the late 1940s and early 1950s, were broad-spectrum antibiotics, active against a wider variety of microorganisms than penicillin. Quinolones were next, first nalidixic acid, then ciprofloxacin, another wide-spectrum antibiotic. The first therapeutically useful antiviral drug was acyclovir, a major contribution to the fight against AIDS. (There are currently eleven anti-AIDS drugs on the U.S. market, which attack the virus by one of two different mechanisms.)

The section on cardiovascular drugs summarizes the history of anticoagulant, antithrombotic, and hypolipemic drugs, emphasizing the mechanism of action of these leading antihypertensive and cholesterol-lowering drugs and their impact on mortality and morbidity.

A section on drugs affecting the central nervous system deals with the discovery, mechanism of action, and medical use of anesthetic, analgesic, anxiolytic, antipsychotic, antidepressant, and antiparkinsonian drugs. The discovery in the 1950s that the psychiatric illnesses (e.g., depression, schizophrenia) respond to drug therapy was largely accidental, made by physicians using drugs for other indications. These early drugs often lacked potency and selectivity, could not be administered orally, and were short acting. Pharmaceutical researchers in industry then optimized these original leads by synthesizing and testing many new but chemically related compounds.

Drugs that affect the peripheral autonomic nervous system, another large class of useful medicines, act largely by either blocking or mimicking the effects of body-own chemicals (neurotransmitters) that are formed and released at the nerve endings. Histamine, a body-own amine involved in the inflammatory response, was discovered at the beginning of the twentieth century; it is released during injury or in response to foreign proteins. Antihistamines, drugs that inhibit responses to histamine, were first discovered in the 1930s and 1940s and marketed as antiallergics. The more recent discovery of a second type of receptors for histamine that control gastric secretion led to drugs (e.g., cimetidine) that control gastric acid secretion.

Hormones, found in the body but used as drugs, are another important class. One of the most important hormones is the antidiabetic insulin. Frederick G. Banting and John J. R. McLeod won the Nobel Prize in physiology or medicine in 1923 for research on its effects in the body. Despite discovery of other drugs, insulin has retained the largest share of the diabetic market. Of the sex hormones, testosterone is used to treat hypogonadism (the failure of the testes to produce testosterone). The estrogens have many indications, but their major uses are in postmenopausal replacement therapy and in oral contraceptives, usually in combination with another sex hormone, progesterone. Other significant hormones are the adrenocorticoid group, isolated in 1934, and cortisone, used to treat rheumatoid arthritis from 1949 on.

Nonsteroidal anti-inflammatory drugs (NSAIDs) include aspirin, ibuprofen, naproxen, and many other drugs that are used not only to suppress inflammation but also to treat minor headache, fever, the common cold, and pain of almost any origin. Aspirin is one of the earliest and most successful, dating from the early 1870s: Felix Hoffman, a chemist at Bayer AG in Germany, used it in 1897 instead of salicylic acid to treat his father, who suffered from arthritis. How it works was not known, however, until the late 1970s, when John Vane found that it acts by inhibiting an enzyme, cyclooxygenase, and therefore prevents the formation of prostaglandins (body-own inflammatory fatty acids). Subsequently many other NSAIDs, including ibuprofen, indomethacin, sulindac, and piroxicam were found to act by the same mechanism.

Two other classes of drugs are increasingly important: immunosuppressants and cancer chemotherapeutics. Immunosuppressants (e.g., azathioprine, cyclosporin A, and tacrolimus) suppress rejection of transplanted organs in humans. A recent advance in cancer chemotherapy was the discovery of paclitaxel, originally isolated from the bark of the Pacific yew; it is used to treat ovarian tumors.

In chapter 3 Scriabine takes a closer look at the birth and development of the biotechnology industry from the 1970s through the 1990s, focusing on the

Table 3. Largest Biotechnology Firms as of 31 December 1998

Company	1998 Annual Product Revenues (in billions of U.S. dollars)
Amgen	2.72
Genentech	1.06
Chiron	0.74
Genzyme	0.67
Alza	0.58
Biogen	0.56
Immunex	0.24
Medimmune	0.20

Data from company annual reports.

history of eight companies (Amgen, Biogen, Centocor, Chiron, Genentech, Genetics Institute, Genzyme, and Immunex). Biotechnology companies were started by scientists and entrepreneurs who applied new biological techniques to manufacture previously known proteins that are useful as drugs, introducing foreign genes into bacterial or cell cultures to produce alteplase, insulin, erythropoietin, growth hormone, and the like on a large scale. Venture capital then spurred the industry's growth. Several of these startups succeeded in developing new products, but as small companies with limited capitalization often operating at a loss, they were eventually taken over by or entered into alliances with large pharmaceutical companies in return for cash infusions.

Originally, this biotechnological revolution was heavily American. In 1997, for example, there were almost 1,400 biotechnology and bioscience companies in the United States compared with about 180 in the United Kingdom, and 100 each in Germany and France. Germany initially lagged behind because of its historical revulsion to the excesses of the Nazi regime, but this attitude has changed and ample funding is now available there. The situation is also favorable in the United Kingdom, France, and Japan. The total world market for biotech firms in 1997 was $13 billion. Yet the total market capitalization of the U.S. biotechnology industry was only about $200 billion in 1998, not far above capitalization for Merck or Pfizer alone. There are still "miles to go" (Table 3).

Chapter 4 is a taxonomy of the drug innovation process over time. Working with physicians, some pharmaceutical companies, and academics, its authors (Ralph Landau, Arthur DeSimone, and Lewis Gasorek) attempted to systematize this complicated technological and industrial story for better comprehension. The chapter demonstrates the different phases of drug discovery: first, a largely empirical and painstaking search for remedies, then a more inventive chemical exploration, which soon blossomed as the insights of biology and physiology were brought to bear. Currently, biotechnology based on molecular biology is advancing with the help of new tools such as computer-based combinatorial chemistry (a computerized system for running virtual chemical reactions en masse), and one can see on the horizon the awesome potential of genetic pharmacology. Scriabine's chapter 2 is closely coordinated with this taxonomy.

The chapter on taxonomy also represents our best judgment as to the "very important" and "important" drugs to be examined and relates them to recent sales figures in the United States, the latest government figures available to us. Most recently, sales volumes of new antiarthritic drugs, such as Celebrex (from Monsanto) and Vioxx (from Merck), have risen rapidly. Drugs that were once important commercially have now faded away or been superseded by superior products, as we indicate. As of 1998 the prescription bestsellers in the United States were as shown in Table 4.

Chapter 5, by Rebecca Henderson and Iain Cockburn, summarizes the econometric studies they have made of major pharmaceutical companies over several years. As the industry matures and more traditional ways of finding and commercializing new drugs become more expensive and difficult, consolidations and other changes occur.[2] Nevertheless, recently developed technologies will restore momentum to well-managed companies, as will reorienting R&D programs toward aggressive goals that will drive new company growth.[3]

In chapter 6 Christopher Flowers and Kenneth Melmon present an intensive study of the tortuous and often serendipitous process of drug discovery. This case history illustrates that the usual straightforward accounts of how discoveries are made rarely tell the real whole story of the incredibly complex search for new chemical entities. Often even the participants are not fully aware of how developments actually occurred! Two recent examples have been publicized: One, a prostate drug (Proscar), turned out to restore hair and was put out in a new version called Propecia by Merck; the other, a failed cardiac drug, unexpectedly counteracted erectile dysfunction and was put out by Pfizer as Viagra. (This last led to a sharp increase in the value of Pfizer stock to nearly equal that of Merck, reduced somewhat more recently.) These are examples also of "lifestyle" drugs, which may find increasing applicability in the future as against medicines strictly for disease treatment.

There is some redundancy in all these sections, but this is deliberate, so as to illustrate the varying dimensions of the innovation process that has contributed so much to human health.

---

[2] As *Chemical and Engineering News* for 23 February 1998 points out.

[3] Henderson, together with Luigi Orsenigo and Gary Pisano, has also studied the changes coming in pharmaceutical research in a forthcoming chapter titled "The Pharmaceutical Industry and the Revolution in Molecular Biology: Exploring the Interactions between Scientific, Institutional, and Organizational Change," to be included in a book from Cambridge University Press by Richard Nelson and David Mowery.

Table 4. Bestseller Prescription Drugs in the United States in 1998

Rank by Year			Drug	Use	Manufacturer	1998 Sales (in millions of U.S. $)
1996	1997	1998				
2	1	1	Prilosec	Anti-ulcerant	Astra Merck	2,933
3	2	2	Prozac	Antidepressant	Eli Lilly	2,181
*	9	3	Claritin	Antihistamine	Schering-Plough	1,848
*	*	4	Lipitor	Cholesterol reducer	Warner-Lambert/Pfizer	1,544
6	3	5	Zocor	Cholesterol reducer	Merck	1,481
4	4	6	Epogen	Anti-anemia	Amgen	1,455
5	5	7	Zoloft	Antidepressant	Pfizer	1,392
*	*	8	Prevacid	Anti-ulcerant	TAP	1,245
*	7	9	Paxil	Antidepressant	SmithKline Beecham	1,190
*	8	10	Norvasc	Calcium blocker	Pfizer	1,086
1	6	*	Zantac	Anti-ulcerant	Glaxo Wellcome	—
8	10	*	Vasotec	Cardiovascular	Merck	—

Data from IMS America.
* Not in top ten for the year.

Why Is the History Important?

This industry is central to human health and welfare. Has it contributed the best solutions possible to these fundamental concerns? Of course, the investors in the industry have benefited hugely, and national wealth has increased. Indeed the general populace has also benefited, particularly in reduced hospitalization and increased longevity. The dramatic reduction (59%) in age-adjusted stroke mortality, occurring in the United States from 1971 to 1994, probably caused by the increased availability and effectiveness of hypertension medicines, provides but one example of such a benefit. In addition a glance at the taxonomy presented in chapter 4 indicates how wide a range of diseases and infirmities have been successfully ameliorated or eliminated altogether by pharmaceuticals. Yet, for all of these benefits, the current cost of pharmaceuticals is only slightly over 10 percent of the overall cost of health care. This seems a small price to pay to improve our quality of life and combat the endless array of new diseases and medical problems that crop up, such as the ongoing problem of AIDS and the still largely unsatisfactory treatments for cancer. Nevertheless, as the tendency to assist in the purchase of prescription drugs for an aging population increases in this country, there seems to be a related rise in the cost of medical care.

Pharmaceutical innovation has also greatly increased longevity. That benefit is a major focus of a recent article by Frank R. Lichtenberg of the Columbia Business School, an "econometric investigation of the contribution of pharmaceutical innovation to mortality reduction and growth in lifetime per capita income."[4] After examining the relationship across diseases between the in-

[4] Frank R. Lichtenberg, "Pharmaceutical Innovation, Mortality Reduction, and Economic Growth," National Bureau of Economic Research, Cambridge, Massachusetts, May 1998.

crease in mean age at death (which is closely related to life expectancy) and rates of introduction of new drugs during two decades, he concluded:

> We perform an econometric investigation of the contribution of pharmaceutical innovation to mortality reduction and growth in lifetime per capita income. In both of the periods studied (1970–80 and 1980–91), there is a highly significant positive relationship across diseases between the increase in mean age at death (which is closely related to life expectancy) and rates of introduction of new, "priority" (as defined by the FDA) drugs. The estimates imply that in the absence of pharmaceutical innovation, there would have been no increase and perhaps even a small decrease in mean age at death, and that new drugs have increased life expectancy, and lifetime income, by about 0.75–1.0% per annum. The drug innovation measures are also strongly positively related to the reduction in life-years lost in both periods. Some of the more conservative estimates imply that a one-time R&D expenditure of about $15 billion subsequently saves 1.6 million life-years per year, whose annual value is about $27 billion. All age groups benefited from the arrival of new drugs in at least one of the two periods. Controlling for growth in inpatient and ambulatory care utilization either has no effect on the drug coefficient or significantly increases it.

In August 1999 the U.S. Centers for Disease Control and Prevention in Atlanta issued a detailed report confirming the improvement in life expectancy and quality of life from new cardiovascular treatment. There are about fifty thousand centenarians in the United States now, whereas there were almost none in 1900.

Yet there are critics of the whole system of medical and pharmaceutical progress. In *Prescriptions for Profits* , for example, Linda Marsa assails all such advances as being predicated on "profits."[5] Of course, that is what a successful capitalist system is designed to do! If the public had not benefited hugely, another system would have been tried. Significantly, there is no record of major breakthroughs in countries ruled by a communist system, where risk taking is penalized. Substantial capital investments are needed for the discovery, development, and marketing of life-saving and life-improving drugs. Not that all is smooth politically for this vital activity even in a capitalist environment: Early in the Clinton administration the pharmaceutical industry was blamed for its high prices as health care costs spiraled up, and the market capitalization of the industry, as mentioned above, dropped by $40 billion. Yet this political attack (along with the administration's entire health care plan) ultimately failed in the face of the industry's real accomplishments.

In *False Hopes: Why America's Quest for Perfect Health Is a Recipe for Failure*, Daniel Callahan has published a somewhat more sophisticated critique.[6] Callahan asks how much health expenditure is prudent in a sustainable economy.

---

[5] Linda Marsa, *Prescriptions for Profits: How the Pharmaceutical Industry Bankrolled the Unholy Marriage between Science and Business* (New York: Scribners, 1997).

[6] Daniel Callahan, *False Hopes: Why America's Quest for Perfect Health Is a Recipe for Failure* (New York: Simon and Schuster, 1998).

Echoing the English Tory politician Enoch Powell's sentiment, "There is virtually no limit to the amount of health care an individual is capable of absorbing," he argues that people should be content to live beyond the biblical three-score-and-ten years and let nature take its course thereafter. Rather than the huge expenditures the sick and the old incur, he urges a focus on healthy living, preventive medicine, and perhaps even enforced behavioral changes toward tobacco elimination, bad diet, and exercise. However, he speaks as a big government exponent in an era when freedom to choose is ever more prized. Certainly the older people he considers superfluous have a different view of the matter. Further, as in the decades studied by Lichtenberg, future pharmaceutical research devoted to chronic diseases and ailments of older people may, after all, if successful, reduce the cost of maladies and medical care in later life and improve the quality of life.[7] Ultimately, the markets will be decisive. And there are still serious large-scale diseases to conquer that affect many individuals besides those over seventy: cancer, AIDS, and the recurring antibiotic-resistant plagues, among others. What is more, prophecies based on current conditions are not always reliable. The classic example is Thomas Malthus, who two hundred years ago predicted doom from overpopulation. In any case, from the standpoint of the patient, taking medicine is better than undergoing surgery, with hospitalization and nursing to follow, even though the need to take medicine over a lifetime may ultimately cost society more. Nevertheless, as *Business Week* of 19 July 1999 shows, some medicines greatly reduce the cost of surgery and hospitalization.

Although this book makes no effort to study these larger issues, we deem it useful to use them as a frame for our main focus: the successes of this industry in treating numerous diseases and in helping to extend human life expectancy (in the United States life expectancy went from about forty-seven years in 1900 to seventy-six years currently). It was not too long ago that plagues struck the European populace and "witches" were burned at the stake to appease the apparently remorseless forces that were devastating the population. At this writing a new era in medicine is already visible, founded on computer-based bioinformation systems that are exploring the vast stores of genomic knowledge and the areas of molecular biology being collected by researchers in many countries who are unraveling the human genome. A new industry of biochemotechnology is also visible, which has as its goal the development of synthetic chemical small-molecule drugs rather than the development or improvement of natural proteins—the goal of the early biotechnology research. In this effort biotechnologists work with chemists to develop chemical drugs that target ex-

[7] Recent studies by the Institute for Clinical Outcomes Research have demonstrated that up-front investments in disease management, including drug therapies, can reduce overall medical costs for some diseases by 30 percent to 50 percent, as *Business Week* for 24 November 1997 points out.

plicitly identified biological targets; nucleic acids (DNA or RNA) enzymes; and surface and intracellular receptors (proteins, lipids, or carbohydrates). This work, as mentioned before, uses the tools of combinatorial chemistry and computer assembly of vast libraries of potential molecules.

Such developments are helping scientists to understand better the disease processes and thus to improve their ability to create better drugs. With this perspective our book may be seen in the future as a record of how science and medicine coped with human ailments in the long era that began with the ancient world and culminated with the science stemming from the fundamental discovery of the double helix by Watson and Crick. This is a highly competitive practice and industry, which faces competition not only from its peers but also from cheap generic drugs when patent protection expires. It is an industry that learned to work closely with government supports, while also being closely regulated by government. It is an example of a well-organized public-private partnership. It is essential that neither partner press its claims beyond a reasonable balance. It is dangerous to tinker with such a successful innovative industry that has done so much for human health.

Chapter One

Innovation in the Pharmaceutical Industry

BASIL ACHILLADELIS

Technological innovation is a dynamic process, perhaps the most dynamic of all industrial activities. With his "gales of creative destruction" Joseph Schumpeter painted a vivid, even romantic, picture of the effects of the diffusion of great technological discoveries and inventions on industry and the world economy. But the dynamism of technological innovation is also expressed in more subtle ways: as a catalyst in the interaction of science and technology that accelerates their otherwise arduous advance; in the technological development and market expansion of industrial sectors; in the research intensity, specialization, and technical and business performance of industrial firms; and in the competitive advantage of national industries. Altogether these effects of technological innovation can be seen and appreciated in the context of the history of research-intensive industrial sectors: chemicals, pharmaceuticals, materials, electricity and its applications, transport, electronics, and computers.

Overview

The pharmaceutical industry is a relatively small research-intensive industry. In the late nineteenth century it was considered to be a special branch of the chemical industry, particularly in Europe where the large chemical companies were also the leading manufacturers of medicines. Only since the 1970s have some pharmaceutical companies become truly large enterprises, comparable to those in chemicals, foods, metals, power, and electronics. And only since the 1980s have pharmaceutical companies become prominent in the manufacturing industry, largely because of the decline in the West of many of the traditional, technologically mature manufacturing sectors. Throughout its history the industry has maintained a close and fruitful relation with institutions of academic research in chemistry, medicine, and the life sciences. Its major innovations have not created waves—maybe not even ripples—of creative destruction in industry or the world economy; but some pharmaceuticals have been household names for nearly a century, and others have deeply affected the nature and morals of society. Indeed, in this latter aspect, perhaps no other industry has had a greater effect.

The history of innovation in the pharmaceutical industry can be divided into two periods (Table 1). During the first period, which began around 1820, scientific methods were adopted for preparing and purifying diverse natural and synthetic products and for detecting and studying their medical and physiological properties. This first period ended around 1930, when this evolutionary process reached a threshold, and pharmaceuticals had to become a truly research-intensive industry. Two generations (clusters) of drugs were introduced during this period: The first was initiated in the 1820s by a cluster of innovations that included the alkaloids and some organic and inorganic chemicals; the second was launched in the 1880s by a cluster of innovations of serums, vaccines, antiprotozoal medicines, antipyretics, analgesics, and hypnotics. The innovators during that period were physicians, academic institutions, public health research laboratories, and the German dyestuffs companies.

The second period began in the 1930s with the introduction of the third generation of medicines—vitamins, sulfonamides, antihistamines, antibiotics, and sex and corticosteroid hormones. This generation revolutionized the structure and business practices of the pharmaceutical industry. Two more generations have since followed. The fourth was introduced in the 1960s and included semisynthetic antibiotics, cardiovascular and central nervous system (CNS) medicines, and nonsteroidal anti-inflammatory drugs (NSAIDs). The fifth generation, launched in the late 1970s and early 1980s, has yet to play itself out. It includes antiulcer and antiviral medicines, enzyme inhibitors, and bioengineered proteins. The scientific basis of medicinal research shifted gradually during this second period from organic chemistry to the life sciences. The innovators have been research-intensive pharmaceutical companies in collaboration with academic and public health research laboratories.

Each of the five generations of medicines was composed of a cluster of technologies introduced within a relatively short period. This phenomenon can be traced to the strong dependence of innovation in the pharmaceutical industry on scientific advance. Extraordinary advances in the physical and life sciences and in medicine led to the near-simultaneous introduction of radical innovations in several therapeutic categories. These radical innovations, which were clinically effective and commercially successful, exerted an orienting effect on subsequent academic and industrial research because the scientific and medical principles on which their activity depended were not completely understood at the time of their introduction. Leading researchers thus strove to discover the missing pieces of knowledge. At the same time the commercial and technological success of these innovations led to their imitation by scores of competing companies eager to enter promising new markets with similar but improved medicines, known as incremental innovations. In a typical cycle knowledge deepens and technology diffuses, making it easier for competitors to introduce incremental innovations, which find ready buyers in markets that expand as the efficacy of the particular class of drugs becomes widely known. Eventually, however, the scientific principles are elucidated, the potential of each technology exhausted, and the commercial opportunities exploited. The researchers then move on to other, more promising subjects. This stage of decline is frequently cut short by the introduction of a new generation of technologies.

THE DYNAMIC EFFECTS OF RADICAL INNOVATION

Technological innovation, defined as the development and commercialization of new products and new manufacturing processes, originates from discoveries (revelations of new knowledge) and inventions (products and processes derived after study and experimentation). Innovation is thus dependent on some combination of luck, serendipity, systematic research, and meticulous development.

Originality is the essential characteristic of all technological innovations, but the degree of originality can vary widely. Classification of innovations according to their degree of originality is an inexact science because of the absence of easily defined discontinuities separating the inspired from the trivial. But innovation can in principle be divided into two types. Highly original innovations are described as radical and are themselves of two kinds. One kind gives rise to new industries or new subsectors of an industry. In the pharmaceutical industry examples include the smallpox vaccine (1796); morphine, the first alkaloid (1806); carbolic acid, the first antiseptic (1860); phenazone (Antipyrin), the first synthetic drug (1884); the antisyphilis drug arsphenamine (Salvarsan), the first chemotherapeutic agent (1911); sulfamidochrysoidine (Prontosil), the first antibacterial (1935); penicillin, the first antibiotic (1942); and the process for recombinant DNA (deoxyribonucleic acid), which opened the field of biotechnology (1975).

The second kind of radical innovation includes those that widen the scope and markets of industrial sectors or subsectors by applying new scientific principles, technology, or materials; that displace products and processes already on the market; and that serve as models for further innovation by imitation. Examples of these in the pharmaceutical industry are barbital (Veronal), the first barbiturate hypnotic (1903), with thirty-two imitations; chlorothiazide (Diuril), the first antihypertensive diuretic (1958), with fifteen imitations; propranolol (Inderal), the first antihypertensive β-blocker (1964), with twenty-four imitations; chlordiazepoxide (Librium), the first benzodiazepine anxiolytic (1960), with thirty-seven imitations; and cimetidine (Tagamet), the first treatment for peptic ulcers (1977), with seven imitations by the mid-1990s.

Innovations are described as incremental when they are designed on the model of existing products or processes with only modest differences in

Table 1. Clusters of Radical Innovations and Technologies in the Pharmaceutical Industry (1820–1990)

Period and Cluster	Technology	First Innovation(s)	Year	Company	Country
One					
First	Alkaloids	Morphine	1806	—	Germany
		Quinine	1820	—	France
	Organic chemicals	Ether	1842	—	United States
One					
Second	Analgesics/antipyretics	Phenazone (Antipyrin)	1884	Hoechst	Germany
		Acetanilide (Antifebrin)	1886	Kalle	Germany
	Hypnotics	Sulfonmethane (Sulphonal)	1888	Bayer	Germany
	(barbiturates)	Barbital (Veronal)	1903	Bayer	Germany
	Biologicals	Anthrax vaccine	1881	—	France
		Diphtheria serum	1890	Hoechst	Germany
	Local anesthetics	Cocaine	1884	—	Germany/Austria
		Orthocaine (Orthoform)	1896	Hoechst	Germany
	Antiprotozoal	Arsphenamine (Salvarsan)	1911	Hoechst	Germany
Two					
Third	Vitamins	Vitamins A, B, C, D	1930	Roche	Switzerland
			1937	Merck	United States
	Sex hormones	Estrone	1931	Schering	Germany
		Testosterone	1931	Organon	Netherlands
	Sulfonamides	Sulfamidochrysoidine (Prontosil)	1935	Bayer	Germany
	Antihistamines	Phenbenzamine (Antergan)	1942	Rhône	France
	Antibiotics	Penicillin	1943	Merck	United States
				Pfizer	United States
	Corticosteroids	Cortisone (Cortone)	1948	Merck	United States

Two					
Fourth	Antihypertensives				
	Diuretics	Chlorothiazide (Diuril)	1958	Merck	United States
	Beta blockers	Propranolol (Inderal)	1964	ICI	United Kingdom
	Central nervous system				
	Tranquilizers	Chlorpromazine (Largactil)	1952	Rhône	France
		Haloperidol (Haldol)	1959	Janssen	Belgium
	Antidepressants	Imipramine (Tofranil)	1959	Geigy	Switzerland
	Anxiolytics	Chlordiazepoxide (Librium)	1960	Roche	Switzerland
	Semisynthetic antibiotics	Phenethicillin (Broxil)	1959	Beecham	United Kingdom
		Cephaloridine, ceporan (Keflin)	1964	Glaxo	United Kingdom
				Lilly	United States
	Nonsteroidal anti-inflammatory drugs	Phenylbutazone (Butazolidin)	1952	Geigy	Switzerland
		Indomethacin (Indocin)	1963	Merck	United States
		Ibuprofen (Brufen)	1964	Boots	United Kingdom
	Oral contraceptives	Mestranol/norethynodrel (Enovid)	1961	Searle	United States
Two					
Fifth	Cardiovascular				
	Calcium channel blockers	Nifedipine (Adalat)	1974	Bayer	Germany
	ACE inhibitors	Captopril (Capoten)	1977	Squibb	United States
	Hypolipidemics	Lovastatin (Mevacor)	1977	Merck	United States
	Antipeptic ulcers	Cimetidine (Tagamet)	1977	SmithKline	United States
	Central nervous system				
	Antinausea	Domperidone (Motilium)	1979	Janssen	Belgium
	Antiparkinson	Bromocriptine (Parlodel)	1984	Sandoz	Switzerland
	Serotonin inhibitor	Femoxetine (Malexil)	1977	Ferrosan	Denmark
	Antiviral	Acyclovir (Zovirax)	1983	Wellcome	United Kingdom
	Antineoplastic	Tamoxifen (Nolvadex)	1977	ICI	United Kingdom
		Cisplatin (Platinol)	1978	Bristol	United States
	Biotechnology	Human insulin (Humulin)	1983	Genentech	United States
				Lilly	United States
		Human growth hormone (Protropin)	1988	Genentech	United States
				Schering	United States

science, technology, materials, composition, and properties that do not provide scope for further innovation by imitation. Incremental innovation is the preeminent vehicle for diffusing technology among competing companies. In some instances incremental innovation can produce very lucrative products, such as the antiulcer drug ranitidine (Zantac, 1982), the antihypertensive angiotensin-converting enzyme (ACE) inhibitor enalapril (Vasotec, 1985), the antibacterial cefaclor (Ceclor, 1979), and the tranquilizer trifluoperazine (Stelazine, 1959).

Radical innovations played a dominant role in the advancement of both medicinal science and technology, in the establishment and growth of therapeutic markets, and in the technological and scientific competencies, market specialization, and business performance of pharmaceutical companies.

THE ADVANCE OF SCIENCE AND TECHNOLOGY

The mechanisms, dynamics, and pace of scientific advance are determined on the one hand by the curiosity of scientists; the strength of their intellect; their ability to design and perform experiments; the precision of their observations; and the depth, scope, and interpretive power of their theories; it is determined on the other hand by the excellence of academic and research institutions, the quality of their leadership, the financial support provided by the public and private sectors, and the prestige accorded the institutions on the part of society at large.

Until the middle of the eighteenth century, scientists—with the exception of mathematicians—studied natural phenomena and classified mineral, plant, and animal species. When inventors began to transform the properties of materials by chemical and physical methods and to introduce new forms of energy, scientists extended their interests and began to study man-made phenomena as well. Since then science, technology, and industry have been inextricably linked; as a result scientific advance since the 1800s has also been influenced by social and economic forces.

The strongest links among science, technology, and industry were forged by those radical innovations that were commercialized when the scientific knowledge on which they were based was only partially understood. The medicinal and therapeutic properties of such radical innovations attract the curiosity of scientists who work to discover the missing scientific principles underlying the properties of the new products or the mechanisms of the new processes. The companies that commercialize these radical innovations want this knowledge so that they can improve the quality of the products, develop better manufacturing processes, and introduce new products to expand their markets. Not content with the normal pace of academic research, the innovating companies—and competitors eager to develop similar drugs—often fund research projects in academic laboratories and also hire outstanding scientists and engineers to develop in-house research expertise in the scientific principles underlying the radical innovation. The synergies thus created by academic and industrial research influence the direction and pace of scientific and technological advance.

RADICAL INNOVATION AND MARKET DEMAND

There is a strong positive relation between originality and the commercial success of individual innovations (Achilladelis, 1993; Achilladelis, Schwarzkopf, and Cines, 1990; see also Klein and Rosenberg, 1986). Many radical innovations open new therapeutic markets; others contribute to the growth of existing markets; and still others displace older, less effective products. It is this relation that encourages company managers to develop in-house research expertise, build networks of cooperation with academic leaders, conduct basic research, and take on the technological and financial uncertainties associated with projects aimed at finding a radical innovation. The alternative would be to follow a competitor's lead, hoping that with relatively modest investments, an improvement to an existing process or product might turn out to be profitable.

Without successful commercialization the interaction of academic and industrial research would be haphazard and indirect. A linear relation between science and technology would most probably ensue, each system obeying its own dynamics with hardly any acceleration caused by the synergies of interaction.

RADICAL INNOVATION AND TECHNOLOGICAL DIFFUSION

A commercially successful radical innovation that offers a robust model for imitation initiates a race for the development of incremental innovations with improved therapeutic properties. This activity is important both for the innovating company that wants to strengthen its position by exploiting its own leads and for competitors that want to get a foot in the door of a promising new market. As the new technology and underlying science diffuses, more companies enter the field and more products are launched. Eventually the technology matures, opportunities for further significant improvement in the therapeutic properties disappear, and the corresponding market becomes saturated with so many products and competitors that new products are unlikely to turn a profit. A new technology may appear at that stage, causing the obsolescence of the older technology and initiating a new cycle of innovation.

There are some exceptions to this general procedure. For technological or therapeutic reasons, some commercially successful radical innovations either cannot be imitated or offer only limited possibilities for imitation. Examples include vitamins, the centrally acting antihypertensives methyldopa (Aldomet) and clonidine (Catapres), and the anticancer drugs tamoxifen (Nolvadex) and cisplatin (Platinol). Sometimes a radical innovation is a process that initiates a new technology and creates opportunities for developing numerous drugs. Such was the case with natural antibiotics, where the processes for screening soil samples to identify metabolites with antibacterial properties and for submerged fermentation made possible the discovery, development, manufacture, and commercialization of more than fifty drugs, including penicillin, streptomycin, tetracycline, and chloramphenicol, none of which are related by chemical structure or composition. Early indications are that biotechnological processes will produce a similar range of diverse drugs.

Distribution patterns over time of both radical and incremental innovations profile the diffusion of technologies. Figure 1 shows the diffusion patterns of some of the most important technologies of the pharmaceutical industry, arranged by the successive generations of pharmaceuticals. The figure shows the concentration of radical innovations at the early stages of each technology, the large numbers of incremental innovations that characterize the maturity of the technology, and the trickle of incremental innovations that accompany its decline.

Figure 2 illustrates the "bandwagon effect"—the distribution over time of the companies that entered a therapeutic market with their own first innovation. The patterns, displayed for four important therapeutic categories, show that at the early stages of a technology when the market is small or uncertain, only a few companies participate. This number increases as the technology matures and the market grows and then drops substantially at the late stages.

To summarize, the introduction of commercially successful, therapeutically effective new drugs provides some robust and versatile models for imitation, and the efforts of competing companies to develop and commercialize similar or improved products and processes determine greatly the rate of diffusion of medicinal technologies and the development of the corresponding therapeutic markets. But this process is not self-regulating. It develops under the influence of a set of driving forces that determine when a new technology is launched, the length of its life, and how widely it diffuses.

FORCES DRIVING TECHNOLOGICAL INNOVATION AND DIFFUSION

Innovation, particularly radical innovation, creates technical and financial uncertainty for the innovating company. What then compels innovating companies to accept these risks?

Figure 3 shows seven social and economic forces that drive technological innovation, sometimes in isolation, but usually in some combination. Six of these forces are "environmental" in character; that is, they affect all research-intensive pharmaceutical companies in a country or the world. These six are discussed immediately below. The seventh is company-specific and is discussed later.

Science-Technology

The pharmaceutical industry has depended throughout its history on advances in scientific disciplines and cross-fertilization among them. Those disciplines,

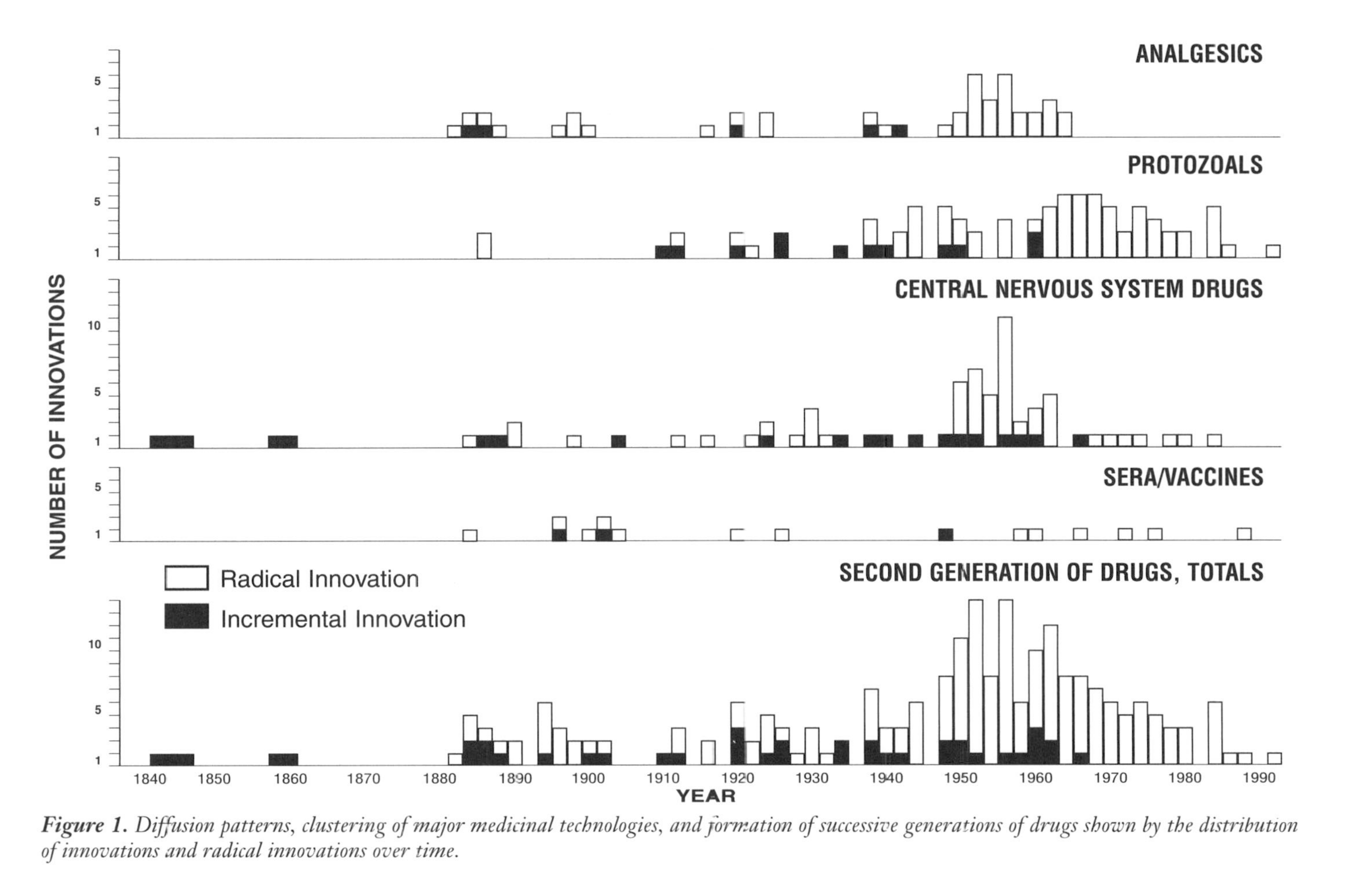

Figure 1. *Diffusion patterns, clustering of major medicinal technologies, and formation of successive generations of drugs shown by the distribution of innovations and radical innovations over time.*

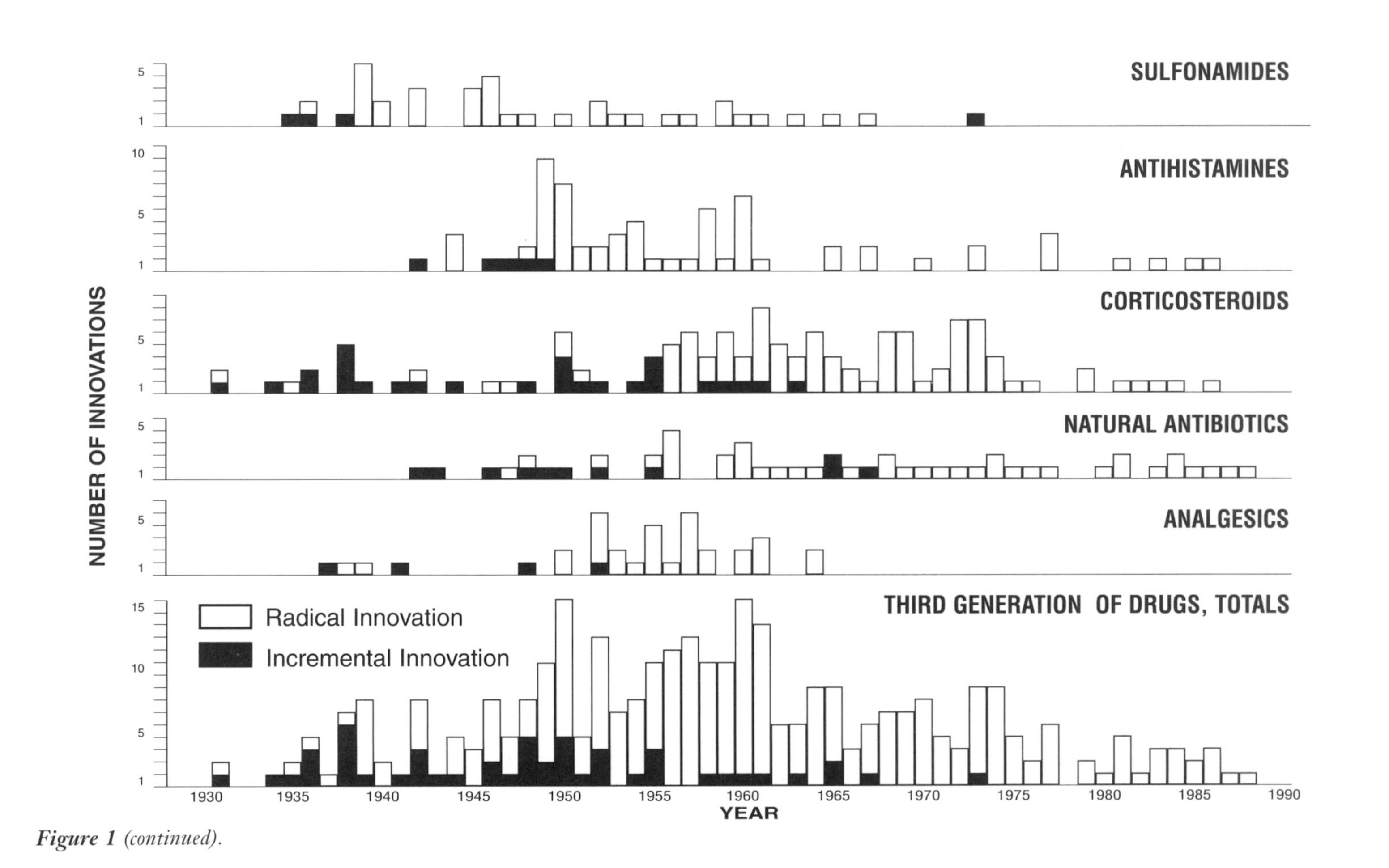

Figure 1 (continued).

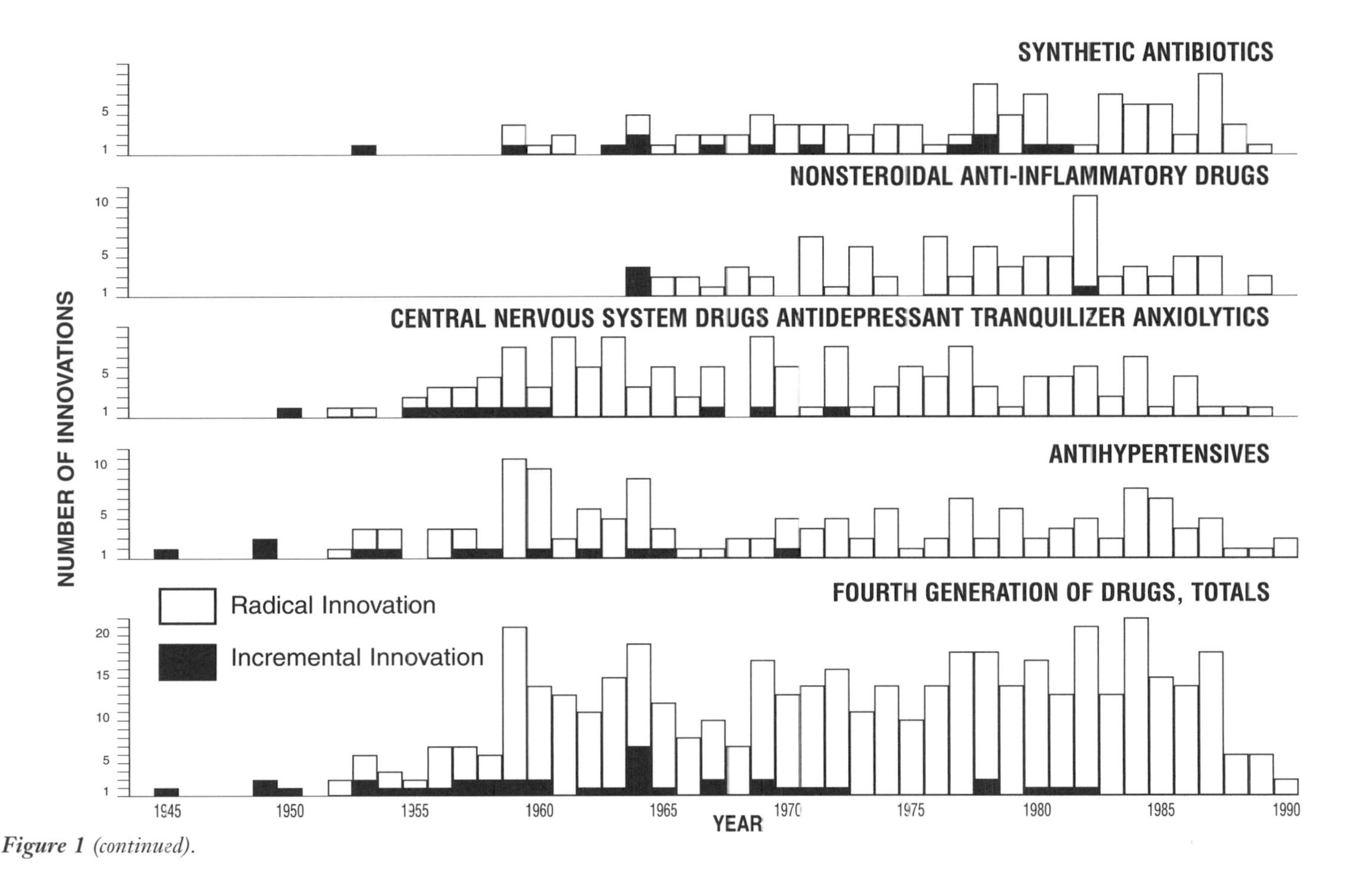

Figure 1 (continued).

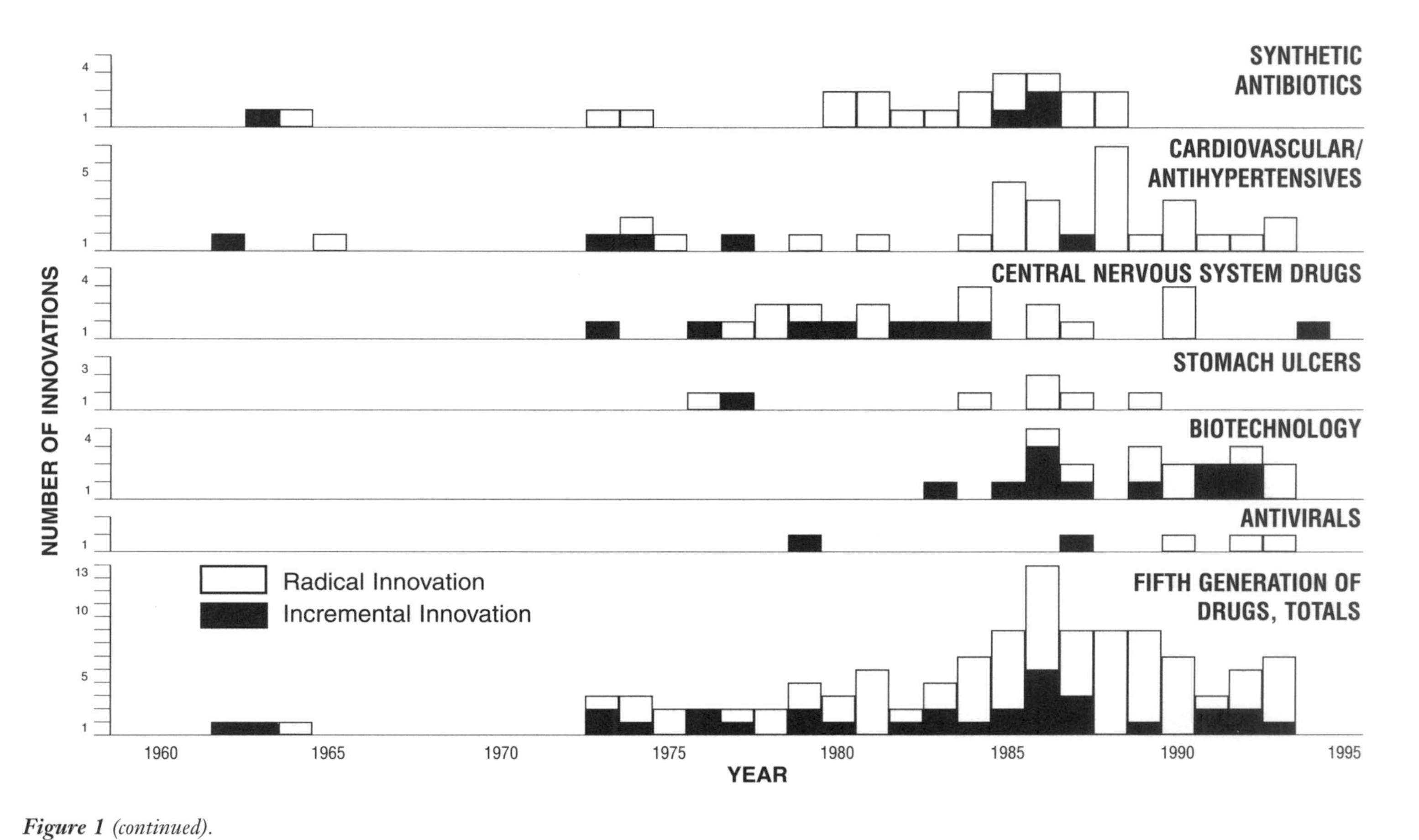

Figure 1 (*continued*).

some of which the industry helped to create, include chemistry, organic chemistry, pharmacology, medicine, biology, bacteriology, enzymology, and molecular biology.

Raw Materials

Since antiquity the discovery and availability of raw materials of mineral, plant, and animal origin have led to the development of technologies that transformed these materials into useful products. The pharmaceutical industry began in the early nineteenth century, when the availability in European countries of colonial tropical medicinal plants led to the manufacture of the first drugs (morphine, quinine, caffeine, nicotine, digoxin, salicylic acid, cocaine) through the isolation of their "active principles."

The availability of thousands of coal-tar chemicals in the second half of the nineteenth century led to the first synthetic drugs—hypnotics, anesthetics, analgesics, antipyretics, and antiseptics—and the emergence of the "modern" pharmaceutical industry. From the 1930s onward thousands of natural products and synthetic chemicals made by the chemical industry are routinely tested by pharmaceutical companies for medicinal properties, an activity that has led to the development of some drugs whose chemical structure gave no indications of their therapeutic properties.

Market Demand

Market demand is one of the strongest driving forces for technological innovation and diffusion, as indicated by the dramatic acceleration of scientific and technological advance and innovation following the spread of capitalism in the mid-1800s, the emergence of the research-intensive company at the turn of the century, and the disparity in innovation in the twentieth century between countries with market economies and those with socialist systems.

The size of individual therapeutic markets has a great effect on the numbers of both pharmaceutical innovations and innovating companies. Although at the early stages of a technology, companies introduce the first radical innovations on the basis of health statistics, therapeutic needs, and anticipated rather than actual demand, once these innovations succeed in the marketplace, rising demand draws many competitors who introduce a host of radical and incremental innovations. Market demand exerts its strongest influence when the technology reaches maturity, and latecomers enter the market with numerous incremental innovations. Such fully developed markets also spur the development of second- and third-generation technologies.

Competition

Competition is a driving force for innovation because it is an essential feature of both academic research and the marketplace. Through their close association with academic institutions and through their own scientists and engineers,

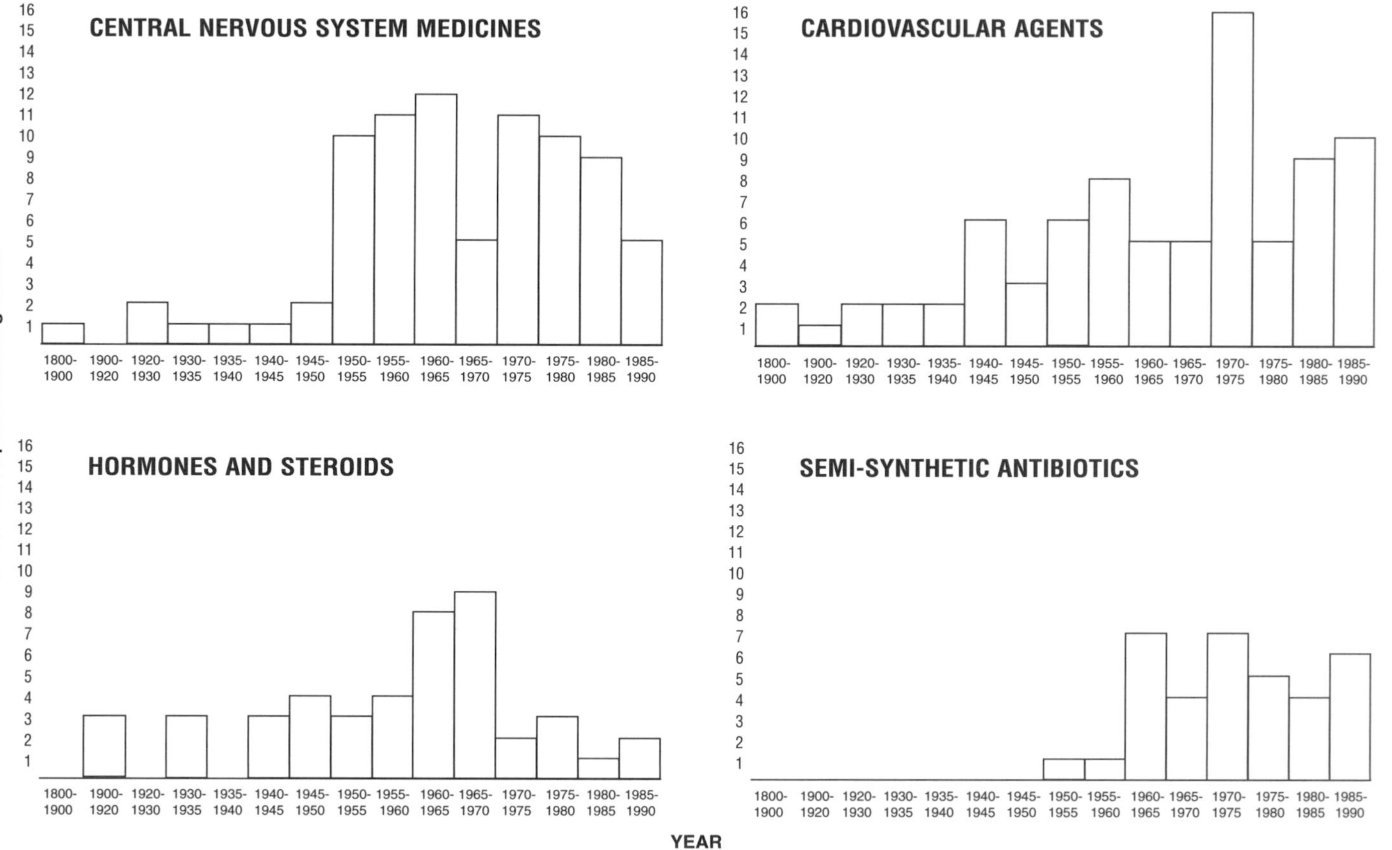
CENTRAL NERVOUS SYSTEM MEDICINES
CARDIOVASCULAR AGENTS
HORMONES AND STEROIDS
SEMI-SYNTHETIC ANTIBIOTICS
Number of Companies Entering Sector
16 15 14 13 12 11 10 9 8 7 6 5 4 3 2 1
1800-1900 1900-1920 1920-1930 1930-1935 1935-1940 1940-1945 1945-1950 1950-1955 1955-1960 1960-1965 1965-1970 1970-1975 1975-1980 1980-1985 1985-1990
YEAR

research-intensive pharmaceutical companies have come to adopt many of the practices and values of the academic community. One of these values is the drive to be first. The introduction of radical innovations whose science and engineering are at the forefront of the relevant discipline adds to the prestige of the company as well as to its profits.

Commercial competition is, of course, far more important in industry than it is in academe. The number of competing companies strongly affects both the number and the pace of innovation. Because of the small number of research-intensive companies in the first half of the twentieth century, the second generation of technologies took sixty years to diffuse, compared with the technologies of the postwar generations, which diffused and declined in about thirty years. The relatively large number of research-intensive companies in the United States after World War II, compared with European countries, contributed to the emergence of the American pharmaceutical industry as the world leader.

Societal Needs

Public and personal health has been a societal expectation since the mid-1800s, when governments responded to the need to contain contagious diseases by building public sanitation facilities and establishing public health agencies and public medical research laboratories. Many of the medicinal innovations of the nineteenth century were introduced by these laboratories, universities, and public hospitals.

A century later "welfare state" legislation, beginning with the British National Health Service in 1947, provided universal or selective health-insurance coverage to the citizens of most industrialized countries. These programs strengthened drug markets and made pharmaceutical companies independent of business cycles, thus encouraging innovation in the pharmaceutical industry in an indirect but extremely important way.

Apart from these general societal forces innovation has also occurred because of specific societal demands. Such cases in the United States include the development of penicillin during World War II under the auspices of the federal government; the development in the 1950s of the polio vaccine, which was totally supported by public funds; and the development in the 1960s of the contraceptive pill, which was supported by a charitable organization. The development of some "orphan drugs" for treating rare diseases and the large public expenditures on research to find cures or treatments for cancer, cardiovascular disease, and AIDS are more recent examples.

Government Legislation

The pharmaceutical industry is perhaps the most regulated of any manufacturing industry. Patent legislation has had the most direct effect on innovation. In countries without patent protection (France before the 1930s and Italy

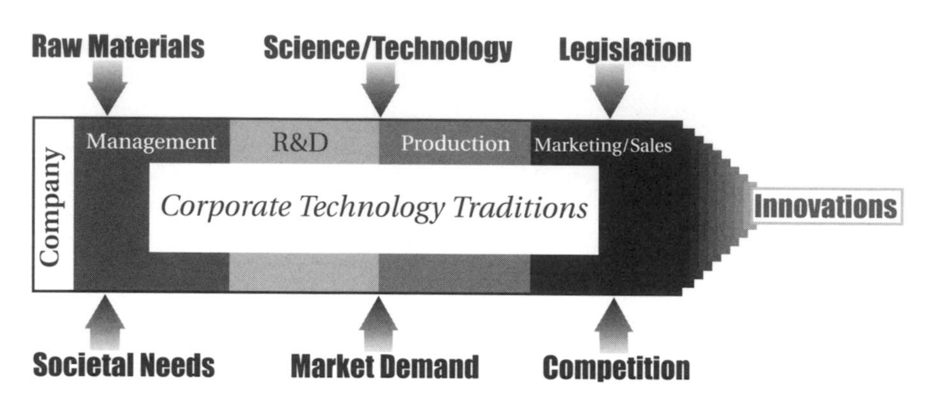

Figure 3. *Driving forces of technological innovation.*

before the 1970s, for example), drug innovation was uneconomic, and research-intensive companies were slow to develop. Strong patent laws in Germany, Switzerland, and the United States had the opposite effect. The early patent protection of natural products (1947) and of recombinant DNA proteins (1985) in the United States contributed to the leadership of the American industry in natural antibiotics and biotechnology.

Government regulation of such industry practices as clinical trials, quality controls, approval of new medicines, and the marketing and sale of drugs has often been criticized by the pharmaceutical industry as adversely affecting innovation. The regulations have increased the cost of developing new drugs, delayed their commercialization, and multiplied the risks associated with research investments. Despite these negative effects on innovation, regulation has also protected the companies and the public from the potentially disastrous consequences associated with the introduction of ineffective or harmful drugs.

Clustering: The Emergence of Successive Generations of Drugs

Since the Industrial Revolution each manufacturing sector has grown from a handful of radical innovations. Improvements in and diffusion of these basic innovations promoted growth, and in most cases that growth was sustained by the introduction of new generations of technologies that led to new products and to revolutionary changes in each industry's markets and structure. Most of these new generations began with the introduction of clusters of radical innovations. Clusters occurred in the 1800s through the 1820s (coal as fuel, the steam engine, railways, textile machinery, cheap cotton fabrics), in the 1870s through the 1880s (petroleum as fuel, the internal combustion engine, automobiles, tractors, cheap steel, electricity), in the 1930s through the 1940s (petrochemicals, plastics, synthetic fibers, cheap automobiles, electronics, commercial aircraft), and in the 1980s through the 1990s (microelectronics, computer hardware and software, biotechnology), with some of the radical innovations giving rise to completely new sectors. The pharmaceutical industry has experienced five clusters of radical innovations, four coinciding with the basic periods and a fifth in the 1960s as described in Table 1.

The clustering of radical innovations, the parallel diffusion of the technologies they generate, and their social and economic consequences have attracted the attention of historians of science and technology, economic historians, and political economists. Several convergent theories have been proposed to account for diverse aspects of this extremely important phenomenon, notably by J. D. Bernal (1954) on science in history and society; T. H. Kuhn (1962) on scientific revolutions; N. D. Kondratiev (1925) on long waves in the world economy; Joseph Schumpeter on "gales of creative destruction"; Nathan Rosenberg, R. Nelson, and G. Dosi on technological paradigms and trajectories; and C. Freeman and C. Perez on technoeconomic paradigms. Our interpretation of the dynamics of technological innovation in the pharmaceutical industry is in agreement with the main conclusions of these theories, but stresses the importance of the driving forces for technological innovation in causing the clustering of radical innovations as well as the parallel evolution and diffusion of the technologies they generate.

Clusters of radical innovations are formed by fluctuations in the intensities of the forces that drive technological innovation. Some of these forces are exceptionally strong during some periods and exert strong pressures on innovating institutions, producing numerous radical innovations, some of which have the commercial appeal, robustness, and versatility that lead to the initiation of new technologies. Among driving forces, those in which intensity can change abruptly—scientific technology, raw materials, and government legislation—are more frequently associated with the clustering of radical innovations.

For the pharmaceutical industry, which thrived on a vigorous interchange with medicine, chemistry, and the life sciences, scientific advance was always responsible for the clustering of radical innovations. Science and technology advance by quantum leaps, followed by periods of less adventuresome steps. These advances are touched off by revolutions of historical dimensions, such as the Chemical Revolution or biotechnology. But they are also sparked by the gradual accumulation of knowledge, which at certain moments in time makes possible the understanding of previously incomprehensible phenomena; by the invention of novel scientific instruments that allow observation and study of previously undetected natural and man-made phenomena; and by the interaction of disciplines, which opens new horizons for study and interpretation. Radical innovations that initiate new technologies are closely associated with the advancing frontier of scientific disciplines; consequently periods of revolutionary scientific advance create many more opportunities for radical innovations than do periods of evolutionary advance. The availability of new raw materials contributed to three of the five generations of radical innovations in drugs: tropical medicinal plants (1820s), organic chemicals derived from coal tar (1870s), and natural products (1930s–1940s). Last, government intervention contributed to the clustering of radical innovations in the 1870s with the creation of public medical research laboratories and in the 1940s with the development of the wartime technologies.

Technologies follow diffusion patterns that are determined by the qualities of the adopted models, by the inherent properties of each technology, and by the forces driving technological innovation, in particular market demand and competition, which become increasingly important at the later stages of diffusion. Thus many of the technologies introduced during the same period are likely to follow parallel diffusion patterns. That is, they are likely to cluster throughout their span because they are exposed to the same set of commercial and competitive driving forces.

The clustering of technologies within the pharmaceutical industry gave rise to successive generations of drugs that dominated the technological and commercial development of the industry. The coincidence of their stages of growth and maturity created waves of radical and especially incremental innovations, rapid growth of market demand, and a powerful bandwagon effect among manufacturers. The industry as a whole expanded and flourished. When one set of technologies reached the stage of decline, a new cluster of radical innovations was introduced, and the industry benefited from a new wave of innovation and commercial expansion. The transition from generation to generation required considerable skill and ingenuity on the part of individual pharmaceutical companies, which were forced to drop traditional products, technologies, and markets to develop new ones and adapt to new competitive settings. Those companies that failed to make these transitions lost their competitive edge and either merged with or were acquired by their more successful competitors.

CORPORATE TECHNOLOGY TRADITIONS

External driving forces are effective only to the extent that they generate internal forces within the institutions that foster innovation: private companies first and foremost and then academic institutions, government research laboratories, and public and private nonprofit research institutions. The research-intensive pharmaceutical firm is the most effective agency for technological innovation because it is sensitive to and able to respond to the stimuli of all the driving forces; it is especially sensitive to market demand.

Research-intensive companies respond to external stimuli by investing capital to develop new products and processes in anticipation of sizable returns from their commercialization. Making profits is the raison d'être of a capitalist enterprise, so however strong the external stimuli and management's conviction that research and innovation are essential for survival, pharmaceutical companies will not support for long research that does not produce profits. Companies may eventually drop a technology or retreat from an industrial sector unless a cash flow is generated from the commercialization of products and processes developed in-house. Corporate research and development is almost always funded at levels commensurate with the size of such cash flows.

Because radical innovations are much more frequently successful in the marketplace and therefore profitable, companies that introduce them can spend more on further research and development than can companies that introduce incremental innovations. The introduction of a radical innovation has another important consequence as well: Although it is a company's response to the stimulus of external driving forces, once launched a radical innovation intensifies rather than satisfies those stimuli. Because the introduction of a commercially successful radical innovation exerts an orienting effect on scientific and technological advance, strengthens market demand, and sharpens competition, the innovating company finds itself under strong pressures to continue to pursue innovation in the same technology and market after the radical innovation had been introduced. This is an important reason, but not the only one, that leads research-intensive companies to specialize in science, technology, and markets.

The strong, postinnovative commitment to additional research and development leads on the one hand to the concentration of the company's finite research resources on one or a few areas, depending on the company's budget and creativity, and on the other hand, to an accumulation of knowledge and experience, including markets and marketing. This expertise creates opportunities for incremental innovations to improve on the original innovation and even for additional radical innovations.

The drive for further innovation is prolonged not only by technical, scientific, and commercial opportunities, but also by the generation of strong corporate forces.

Scientific Leadership

The science underlying a radical innovation introduced at the early stage of a technology is usually close to the advancing frontier of the relevant scientific disciplines. Company researchers, encouraged by their success, deepen and expand their knowledge and often come to be recognized as authorities in their disciplines. This distinction is appreciated by both researchers and managers who project it as proof of the company's technological leadership over its competitors and who are eager to nurse and maintain that leadership for as long as possible.

Influence on Top Management

Innovators and managers associated with an original and lucrative innovation are usually promoted to positions of influence from which they steer the company's innovative efforts toward technologies and markets with which they are most familiar.

Time Lag

An average of eight to ten years elapses between the beginning of research and development and the commercialization of a radical innovation. Companies are thus eager to consolidate their position by introducing incremental innovations with improved properties before their competitors cover that lag and begin to work on their own products.

Familiarity with Markets

Early commercialization of a radical innovation requires expansion of existing markets or opening of new ones. The effort expended in either case leads to the familiarization and specialization of the company's sales and marketing departments in these markets. Feedback from doctors, hospitals, and patients often leads to incremental innovations that further expand the market.

Economies of Size and Scope

The cash flow created by a commercially successful radical innovation leads to further investments in related research and development, manufacturing capacity, and marketing and sales efforts, all of which create economies of size and scope. Subsequent research projects that can take advantage of accumulated knowledge, research facilities, and the company's familiarity with its markets can save on costs and lead to further specialization.

The "Invented Here" Syndrome

Companies show a strong tendency to pursue their own leads rather than those of their competitors. As a result a company's name often is identified with a particular technology and its market, a position that managements are keen to preserve.

The cumulative effect of these driving forces leads to the establishment of corporate technology traditions (CTTs), which act as powerful internal driving forces that strongly influence the innovating company's subsequent research, innovation, and markets. These traditions acquire their own momentum, so that over a twenty- to forty-year period the same technology forms the basis for a disproportionate number of innovations. Not all of these innovations are highly original, of course, but some succeed in the market and help perpetuate and reinforce the CTT. In many cases CTTs become so strong that companies become less sensitive to external driving forces. This tendency can be seen in companies that doggedly pursue work in a particular technology even as it declines and grows obsolete. The CTT in these cases has become so deeply embedded in the corporate psyche that it actually undermines the company's technological and market position.

Leading pharmaceutical companies have all initiated therapeutic technologies and created long-lasting CTTs; the largest and most successful companies have created CTTs in more than one therapeutic market. Table 2 presents some CTTs for several therapeutic markets.

COMPETITIVE ADVANTAGES OF NATIONAL INDUSTRIES

Despite the universal need and demand for medicines the pharmaceutical industry shows a high degree of concentration in a few countries, and the concentration of the innovating segment of the industry is even higher. Five countries—France, Germany, Switzerland, the United Kingdom, and the United States—have introduced about 80 percent of all the new drugs marketed worldwide in seven of the most important therapeutic markets (Table 3). The pharmaceutical industries of these five countries also accounted for 60 percent of the world's drug exports between 1975 and 1988. The next five most innovative countries (Belgium, Denmark, Italy, the Netherlands, and Sweden) accounted for another 19 percent (Table 4). It should be noted that Japan, which produces 22 percent of world output, accounted in 1988 for less than 3 percent of exports, largely because the Japanese companies, at least through the mid-1970s, introduced mainly imitative incremental innovations and confined their activities to their very large and protected domestic market. The strong relation between export markets and innovation record indicates that the competitive advantage of national pharmaceutical industries stems from their ability to develop and commercialize new drugs for the world market.

The historical development of the pharmaceutical industry shows that each country's environment helps to determine the intensity of the forces driving technological advance. The excellence and specialization of universities and public research institutions, the availability and kind of raw materials, the size of the home markets, the number of research-intensive companies and the competition among them, the perception of public needs and government intervention all are strongly dependent on each country's particular environment. Conditions favorable for technological innovation were created for individual national pharmaceutical industries during periods when most of the driving forces in a country happened to be exceptionally strong. This was, for example, the cause of Germany's predominance in the second generation and of the United States' leadership in the third generation of pharmaceuticals. Between 1880 and 1930 German universities were leading in chemistry and pharmacology. Bismarck's government passed an effective patent law, built city sanitation facilities, and established medical research laboratories. The dyestuff industry produced a wealth of new organic chemicals, some of which had medicinal properties and often were used as intermediates for the synthesis of new drugs. The German dyestuff companies were the only ones with in-house research capabilities and the German pharmaceutical industry that emerged had a near monopoly in the world market of "modern" drugs. Between 1930 and 1960 American universities led in natural products chemistry, physiological chemistry, and spectroscopic methods. The federal government supported research projects on penicillin, corticosteroids, antiprotozoal medicines, and the polio vaccine. The American market was by far the largest, while overseas markets were also open to American medicines. About fifteen R&D–intensive companies competed for the domestic and overseas markets.

Table 2. Some Corporate Technology Traditions and Their Characteristics

Company	Number of Innovations (Percentage of Total)*	RIs†	MSs‡	Period of Commitment			Main Technology (Number of Products)	Radical Innovation/Model
				First Product	Last Product	Total Years		
Central nervous system drugs								
Roche	19 (8.2%)	4	6	1950	1989	39	Benzodiazepines (9)	Librium (1960)
Janssen	15 (6.2%)	3	3	1958	1989	31	Butyrophenones (12)	Haldol (1958)
Abbott	11 (4.5%)	1	4	1922	1977	55	Barbiturates (6)	Neonal (1922)
Antibacterial agents								
Bayer	18 (8.0%)	4	6	1935	1986	51	Sulfonamides (11)	Prontosil (1935)
Pfizer	16 (7.2%)	6	6	1942	1986	44	Natural antibiotics (8)	Penicillin (1942)
							Tetracyclines (4)	Terramycin (1950)
Lilly	15 (6.8%)	8	7	1952	1983	31	Natural antibiotics (5)	Ilotycin (1952)
							Cephalosporins (9)	Kefspor (1964)
Beecham	15 (6.8%)	4	3	1959	1985	26	Semisynthetic penicillins (15)	Broxil (1959)
Bristol	15 (6.8%)	1	3	1956	1987	31	Semisynthetic penicillins (13)	Syncillin (1959)
Cardiovascular agents								
Ciba	21 (7.0%)	6	5	1941	1990	49	Antihypertensives (13)	Apresoline (1942)
Merck	16 (5.5%)	7	9	1955	1988	33	Antihypertensives (10)	Inversine (1955)
							Diuretics (6)	Diuril (1958)

Table 2. (continued)

Company	Number of Innovations (Percentage of Total)*	RIs†	MSs‡	Period of Commitment			Main Technology (Number of Products)	Radical Innovation/Model
				First Product	Last Product	Total Years		
Sex hormones/corticosteroids								
Syntex	19 (12%)	4	7	1942	1981	39	Sex hormones (11)	Progesterone from diosgenin process (1942)
							Corticosteroids (8)	Deoxycortone from diosgenin process (1947)
Schering AG	17 (11%)	7	6	1931	1977	46	Sex hormones (14)	Estrone (1931)
Ciba	15 (10%)	7	6	1913	1969	56	Sex hormones (5)	Ovarian extract (1913)
							Corticosteroids (9)	Deoxycortone (1939)
Upjohn	14 (8%)	2	3	1935	1976	41	Hormones (5)	Adrenocortical extract (ACE) (1935)
							Corticosteroids (9)	Cortisone from progesterone process (1952)
Analgesics/antipyretics								
Hoechst	11 (8.0%)	4	5	1882	1985	103	Antipyretics/Analgesics (11)	Antipyrin (1884)
Janssen	7 (5.0%)	1	2	1958	1984	26	Pethidine-related analgesics (7)	Palfium (1958)

* Number of innovations as percentage of all the new drugs of the corresponding therapeutic sector.

† RI: radical innovation.

‡ MS: commercially successful innovation.

Table 3. Relative Strength in Innovation of National Pharmaceutical Industries (1820–1990)

Sector	United States	Germany	Switzerland	United Kingdom	France	Japan	Other	Total
Antiprotozoal agents	34	15	12	12	9	—	6	88
Antibacterial agents	111	35	14	35	17	26	9	247
Central nervous system medicines	89	28	42	19	23	12	34	247
Antihistamines	27	5	12	6	11	1	8	70
Cardiovascular agents	121	48	37	27	23	3	28	287
Hormones/ corticosteroids	88	18	22	14	8	—	25	175
Analgesics/anti-inflammatories	45	27	16	11	12	4	16	131
Total	**515**	**176**	**155**	**124**	**103**	**46**	**126**	**1,245**
Percentage	**41**	**14**	**13**	**10**	**8**	**4**	**10**	

The influence of national environments on driving forces for technological innovation was particularly strong from 1800 to about 1950, when companies operated strictly within a national framework of rules and economic policies. (That was true even for Germany, which had significant export markets before World War I.) The small size of both manufacturers and markets in the 1800s, the upheaval created by World War I, the economic recession of the 1920s, the Great Depression of the early 1930s, and the onset of World War II forced governments to protect their manufacturing industries by raising barriers to international trade. Moreover, technology was rarely transferred across borders because companies seldom licensed their innovative products and processes to third parties, while the unstable political situation limited exchanges between scientists and engineers of various countries, thus slowing the diffusion of scientific advance.

It was during this period of relative national insularity that most of today's leading pharmaceutical companies introduced their first radical innovations, became research-intensive, and initiated their first CTTs by means of which they were able to prolong their technological and commercial prominence. The geography of technological advance in the pharmaceutical industry was largely determined in this early period.

From 1950 to 1980 governments still protected their domestic markets and encouraged their home industries, supported scientific research and began funding national health-insurance systems. But the influence of the national environment on the forces driving innovation was significantly reduced by the Bretton Woods Agreement, the Marshall Plan, multilateral trade agreements, and the European Common Market, all of which fostered economic growth, liberalized world trade, and created a new competitive environment based on international rather than national rules. These measures enabled the large research-intensive pharmaceutical companies to invest overseas and to become multinational by spreading their marketing and sales organizations and manufacturing facilities across national frontiers to tap a largely homogenized world market. Although most companies kept their research departments in their home countries, technology transfer accelerated dramatically because of the multinational character of the large companies, the widespread licensing of innovative technologies, and the easy movement of scientists and engineers across national borders.

From 1980 onward the influence of national environments on the pharmaceutical industry all but disappeared as a result of the globalization of the operations and markets of the large companies by means of overseas investments and cross-national mergers. The new competitive environment is shaped by the globalization of financial markets, the strengthening of international trade agreements, the expansion of the European Union into eastern Europe and beyond, the dislocation of manufacturing sectors with mature technologies going to developing countries, and the large investments in the high-technology industries of the United States, Japan, and Europe. Only academic institutions have kept some of their national characteristics through scientific specialization, as the prominence of the United States in biotechnology illustrates. But multinational companies have realized the importance of tapping scientific knowledge wherever it is produced and have opened research departments in foreign countries and forged ties with foreign universities. As a result national characteristics have lost much of their significance. Today's pharmaceutical industry has come to depend on the technological and commercial performance of companies that are essentially global.

Table 4. Value of Drug Exports of Most Innovative Pharmaceutical Industries

Country	1975	1980	1988	Percentage of Exports
	(in millions of current U.S. dollars)			
Germany	1,010	2,148	3,996	15.1
United States	725	1,772	3,541	13.4
Switzerland	860	1,629	3,172	12.0
United Kingdom	744	1,539	2,793	10.5
France	573	1,387	2,345	8.8
Italy	384	701	1,282	4.8
Netherlands	332	589	1,038	4.0
Belgium	303	640	1,026	3.9
Denmark	145	315	839	3.2
Sweden	112	309	829	3.1
Hungary	214	322	739	2.8
Japan	124	296	721	2.7

Data from R. Ballance, J. Pegany, H. Forstner, *The World's Pharmaceutical Industries* (Aldershot, Hampshire, U.K.: Edward Elgar, 1992).

Origins of an Industry (1820–1880)

The modern pharmaceutical industry had its origins in two revolutions of the eighteenth and nineteenth centuries—the intellectual revolution, which led to the development of the modern physical and life sciences, including chemistry, physiology, bacteriology, and pharmacology; and the Industrial Revolution, which catalyzed the technology that enabled scientists and engineers to unlock and harness the mysteries of the natural world. The pharmaceutical innovators during this early period were academic researchers and physicians, primarily in Europe, who began to discover and explain the physiological processes of the human body and the nature of disease. It was these early researchers who made the first major breakthroughs in therapeutic drugs: the isolation of the "active principles" of certain plants and the discovery of medicinal properties in some simple organic chemicals. It was not until the 1880s, however, when the academic researchers and physicans began working with the chemical companies, that a second generation of drugs was born and the modern pharmaceutical industry began to take shape.

The Industrial Revolution fostered much of the technology that would eventually lead to the manufacture of drugs. But it also created the unsanitary conditions that forced governments to play a leading role in improving and protecting public health, in stimulating scientists to investigate the causes of illnesses and find remedies for them, and in giving companies economic incentives to accelerate the pace of scientific advance by investing in research of their own and applying their findings to the manufacture and commercialization of drugs.

What might be conceived as the first generation of modern drugs stretched from about 1820 to 1880 and encompassed these early scientific advances. The discoveries that natural products had "active principles" of therapeutic value that could be manipulated to produce effective medicines and that some simple organic chemicals derived from coal tar also had medicinal properties were probably the two findings of greatest significance for the emerging pharmaceutical industry. The second generation of drugs, which ran from 1880 to 1930, saw not only the development of a range of new drug therapies, but the successful commercialization of these products both in Europe and the United States. The most innovative new pharmaceutical companies relied on academic researchers and on their own scientists and engineers to develop new products and processes for manufacturing them.

ISOLATION OF "ACTIVE PRINCIPLES"

Before the first modern drugs were discovered in the early 1800s—and for several decades afterward—most medicines were natural products of plant origin, organic and inorganic salts, formulated into syrups, liquid extracts, and powders of uneven dosage and often of uncertain effectiveness. Although physicians knew that certain medicines cured or eased certain maladies, little was known about how these medicines actually worked.

The first person to argue that certain substances contained an active principle was Paracelsus (1493–1541), who searched for the "arcanum," the healing essence within all medicinal preparations irrespective of their origin. Paracelsus was also the first to promote and to use chemicals as medicines. He achieved notable results by using mercury salts orally to treat syphilis. Until the 1800s mercury salts was one of the few known medicines that was truly effective, along with cinchona bark (quinine) for treating malaria, tartar emetic and ipecacuanha (emetine) for treating dysentery, and digitalis for treating cardiac edema.

Not until the development of modern chemistry three centuries later, however, was Paracelsus's theory of active principles proved correct. Of all the physical sciences chemistry was considered to be an "applied science" and became intimately associated with agriculture, industry, and medicine. The first experimental studies in chemistry were carried out in the mid-1700s by the British chemists Joseph Priestley (1733–1804), Joseph Black (1728–1799), and Henry Cavendish (1731–1810), and the Swede Karl Wilhelm Scheele (1742–1786). Inspired by the steam engine, they studied combustion and the chemistry of gases. Priestley and Scheele independently isolated oxygen, Black discovered carbon dioxide, and Cavendish isolated hydrogen. All four scientists promoted the theory of "phlogiston," a hypothetical substance that they thought caused the production of heat when it was liberated during combustion. As was usually the case, new materials—gases in this instance—were tested as possible cures for illnesses, and patients suffering from a score of diseases were made to inhale oxygen or carbon dioxide. The only valuable result of these efforts was Humphry Davy's (1778 1829) discovery of the euphoriant properties of nitrous oxide (1793), which, fifty years later, was found to be an excellent anesthetic.

It was the French chemist Antoine Lavoisier (1743–1794), the founder of modern chemistry, who discovered the role of oxygen and carbon dioxide in combustion and thus disproved the phlogiston theory. Then, together with Armand Seguin, Lavoisier applied his findings in physiology and elucidated the role of these two gases in respiration and the generation of body heat. Lavoisier was the most illustrious member of a school of famous French chemists that dominated chemistry in the late 1700s and early 1800s. The French School emphasized experimental work, including both qualitative and quantitative analysis, and had a strong bent toward the practical application of its findings. Jean Baptiste Dumas wrote *Traité de Chimie Appliquée aux Arts* (*A treatise on the application of chemistry in manufacturing*); Jean Antoine Chaptal, who eventually became minister of the interior under Bonaparte, wrote the first book on industrial chemistry; Nicolas Leblanc, a physician turned chemist, introduced the process for manufacturing soda ash (sodium carbonate) that bears his name; Claude-Louis Berthollet discovered the bleaching and disinfectant properties of chlorine, and Antoine Fourcroy, a medical student turned chemist, was the first to use chemical methods to determine the composition of mineral waters and plants.

In 1793 the French revolutionary National Convention decided to suppress the academic and professional institutions, including the Académie des Sciences, that had enjoyed privilege under the monarchy and to establish new institutions of higher education throughout France. The new institutions for medicinal research in France were the Société de Pharmacie and the École Superieur de Pharmacie (1803). The latter's first director, Nicolas Vaquelin, an analytical chemist and a political figure of the new regime, encouraged the application of chemical methods in pharmacy and focused research on isolating the active principles of medicinal plants. This research was badly needed. Not only were medicines unpalatable, but because they were dispensed as powdered dried leaves, syrups, and other concoctions, the concentrations of the active ingredients varied from dosage to dosage, making their therapeutic effects difficult to predict. Moreover, many of these medicines were toxic when taken in stronger doses, while some, such as digitalis, were effective only at near-fatal doses.

DISCOVERY OF MORPHINE

The great discovery that launched the first generation of drugs was the isolation in 1806 of pure morphine from opium by the twenty-year-old German apothecary Friedrich Wilhelm Sertürner. Sertürner was the first researcher to use chemical methods to analyze an active principle, in this case the alkaloid morphine, and to use "pharmacological" tests on his dogs to determine its properties. Not being an academic, he buried his findings in an obscure journal where they lay unnoticed for ten years. The French chemist and physicist Joseph Gay-Lussac (1778–1850) discovered them, recognized their importance, and republished Sertürner's paper in 1816 in *Annales de Chimie*. He also advised French chemists to pursue Sertürner's lead and to search for alkaloids in other plants. The results were dramatic: Between 1820 and 1840 hundreds of alkaloids and other chemicals of plant origin were isolated and evaluated for medicinal use.

Next to the discovery of morphine the most important discovery in alkaloids was the isolation in 1820 of pure quinine from cinchona bark by Pierre Joseph Pelletier and Joseph Caventou. Quinine was particularly important because it was used to treat not only malaria but also other common illnesses that caused fevers.

THE DEVELOPMENT OF PHARMACOLOGY

The isolation of pure alkaloids and other medicinal chemicals present in natural products gave physicians and physiologists the opportunity to examine systematically their physiological effects using accurately measured quantities, a process that eventually led to the establishment of pharmacology as a separate scientific discipline. The beginnings of pharmacology can be traced to François Magendie and his erstwhile student Claude Bernard; both men were professors of physiology at the École de France and made their historic discoveries in that discipline. Convinced of the importance of the experimental method and influenced by the advances in organic chemistry, they extended their work to studying the physiological effects of certain alkaloids. Their teachings and research thus helped identify pharmacology as a distinct—if not an independent—discipline.

Magendie was the first physiologist to use pure alkaloids to treat disease. Together with Pelletier he worked with ipecacuanha, which was used to treat dysentery, eventually isolating a crystalline mixture of alkaloids that caused vomiting. The two men also conducted a systematic quantitative study of the dosages of this mixture, which they called emetine. In 1821 Magendie published a formulary for the preparation and use of numerous medicines that was translated into many languages and became a classic. Bernard carried out the historic study of the effect of curare on the nervous system, which led to the discovery of the independence of the motor and the sensory nerves. He came to regard pharmacology as a potent research tool for studies in physiology, and by advocating this cross-fertilization, he helped to advance both disciplines.

Just as Justus von Liebig was instrumental in relocating the center of chemistry from France to Germany, Rudolf Buchheim and Oswald Schmiedeberg did the same for pharmacology. Buchheim was a physiologist, who in 1846 was offered the first chair of pharmacology ever established, at the University of Dorpat (now Tartu in Estonia). Buchheim created the discipline of experimental pharmacology and gained its recognition as an important branch of medicine. In 1860 he organized the first research laboratory in pharmacology. In 1869 he and Oscar Liebreich discovered the hypnotic properties of chloral hydrate. As early as 1856 he published his *Lehrbuch der Arzneimittellehre* (Textbook of pharmacology), which put forth a new classification system of medicines.

Schmiedeberg, a physician who studied under Buchheim and succeeded him in the pharmacology chair at Dorpat, is generally recognized as the father of modern pharmacology. Schmiedeberg defined pharmacology as "an independent and purely biological science concerned with the action of pharmacological agents regardless of their practical importance. Pharmacology should investigate the effects of chemically reactive substances under physiological conditions, but the results might prove to be of toxicological, therapeutic, or purely physiological importance." In 1872 he moved to the University of Strassburg, where he established and led the Pharmacological Institute in which most of the outstanding researchers of the following fifty years trained. Many of these men became professors in newly established chairs of pharmacology and spread their teacher's influence throughout the medical world. His research covered the whole range of medicines and many metabolic processes, and his textbook, *Grundriss der Arzneimittellehre* (Outline of pharmacology), published in 1883, became a classic.

INFLUENCE OF THE INDUSTRIAL REVOLUTION

The beginnings of modern pharmaceuticals were set against the backdrop of the Industrial Revolution, with the devastating illnesses that industrialization was causing. During the first half of the nineteenth century public health was deteriorating in the fast-growing industrial cities of northern Europe and the United States to which large segments of the rural population and immigrants from poor countries were moving in waves. Narrow streets, multistory buildings, and lack of running water and sanitation facilities were the common characteristics of working-class boroughs. Factories in which workers spent twelve hours or more daily were equally filthy and overcrowded. Poverty, malnutrition, and overcrowding in badly aerated space caused illness, death, and the spread of contagious diseases, which were blamed on the "bad air" of the dwellings and of open spaces, such as swamps. The most common endemic diseases of the nineteenth century were fevers and dysenteries, including malaria, typhus, typhoid fever, relapsing fever, and tuberculosis.

The conditions in hospitals, prisons, and asylums were even worse. The oldest "modern" hospitals in Europe and the United States were built in the eighteenth century, had no running water and hardly any sanitation facilities, and were themselves sources of infection, sepsis, and gangrene. Mortality among surgical patients in hospitals was four times higher than among those operated on privately.

Some improvements in personal and public hygiene practices began to take hold in the first half of the nineteenth century; these included bathing, use of soap and water, and fumigation and ventilation of living spaces and public buildings. Medical advice reduced mortality among infants and children. For example, mortality during childbirth at a Vienna hospital fell from 12 percent to 1 percent in 1847 after doctors and midwives followed Ignaz Semmelweis's advice to wash their hands before delivering babies.

Other improvements resulted from industrialization itself. Increased yields from agriculture and better transport made food more plentiful, varied, and above all, more certain, and the availability of cheap, washable cotton clothing dramatically improved personal cleanliness. Leprosy and plague were virtually eradicated from western Europe by segregating the sick and imposing strict quarantines at ports of entry, while the incidence of malaria was diminished when marshes were drained in land-reclamation efforts. (It would be several decades before scientists discovered that malaria was transmitted by mosquitoes and not "bad air.")

Compared with endemic illnesses, epidemic contagious diseases produced a stronger stir among the population because they came suddenly and forcefully and killed not only the very young, the weak, and the old, but also middle-aged working people, creating more personal, social, and economic hardship for their families. Typhus killed 12 percent of the Irish population between 1816 and 1819. A very serious epidemic of influenza occurred in 1833 in Europe, while in the United States the disease was nearly endemic. Measles and whooping cough became increasingly more serious, even lethal. Diphtheria, which was unknown in Europe before 1860, became pandemic and the leading killer of infants and children.

The most catastrophic outbreak was a cholera epidemic that struck both Europe and America between 1830 and 1832. It was this epidemic that stirred governments to take steps to improve public health. Authorities in Britain, France, Germany, and the United States realized that the contagion had struck more forcefully in the cities than in the countryside and had particularly affected the poor districts of the cities. Although the medium by which cholera is transmitted—contaminated water—had not yet been identified, the statistical evidence indicated that filth and overcrowding played at least an indirect role in the spread of the disease. Rudolph Virchow, a prominent German physician influential in formulating German public health policy, drew a similar conclusion on a visit to Silesia in 1848, where, after a serious outbreak of typhus, he saw the unsanitary conditions in which textile workers lived.

The movement to improve public health became entangled with contemporary movements to improve living and working conditions for industrial workers and eventually led to the passage of the first modern legislation on public health. The British Public Health Act of 1848 created a central authority, the General Board of Health, for this purpose. Most industrialized countries passed similar laws and established similar public health authorities.

ADVANCES IN THE LIFE SCIENCES

All of these initial government actions focused on improving the environment in which people lived. They did not address individual diseases. The search for the causes and cures of the illnesses was left to doctors and scientists. Many of the revolutionary advances made in the physical and biological sciences as well as in medicine came after the development in the early 1800s of the achromatic microscope, which allowed researchers to see smaller units than ever before. (One of the major contributors in the evolution of this microscope was Joseph Jackson Lister, father of Joseph Lister, who later in the century would discover the importance of antiseptics in surgery.) The revelations made possible by the achromatic microscope led to a fundamental change in thinking that has affected medical research and practice ever since: the conception of the living body as a vast organization of millions of cells.

Virchow was one of the initial and leading exponents of this theory. In his *Die Zelluläre Pathologie* (Cellular pathology), published in 1858, he argued that all cells arise from other cells and that every diseased tissue consists of cells that are the offspring of preexisting cells. This doctrine proved extraordinarily fertile because it made the microscope an essential piece of equipment for every physician and helped not only in the diagnosis but also in the discovery of the fundamental nature of disease.

The French chemist Louis Pasteur (1822–1895) was the first to propound the theory that infections were caused by germs. He found that fermentation depended on the presence of "living globules" (microorganisms) that were specific to different types of fermentation and that were carried in the air in the form of very robust particles (spores). These microorganisms could be excluded from the air by filtration and from liquids by heating. These observations led Pasteur to hypothesize that putrefaction was a type of fermentation caused by the presence of such microorganisms and not, as previous theories had claimed, that they were spontaneously generated from dead materials under suitable circumstances.

Joseph Lister, a Scottish surgeon appalled by the high mortality of patients recuperating in hospitals after otherwise successful surgery, decided to test Pasteur's hypothesis and sought materials that would protect healing wounds from contact with air. Following a lead from France, he applied a mixture of phenol (carbolic acid) and plaster to cover wounds and found that healing proceeded without further complications. Thus Lister proved that Pasteur's hypothesis was correct. He and other researchers subsequently pursued the search for other organic chemicals with antiseptic properties, which, like phenol, were produced by the emerging coal-tar industry.

Drugs from Coal Tar

The discovery that some simple organic chemicals derived or synthesized from coal tar possessed medicinal properties would become, like the discovery of "active principles," one of the building blocks of the pharmaceutical industry. Coal tar is the waste material left when coal is made into coke. First produced in 1620, coke gradually replaced charcoal for all its industrial uses, including the manufacture of cast iron. Even more coal tar was produced after 1792, when coal gas was initially trapped and used for illumination, first in London and later in other European cities. By the early 1820s coking operations had expanded so much that the accumulating coal tar was beginning to pose a serious environmental problem, and people began searching for ways to use the substance rather than simply discarding it. In Germany, for example, coal tar was used to waterproof roofs.

Two Germans working independently of each other began to unlock the medicinal potential of organic chemicals derived from coal tar. In 1828 Friedrich Wöhler, working at the University of Göttingen, produced urea, a known constituent of urine, by the evaporation of ammonium cyanate, an inorganic compound. Wöhler's achievement laid to rest the theory that living organisms produced organic chemicals through the agency of a "vital force" and thus could not be synthesized in the university or industrial laboratory. A few years later, in 1832, Friedlib Runge, a physician turned chemist, was appointed chief chemist at a chemical plant in Oranienburg, a small town outside Berlin that received the coal-tar residues of that city's coal gasworks. Runge, who had isolated the alkaloid caffeine while still a student at the University of Jena, decided to try to isolate useful substances from coal tar. Within a few years he had used fractional distillation to isolate several of the chemicals, such as methanol, aniline, and carbolic acid, that would become the building blocks for the chemical, dyestuff, and pharmaceutical industries; his work drew the attention of academic and industrial researchers to this subject. Together, Wöhler's academic and Runge's industrial discoveries inspired the search for organic chemicals with medicinal properties among the thousands that were isolated from coal tar or synthesized from its distillation products. The most important discoveries in the first half of the nineteenth century, however, were the discoveries of the anesthetic properties of ether by the American physician Crawford Long and the use of chloroform as a hypnotic by James Young Simpson, professor of midwifery at the University of Edinburgh.

Discovery of Synthetic Dyes

The next significant advances in drugs were dependent on advances in industrial organic chemistry, which were inextricably woven in with the development of the dyestuffs industry in the second half of the nineteenth century. That began in 1856 with the serendipitous discovery of mauvein, the first synthetic dye, by William Henry Perkin, who was trying to synthesize quinine from aniline. The immense commercial success of this synthetic dye led to the introduction of numerous aniline dyes and the establishment of many companies near London and in the textile-producing areas of Great Britain. The synthesis of new dyes called for further advances in organic chemistry because the molecules were structurally complex and little was known about their composition and chemistry. The simultaneous advance of science and technology required cooperation between academic institutions and industrial enterprises, which proved most successful in Germany where the predominantly German entrepreneurs of the British companies returned after the reunification of the German states in 1870, establishing the great dyestuffs-chemical companies, such as BASF, Hoechst, Bayer, and Agfa. Chemical companies soon found it necessary to establish their own research laboratories and to employ graduate scientists and engineers who could cooperate with academic researchers to develop new dyes that could be patented.

The advance of industrial organic chemistry, particularly after the discovery of the synthetic dyes, was pivotal in the establishment of the pharmaceutical industry for many reasons. First, it made available many pure organic chemicals from coal tar, fermentations, and cellulose, some of which were found to be useful in medicine. These included chloral hydrate and amyl alcohol as anesthetics or hypnotics; amyl nitrite and nitroglycerin for treating angina pectoris; and phenacetin, salicylate, and their derivatives as analgesics and antipyretics. Second, many of the synthetic dyes were used to stain bacteria and other microorganisms, allowing scientists to characterize them and study their life cycles. Finally, organic chemistry led to the development of laboratory methods for synthesizing both simple molecules, such as solvents, and complex ones, such as dyestuffs; for analyzing chemicals qualitatively, quantitatively, and structurally; and for investigating the relations between chemical structure and physicochemical properties.

Early Vaccines and Serums

Even as academic researchers and chemical companies were beginning to explore the medicinal properties of organic chemicals, other researchers began to collaborate with public health authorities to produce vaccines and serums against some of the more virulent diseases of the time.

Since ancient times it had been recognized that survivors of contagious diseases became immune to further attack. Inoculation against smallpox was introduced in western Europe in the seventeenth century, but the method used at the time frequently led to illness and death rather than to immunity. In 1798 Edward Jenner noted that milkmaids who had contracted cowpox, a mild form of smallpox, became immune to the lethal strain. He developed a cowpox vaccine, which virtually eradicated the contagious disease, although no one understood for several decades precisely how vaccination worked.

In 1838 Casimir Joseph Davaine, a French pathologist, observed under a microscope rodlike microorganisms in the blood of animals that had died from anthrax. In 1863 he found that the virulence of the disease depended on the concentration of these microorganisms, which he named "bacteria." Inspired by Pasteur's germ hypothesis, Davaine began to inoculate rabbits with contaminated blood in an effort to reproduce the disease and prove that these microorganisms were its cause. Davaine was thus the first to identify pathogenic microorganisms of a specific disease.

From the 1870s onward Pasteur, Robert Koch, and their students identified numerous pathogenic microorganisms, using microscopes and the newly discovered synthetic dyestuffs for staining, a method introduced by Carl Weigert in 1871 and perfected by his cousin Paul Ehrlich. Techniques were developed for studying the life cycle of bacteria, producing them in liquid or solid media, reproducing the diseases in animals to ascertain the relation between specific bacteria and diseases, and attenuating the virulence of pathogenic bacteria before using them for vaccination. Table 5 lists the major pathogens discovered during this period.

The innovations of the 1880s were Pasteur's development of effective vaccines to prevent anthrax and smallpox (1883) and to treat rabies (1885) and Emil Adolf von Behring's development of the diphtheria serum antitoxin (1891),

Table 5. Discoveries of Major Pathogens

Pathogen	Disease	Year
Borrelia	Relapsing fever	1873
Mycobacterium leprae	Leprosy	1874
Entamoeba histolytica	Dysentery, amoebiasis	1875
Bacillus anthracis	Anthrax	1876
Neisseria gonorrhoeae	Gonorrhea	1879
Trypanosoma brucei	Sleeping sickness	1880
Salmonella typhi	Typhus	1880
Streptococcus pneumoniae	Pneumonia	1880
Staphylococcus pyogenes	Septicemia	1880
Mycobacterium tuberculosis	Tuberculosis	1882
Vibrio cholerae	Cholera	1883
Klebsiella pneumoniae	Pneumonia	1883
Clostridium tetani	Tetanus	1884
Brucella melitensis	Brucellosis	1887
Neisseria meningitidis	Meningitis	1887
Babesia eigena	Texas cattle fever	1889
Salmonella paratyphi	Paratyphus	1893
Pasteurella pestis	Bubonic plague	1894
Clostridium botulinum	Botulism	1896
Shigella dysenteriae	Dysentery	1897
Leishmania donovani	Leishmaniosis	1900
Trypanosoma pallidum	Syphilis	1905
Bordetella pertussis	Whooping cough	1906

which was used both as a preventive vaccine and as a treatment for the disease in its early stages. These advances were made by university and public research laboratories, and institutions were needed to manufacture them in quantities sufficient to meet the great demand.

The Pasteur Institute, established by public subscription in Paris in 1888, and the Lister Institute, established in London in 1891, were among the first to manufacture the diphtheria antitoxin. The New York State Board of Health was the first to produce the antitoxin outside of Europe (1894), followed by the U.S. Marine Hospital Service in Washington, D.C., and the city health departments of Philadelphia, Boston, Buffalo, Rochester, Pittsburgh, and St. Louis.

Several public health authorities soon went a step further and began to develop and produce vaccines and serums for other diseases. Thus, the Imperial Health Office in Berlin set up the Research Institute for Public Hygiene and later the Berlin Institute for Infectious Diseases, which was initially headed by Koch, and the Prussian State Institute for the Investigation and Control of Serums, led by Paul Ehrlich beginning in 1896. The Kitasato Institute in Tokyo, the Rockefeller Institute in New York, and the Frankfurt Institute for Experimental Therapy were also set up during this period.

The public sector's involvement with medicinal research came at a time when few private companies were in a position to undertake the task. Public

Table 6. Medicinal Innovations by Nationality of the Innovating Institution (1800–1930)

Period	Germany		France		Switzerland	
	Academic	Industrial	Academic	Industrial	Academic	Industrial
Up to 1800	—	—	1	—	2	—
1801–1820	2	—	9	—	—	—
1821–1840	3	—	2	—	—	—
1841–1860	2	1	1	—	—	—
1861–1880	1	1	1	—	1	—
1881–1890	8	9	4	—	1	—
1891–1900	2	9	—	—	—	2
1901–1910	2	5	1	1	—	—
1911–1920	0	11	—	1	—	2
1921–1930	0	9	1	1	—	2
Totals	20	45	20	3	4	6
	65 (44%)		23 (16%)		10 (7%)	

and nonprofit research laboratories continue to contribute handsomely to medical and medicinal research in cooperation with universities and research-intensive pharmaceutical companies.

The Emergence of the Modern Pharmaceutical Industry (1880–1930)

For most of the nineteenth century medicines were dispensed by apothecaries who traded in plant and animal extracts and inorganic salts. Because they were natural products or simple chemicals, these substances were generally not patentable, so technological competition was restricted to the formulation of more palatable forms and the standardization in content of the active principles. These companies did not employ scientists who were able to follow the leads of academic researchers and therefore did not contribute much to the discovery of the first generation of drugs. That was essentially the work of academic researchers and physicians.

In 1826 the two discoverers of quinine, Pierre Joseph Pelletier and Joseph Caventou, established the first "modern" entrepreneurial pharmaceutical company, producing pure quinine from imported cinchona bark. In 1848 Georg Merck discovered the alkaloid papaverine. This led him to transform his family's seventeenth-century Angel Apothecary in Darmstadt, established by Emmanuel Merck, into the manufacturing apothecary E. Merck, which produced alkaloids and medicinal chemicals. As the rather simple technology diffused, some apothecaries-manufacturers began to produce alkaloids and other "active principles" of plant origin, extending their long lists of available powders, pills, and syrups.

United Kingdom		United States		Other		Total for
Academic	Industrial	Academic	Industrial	Academic	Industrial	Period
3	—	—	—	—	—	6
—	—	—	—	—	—	11
2	—	—	—	—	—	7
3	—	2	—	—	—	9
5	—	—	—	—	—	9
—	—	—	—	1	—	24
—	1	—	1	—	—	19
—	1	—	1	—	—	13
2	—	3	2	2	—	20
—	1	1	11	3	—	30
15	3	6	15	6	—	148
18		21		6		148
(12%)		(14%)		(4%)		100%

During much of the nineteenth century chemical companies dealt with the manufacture of inorganic chemicals in bulk for the textile, soap, and glass industries. Commercial applications for organic chemicals produced from coal tar had not yet been found. Even Runge's company did not take advantage of his many discoveries.

The 1880s were a watershed in the development of the pharmaceutical industry—the period when collaboration among the chemical companies, academic scientists, and public health institutes led to the second generation of drugs and to the formation of the research-intensive pharmaceutical company. Until World War I the most innovative companies were in Germany; a few companies in Great Britain, Switzerland, and the United States were also notable innovators (e.g., Burroughs Wellcome, Roche, Ciba, and Parke-Davis). The French company Rhône-Poulenc did not enter the pharmaceutical market until 1920. And the American companies, which would dominate the worldwide industry after World War II, did not become a force in the pharmaceutical market until after World War I.

Table 6 shows the distribution of 148 major innovations among national industries, academia, and industry from 1800 to 1930. Up to 1880 nearly all the innovations were introduced by academics and physicians, but after 1880 the contribution of industrial companies gradually became more important, especially in Germany after 1880 and in the United States after World War I. French researchers discovered most of the earliest medicines, but Germany soon became the leading innovator, introducing 65 (44 percent) of the 148 medicinal innovations in the 130-year period. The five leading national industries in that early period—Germany, France, Switzerland, the United

Kingdom, and the United States—remain to this day the most innovative among national pharmaceutical industries despite changes in their order of importance.

THE GERMAN INDUSTRY

In Germany innovation in pharmaceuticals was dominated by Bayer and Hoechst, both research-intensive chemical companies established in the 1860s and primarily engaged in the synthesis of dyestuffs. For the German dyestuffs industry the historic development occurred in the 1870s, with the synthesis of alizarin, the natural coloring substance of madder, introduced simultaneously in 1870 by Perkin in England, Carl Graebbe of the Berlin Polytechnic, and Carl Lieberman of BASF. Ten years later Adolf von Baeyer at the University of Munich synthesized the coloring matter of indigo, the most important of the natural dyes. Although it took BASF and Hoechst fifteen years of research to develop a commercial process, the trend in replacing natural products by synthesizing them from coal-tar chemicals was set.

By the early 1880s the German companies had left behind the British and French and were well on their way to building their near-monopoly of dyestuffs in the world market. They had forged close relationships with university professors and their students and had established in-house research and development laboratories. As success followed success, the companies came to depend on their ability to innovate—and it was this ability, together with the relations they had forged with the academic community, that helped them to create the modern pharmaceutical industry.

Hoechst

The first company to develop synthetic drugs was Farbwerke Hoechst, which had been established in 1862 as a manufacturer of aniline dyes. Eugen Lucius, a chemist and one of the company's founders, saw a commercial opportunity in synthesizing quinine, the most effective antipyretic (fever reducer) known at the time. Like natural dyestuffs, quinine was imported from the Tropics, but Germany lacked a significant colonial presence in tropical regions of the world and so had no easy access to quinine. Lucius was not the first to think about synthesizing quinine. In 1856 August Wilhelm Hofmann and William Perkin attempted to make quinine from aniline, but their experiments led them instead to mauvein and synthetic dyestuffs.

Like most alkaloids, quinine is a large and complex molecule; in the mid-1800s its structure was only partially known, which made attempts at its synthesis haphazard and virtually doomed to failure. To overcome this difficulty, chemists turned to breaking the quinine molecule into smaller fragments, reasoning that synthesis would be easier if any of the fragments were found to contain antipyretic properties. Wilhelm Koenig and Otto Fischer at the University of Munich synthesized a variety of such products, which were evaluated by Wilhelm Filehne, a Strassburg graduate who had been appointed professor of pharmacology at the University of Erlangen. Filehne found that two of these products possessed antipyretic properties, and Hoechst put them on the market in 1882. But they proved toxic to humans and were immediately abandoned.

Ludwig Knorr, an assistant to Emil Fischer, professor of chemistry at the University of Erlangen, applied a novel method of synthesis and, in cooperation with Filehne, produced phenazone, which was found to be an effective antipyretic far more palatable than quinine. Knorr granted the patent rights to Hoechst, which marketed phenazone in 1884 under the name Antipyrin. Only after the drug reached the market did Knorr discover that he had mistaken its structure and that phenazone was unrelated to any quinine fragment. For the next twelve years, until aspirin was introduced in 1899, Antipyrin was the most widely sold drug in the world. At the turn of the century Hoechst was producing 14,000 kilograms of it annually. Antipyrin, perhaps more than any other drug, proved that the development of synthetic drugs could be a profitable business and led Hoechst and its competitors to intensify their efforts in this field. Antipyrin also initiated a CTT in antipyretics and analgesics at Hoechst that lasted until the 1940s. The collaboration among Knorr, Filehne, and Hoechst also lasted for several more years. In 1896 they produced a new antipyretic, aminopyrine (Pyramidon), which was three times more potent than Antipyrin and was the best-selling drug in Europe up to 1934.

At about the same time that Antipyrin was being developed, chemists at the University of Strassburg were conducting clinical trials on the use of naphthalene as an antiseptic and vermicide. One of the investigators accidentally pulled the wrong bottle from the shelf and, instead of naphthalene, administered acetanilide, a simple chemical made from aniline. The group noticed an unexpected drop in the patient's fever, traced this phenomenon back to the error, and after more experiments published a paper describing the discovery of a new antipyretic. In 1886 a small chemical company, Kalle and Company, marketed it under the trade name Antifebrin. The new antipyretic soon caught the attention of the larger dyestuffs companies because, unlike the chemically complex alkaloids, acetanilide was a very simple compound made from coal-tar chemicals. Kalle grew considerably in size, primarily because of the commercial success of Antifebrin, and in 1908 it was bought by Hoechst.

During this same period Hoechst also became the first commercial manufacturer of vaccines, serums, and toxins (biologicals). That aspect of its business began in 1883 when the company named August Laubenheimer its research director. Laubenheimer was particularly interested in bacteriology, and in 1891 he invited the prominent physicians Emil Adolf von Behring and Paul Ehrlich to develop for Hoechst a commercial process for manufacturing diphtheria antitoxin, which von Behring had discovered the previous year. The process they developed used horses rather than smaller animals as "factories" for the antitoxin. This was the first time this method, introduced a few years earlier by Pierre Emile Roux at the Pasteur Institute, was used on an industrial scale. Ehrlich and von Behring's collaboration with the chemical company led to the commercialization not only of diphtheria antitoxin (1892), but also of the tetanus antitoxin (which von Behring also discovered) and serums against numerous diseases including foot-and-mouth disease, dysentery, and anthrax.

The third historic partnership was Hoechst's work with Ehrlich in the manufacture of arsphenamine (Salvarsan, 1910), the antisyphilis drug that opened the field of chemotherapy. While developing methods in the 1880s and 1890s for staining bacteria with synthetic dyes to make them visible under the microscope, Ehrlich noticed that methylene blue, one of the dyes he used, selectively attached to nerve fibers. From this observation Ehrlich reasoned that drugs might be developed that could selectively attack and destroy pathogenic microorganisms without harming the host organism. The antipyretics, hypnotics, anesthetics, and sedatives that had been discovered in the preceding decades did not cure disease, Ehrlich noted, but simply relieved its symptoms. Furthermore, he observed, pharmacologists mainly studied the toxicity of drugs, experimenting on healthy animals or sick people, and thus failed to elucidate the mechanisms by which a specific drug cures a specific disease. Ehrlich suggested that methods be developed to infect laboratory animals with a disease in a way that could be reproduced and then to study the effects of various substances on the infected organism. He called this method *experimental therapeutics.* Ehrlich proposed that serums might be the best substances to investigate because they are physiological products that cannot harm the host organism. For those diseases that serums could not cure (notably malaria, trypanosomiasis, and other ailments caused by protozoa), synthetic chemicals would have to be developed. But because chemicals are poisons that affect the body as well as the parasite, the chemicals used to treat these diseases would have to have a high affinity and high lethal potency for the parasites and a low toxicity for the host. To find these chemicals, Ehrlich postulated, it was essential to synthesize and test as many derivatives of promising chemicals as possible. This process he called *chemotherapy.*

To prove the soundness of his ideas, Ehrlich focused his initial investigations on one of the most obvious targets, the trypanosomes—protozoa that caused many diseases, among them the dreaded sleeping sickness of the Tropics that attacked both humans and animals. (The colonial powers of the nineteenth century were particularly interested in cures for tropical diseases because illness among workers in the tropical colonies slowed development of the agricultural and mineral resources. The colonialists needed to supply their expanding factories with raw materials. Many university schools of tropical medicine were established during this period.)

The substances Ehrlich chose to work with were chemicals containing arsenic, which had been used since the 1700s to treat malaria, syphilis, and sleeping sickness. In 1868 M. A. Béchamp synthesized sodium arsanilate, which H. W. Thomas of the Liverpool School of Hygiene and Tropical Medicine found effective against trypanosomes in mice. This compound was introduced as a therapeutic agent in the 1880s by the German firm Vereinigte Chemische Werke under the trade name Atoxyl. Robert Koch, in his research at the Berlin Institute for Infectious Diseases, found that Atoxyl was effective against sleeping sickness in cattle but that it caused blindness in humans. In 1903 Ehrlich began to synthesize hundreds of chemicals related to sodium arsanilate in search of a compound that would be more effective against trypanosomes and safe for humans. Cassella, a small chemical company in Frankfurt, agreed to support his research in exchange for the patent rights to any drug he discovered. Hoechst acquired Cassella in 1909 and continued to support Ehrlich's work.

A breakthrough came in 1906 when the syphilis spirochete was identified at the Hamburg School of Tropical Medicine. Ehrlich decided to add that disease to his target illnesses. With the help of Sacachiro Hata, a biochemist at the Kitasato Institute who had developed methods to infect rabbits with syphilis, Ehrlich tested various arsenic compounds on the infected rabbits. One of these, arsphenamine, was found to cure syphilis. Hoechst developed the extremely complicated industrial process for its commercial manufacture and launched it in 1910 under the trade name Salvarsan. Ehrlich's success lent credibility to his ideas about experimental therapeutics and chemotherapy, which have mapped the course of pharmaceutical research ever since.

Hoechst's other major achievements at the turn of the century were the development of a commercial process for manufacturing synthetic epinephrine. Epinephrine (Adrenalin) was the first hormone isolated in pure form; John Jacob Abel, a professor of pharmacology at Johns Hopkins University, and a Japanese chemist, Jokichi Takamine, produced it by extraction and purification from adrenal glands. A key messenger of nervous impulses, epinephrine was used as a vasoconstrictor to reverse a severe drop in blood pressure during surgery and as a hemostatic to control bleeding. Another Hoechst achievement was the development of procaine (Novocain, 1903), the local anesthetic that dominated world markets for the next fifty years. Novocain was developed by Albert Einhorn, a professor of chemistry at the University of Munich, who thought that fragments of the cocaine molecule, which was isolated in 1860 and whose topical anesthetic properties were discovered in 1894, might also possess anesthetic properties. Other local anesthetics discovered by Einhorn and produced synthetically by Hoechst were orthocaine (Orthoform [1896], Nirvanin [1899]), and benzocaine (Anesthesin [1902]).

After the German chemical companies formed a cartel, first in the framework of the small I.G. (1916) and later as I.G. Farben (1925), pharmaceutical research at Hoechst declined, largely because Bayer's Elberfeld Research and Development Laboratory was chosen by the I.G. Central Administration as the main pharmaceutical research center of the conglomerate. Thus Hoechst produced fewer medicines in the 1920s and 1930s than did Bayer. Hoechst, however, pursued its CTT in analgesics and antipyretics; in 1937 it introduced pethidine in Europe (meperidine [Dolantin] in the United States), a narcotic analgesic that was widely used to relieve pain during childbirth, and in 1940 methadone (Amidon), the first synthetic analgesic alternative to morphine that causes less sedation and respiratory depression. Both compounds were structurally simple compared with morphine and were marketed and imitated extensively after World War II, when their patents were abrogated by the Allies.

Bayer

Bayer's entry into the pharmaceutical sector began with the search for other antipyretics among simple organic chemicals rather than the highly complex alkaloids. For Bayer the crucial year was 1888, when it commercialized the antipyretic acetophenetidine (Phenacetin). Bayer chemists had synthesized this drug, which had a composition similar to that of acetanilide, while searching for uses of nitrobenzene, a by-product of the manufacture of dyestuffs. Acetophenetidine was evaluated clinically by Alfred Kast, professor of pharmacology at the University of Freiburg. A major commercial success, Phenacetin was popular for ninety years, after which it was gradually replaced by nonsteroidal anti-inflammatory drugs. Bayer's research on antipyretics and analgesics was crowned—and indeed terminated—by the introduction in 1899 of acetylsalicylic acid (Aspirin), a product that could not be patented because its synthesis had been published much earlier. It was launched as a proprietary product under a registered name after an extensive and vigorous marketing campaign aimed at physicians. This was the first instance of the modern method of marketing pharmaceuticals.

Bayer's other strong and profitable CTT was in hypnotics, which it initiated in 1888 when it commercialized sulfonmethane (Sulfonal), which was discovered by Eugen Bauman, professor of chemistry at the University of Freiburg, and evaluated by Kast. Sulfonal was palatable and did not irritate the gastrointestinal tract or disturb the cardiovascular system as chloroform and chloral hydrate did. Sulfonal's profitability encouraged Bayer to pursue research in pharmaceuticals. Its collaboration with Emil Fischer and Josef von Mering, a student of Schmiedeberg's who had specialized in hypnotics, proved extremely fruitful, producing sulfonethylmethane (Trional [1889]) and isopropylurethane (Hedonal [1899]). By the turn of the century von Mehring had come to recognize some of the relations between chemical structure and hypnotic properties, which facilitated his choice of candidate drugs. In 1903 he and Fischer synthesized barbital (Veronal), the first barbiturate hypnotic, whose superior properties rendered obsolete all other hypnotics then in use. This series of lucrative products reached its apex in 1911, when Fischer discovered phenobarbital (Luminal), which was not only a hypnotic, but also an anticonvulsant.

These innovations, most of which were extremely successful commercially, created the profits by which further research could be amply supported. Furthermore, as Lucius had at Hoechst, Carl Duisberg, the chairman of Bayer, became a champion of the pharmaceutical business. A chemist and codeveloper of Phenacetin, Duisberg established the pharmaceuticals department in 1888. In its early years the department depended on suggestions from academics for the development of new drugs, while company chemists dealt with their formulation and production. In 1896 Duisberg set up a pharmaceutical research unit independent of the production department and hired outstanding scientists to lead it. Heinrich Hörlein, a chemist and student of Ludwig Knorr, became the director of pharmaceutical research, organizing laboratories not only in synthetic chemistry but also in pharmacology and chemotherapy so that Bayer could extend its research to cover in-vitro and in-vivo evaluations of the compounds it synthesized. After numerous pathogens were discovered in the 1880s, the company embarked on research in antiprotozoal drugs, testing dyes as possible selective antibacterial agents. It was to be a long and arduous effort. Trypan red (1890) and trypan blue (1906) showed some effects against trypanosomes but with unacceptable side effects. In search of academic support Bayer crossed the border and forged ties with the Pasteur Institute and the Lister Institute, while Hoechst monopolized Ehrlich, von Behring, and the Berlin Medical Research Institute.

More than fifteen years of research were necessary before suramin (Germanin, Bayer 205 [1920]), the first antitrypanosomal and a cure for sleeping sickness, was launched. The drug was used for the next sixty-five years. A purely in-house innovation, suramin initiated another CTT and was followed by the three antimalarials: pamaquine (Plasmoquine [1925]), quinacrine (Atabrine [1932]), and chloroquine (Aralen [1934]). This sequence of innovations led to Bayer's historic achievement—the first sulfonamide antibacterial sulfamidochrysoidine (Prontosil [1935]), which introduced the third generation of pharmaceuticals.

Bayer and Hoechst created the modern pharmaceutical industry, but their great successes were achieved only through the companies' extremely close cooperation with academic researchers and universities. In contrast to other science- and technology-based sectors of manufacturing, in which the gifted scientist-entrepreneur was the decisive agent, the interaction, cooperation, and cross-fertilization in organic chemicals and pharmaceuticals was institutional rather than personal because the great advances made in organic chemistry, the life sciences, and medicine over more than a century created an abundance of opportunities for profitable exploitation. The ties between chemical company and university were strengthened by the founders of these companies and their successors, who were themselves renowned scientists—many lured from academic posts—who could forge effective links between industry and the academic institutions. In-house research laboratories gave the companies the ability to pursue the leads made in academia, to expand on them, and to create CTTs, which led to the commercial successes that kept the companies ahead of their competitors.

In contrast to the chemical companies the traditional German pharmaceutical companies, such as E. Merck, A. Knoll, and Schering AG, were much slower to shift their emphasis from extraction of natural products to the synthesis of new drugs. In the 1920s Merck and Knoll, for example, were still introducing pure alkaloids and other pure chemicals, while in the early 1930s Schering, through its cooperation with the University of Göttingen, began to work on hormones by extracting them from physiological products. In that sense they resembled the most progressive of their counterparts in Britain, France, and the United States.

THE SWISS COMPANIES

Three of the four major Swiss chemical-pharmaceutical companies—Ciba, Geigy, and Sandoz—were established in the 1880s as dyestuffs manufacturers in the German-speaking part of the country (around Basel) and shared with their German counterparts a common cultural, educational, and scientific heritage. Hoffmann–La Roche was established in 1894 as a pharmaceutical company. In the years leading up to World War I, the Swiss companies were second only to their German competitors, and when the war began, they took advantage of the embargo against German products, displacing German dyestuffs from their overseas markets.

This situation did not last long: The formation of the German cartel, I.G. Farben, in 1925, its prodigious research effort, and its horizontal diversification into fertilizers, organic chemical intermediates, plastics, detergents, and synthetic fibers made Germany once again the undisputed leader in the chemical industry. At the same time mergers of smaller companies in Britain and the United States created such formidable competitors as Imperial Chemical Industries (ICI), DuPont, Union Carbide, Dow Chemical, Monsanto, and American Cyanamid. Switzerland's small domestic market, geography, and lack of raw materials put the Swiss companies at a disadvantage, particularly in the manufacture of chemicals sold in tons rather than pounds because of the forbidding cost of transporting raw materials, intermediates, and final products in and out of the landlocked country.

Ciba, Geigy, and Sandoz sought to strengthen their position by forming the Dyestuffs Cartel in 1918. The cartel eased competition among them through price and market share agreements and through shared manufacturing facilities in Britain (Manchester) and the United States (the Toms River plant in New Jersey). These facilities allowed the three companies to produce such bulky chemicals as plastics and solvents and to operate behind the high tariff walls that were raised in the 1920s and 1930s in both Europe and the United States. Despite these measures the companies understood that their long-term viability depended on using their skills in synthetic organic chemistry to develop and manufacture low-volume, high-value-added products (e.g., dyestuffs, drugs, specialty chemicals, and eventually pesticides).

The Swiss companies had entered the pharmaceutical market very early with products based on synthetic organic chemistry. In 1896 Hoffmann–La Roche introduced the expectorant guaiacolsulfonic acid (Thiocol), which proved to be a great commercial success up to World War II. Ciba followed at the turn of the century with iodochlorhydroxyquin (Vioform, Entero-Vioform), an antiseptic for topical and intestinal use. Sandoz focused its research on natural products, particularly alkaloids. Its first product was ergotamine (Gynergen [1918]), the first ergot alkaloid isolated in pure form, which functioned as a uterine, arterial, and venous stimulant and as an epinephrine antagonist. The discovery of ergotamine created a strong CTT, which lasted at least until the 1980s when Sandoz introduced synthetic ergot medicines for the treatment of Parkinson's disease. Geigy is the relative newcomer; it did not enter the pharmaceutical industry until 1938.

Up to the late 1920s Swiss contributions to pharmaceuticals were modest for three reasons. Pharmaceutical innovation was slowing in general; the Swiss companies already had a thriving business in dyestuffs; and they did not contribute at all to innovation in serums, vaccines, or antiprotozoal drugs. Ciba, which was the most strongly committed to pharmaceuticals, manufactured medicinal chemicals, which it sold in bulk to other pharmaceutical companies. In essence Ciba operated like the chemical company it was. In the 1920s and 1930s, however, Ciba introduced some trademarked products, notably the local anesthetic dibucaine (Nupercaine), and, most important, the respiratory stimulant nikethamide (Coramine), a major commercial success that preceded Ciba's CTT in cardiovascular medicines.

THE BRITISH COMPANIES

The contributions of British physicians, universities, and teaching hospitals to the advance of medicine and the discovery of medicinal properties of natural products in the nineteenth century were at least equally important to those made by the French and Germans (Table 6). This great tradition of academic research in medicine and the life sciences, which was strengthened in the 1920s by the establishment of the Medical Research Council and its laboratory, was responsible for Britain's outstanding innovations in pharmaceuticals in the second half of the twentieth century. British companies, however, made only modest contributions to the development of "modern" drugs during the period from 1820 to 1930.

Because Britain was the leading trading nation, its manufacturing apothecaries, some of which were established in the 1600s, had a long tradition in the lucrative trade and wholesale business in tropical and subtropical medicinal substances. They became major producers of alkaloid products in the 1800s because they controlled the trade in quinine, opium, and other medicinal plants.

With the exception of Burroughs Wellcome, the transformation of the manufacturing apothecaries into research-intensive companies was very slow. Nor did British chemical companies show much interest in pharmaceuticals until the 1930s, in part because of the decline of their dyestuffs businesses after 1870 and because they had created thriving businesses in inorganic chemicals. Neither the manufacturing apothecaries nor the chemical companies had developed the alliances with academic researchers that would allow them to participate actively in the emerging pharmaceutical industry.

The exception was Burroughs Wellcome. It was formed as a manufacturing apothecary in 1880 in London by two Americans, graduates of the Philadelphia School of Pharmacy. They focused their business on producing accurately formulated medicines, mainly in tablet form. Henry S. Wellcome was one of the first industrialists to realize that the future of the pharmaceutical industry lay in fundamental scientific research. In 1894 he set up the Physiological Research Laboratory, the first of its kind in Britain, receiving the first British license granted to a non-physician to conduct experiments upon animals. The following year Silas M. Burroughs died, leaving Wellcome the sole owner. The company entered the biologicals sector in 1900 when it began to manufacture diphtheria vaccine.

But its real history as a research-intensive company began in 1904, when it hired Henry Dale as its pharmacologist. Dale, a graduate of Cambridge University who had worked with the famous British physiologists J. M. Langley and E. H. Starling, and for a short period with Paul Ehrlich, soon became the director of research at Burroughs Wellcome and the doyen of the British pharmaceutical industry. Dale, who won the Nobel Prize in medicine or physiology for his work on the chemical transmission of nervous impulses, influenced the course of medical research in Britain first as an academic researcher, then as director of a research-intensive pharmaceutical company, and finally as the first director of the National Institute for Medical Research, which was created in 1914.

Wellcome was interested in the action of ergot, a natural product used since the mid-nineteenth century to stimulate the uterus during childbirth. Dale's work on ergot, which he undertook only reluctantly, led to numerous advances in physiology and medicine and a score of innovations. The isolation of tyramine (marketed by Burroughs Wellcome in 1908 to treat precipitous drops in blood pressure), of histamine (and the recognition of its role in anaphylactic shock), and of acetylcholine (and the investigation of its physiological properties) led to the study of the physiological actions of sympathomimetic amines and to the elucidation of the role of chemical transmitters in the autonomous nervous system. The creation and diffusion of this expertise, largely by Dale and his coworkers, who moved among industry, government laboratories, and academia, created a British technology tradition leading to the introduction in the late 1940s of anticholinergic drugs and the early antihypertensives by Burroughs Wellcome and other companies and eventually in 1964 to the beta-blockers introduced by the chemical company Imperial Chemical Industries (ICI).

In the meantime, in 1923, Burroughs Wellcome produced insulin; in 1930 extracted and manufactured digoxin, the most active component of digitalis; and in 1935 extracted and produced ergometrine, the strongest component of ergot. Wellcome, who was elected a Fellow of the Royal Society for his contributions to medical advance, died in 1935, leaving the ownership of the company to a charitable trust with instructions that all Burroughs Wellcome profits be used to support medical and allied research in universities and teaching hospitals around the world. As the company prospered, the Wellcome Foundation Ltd. became one of the largest philanthropic supporters of medical research in the world. Burroughs Wellcome continued its strong ties to academic research, becoming one of the most innovative pharmaceutical companies, particularly in antineoplastic and antiviral drugs. In 1992 the trustees decided to sell the company to the public, and in 1995 Burroughs Wellcome merged with Glaxo to form Glaxo Wellcome, then the largest pharmaceutical company in the world. Glaxo, a manufacturer of powdered milk and infant foods, entered the pharmaceutical business in the late 1920s with innovations in thyroid hormones and vitamin D.

The other large British pharmaceutical companies also entered the business relatively late. Beecham manufactured proprietary products, such as toothpaste and toiletries, and entered the pharmaceutical industry only in the 1950s with the discovery and development of semisynthetic penicillins. Two other chemical companies, May and Baker and ICI, entered in the late 1930s, with innovations in sulfonamides and antimalarials.

THE FRENCH INDUSTRY

The French pharmaceutical industry has been centered in a single company, which had its origins in 1858, when Etienne Poulenc, an apothecary, began to manufacture photographic chemicals (e.g., collodion, gelatine, sodium thiosulfate, silver iodide, and silver bromide). In 1900 the company became Les Etablissements Poulenc and added such medicinals as bismuth salts, citrates, and phosphates to its list of products.

In 1906 Poulenc hired Ernest Fourneau, a chemist who had studied with Emil Fischer and Richard Willstätter. Fourneau was convinced that France could develop its own pharmaceutical industry and persuaded Poulenc to establish a pharmaceutical research laboratory. He made France's first original synthetic drug, a topical anesthetic, amylocaine (Stovaine), which became a commercial success and gave the company the confidence to pursue research on other potential new drugs. Fourneau's role in the French industry is strikingly similar to that of Dale's in the British industry. In the early 1920s Fourneau became director of the Pasteur Institute, and from this position he nursed cooperative efforts between the institute and Poulenc and contributed to many of the innovations of the French pharmaceutical industry in the 1920s and 1930s.

In the late 1920s Les Etablissements Poulenc bought the British company May and Baker, a producer of specialty chemicals and pharmaceuticals and merged with the Société Chimique des Usines du Rhône, a larger company established in the nineteenth century as a dyestuffs manufacturer. Faced with stiff German competition, Rhône soon abandoned dyestuffs and became a manufacturer of specialty photographic chemicals and pharmaceuticals. Once Rhône and Poulenc merged, the new company formed a wholly owned subsidiary, SPECIA, for pharmaceuticals.

THE AMERICAN COMPANIES

The American pharmaceutical companies have dominated world markets since the end of World War II, both in innovation and in market share, but they were responsible for only a few of the new drugs that were introduced before 1920. Their climb to the top after 1945 resulted in part from the outcome of the war, which left the American companies nearly the sole survivors in an expanding market. But the foundation of their success was laid in the interwar years when many of the American pharmaceutical companies strengthened their scientific base and built strong ties with U.S. academic institutions. Since the beginning of the century such institutions as Johns Hopkins, Harvard, Columbia, University of Wisconsin, and Washington University of St. Louis had made outstanding advances in medicine and the physical and life sciences, becoming leaders in physiological chemistry. These ties helped the companies to focus their research and development efforts and led to some of the major innovations of the 1920s and 1930s. Last but not least, federal legislation such as the 1902 Licensing Act and the 1906 Pure Food and Drug Act imposed quality standards on the companies, which forced them to improve their science and technology, while the 1917 Trading with the Enemy Act, which abrogated German patents, gave the American companies the opportunity to develop manufacturing processes for products with well-established markets.

The Early Years

Most of the modern American pharmaceutical companies can trace their history back to the 1800s, when they began to cater to the medicinal needs of a fast-growing population. In the United States the antecedents of the research-intensive pharmaceutical companies were not the chemical companies but the manufacturing apothecaries. And it was a war, the War of 1812, which disrupted U.S. trade with England, that was largely responsible for the establishment of the first American manufacturing apothecaries. These companies, located primarily in Philadelphia and New York, sought to replace the medicines and the raw materials for their manufacture that had been previously imported from England by trading directly with tropical and subtropical countries, particularly in Latin America. The American companies also sought to take advantage of the advances French and German scientists had made in the isolation of pure alkaloids, which had dramatically improved the quality of the few but most effective drugs of the period.

Between 1818 and 1822 six Philadelphia apothecaries became manufacturers. For example, the English pharmacist John Farr and the Swiss pharmacist Abraham Kunzi began to manufacture medicinal chemicals and Seidlitz Powders, a digestive already popular throughout Europe. Their enterprise evolved in 1836 into the Powers and Weightman Company, which in 1905 merged with Rosengarten and Company, a Philadelphia business that had similar beginnings. In 1927 Powers, Weightman and Rosengarten merged with Merck and Company. John K. Smith and John Gilbert established a manufacturing apothecary in 1835, which in 1891 merged with the French and Richard Company to become Smith Kline and French, the leading Philadelphia pharmaceutical company. A few companies were established as manufacturers of chemical specialties with medicinal applications: Robert Shoemaker (Philadelphia, 1838) made santonin, borax, iodides, and bromides, and E. R. Squibb (New York, 1859) made high-purity ether for anesthesia.

The Civil War boosted the business of these manufacturing apothecaries, but it was during the later westward expansion and the development of the industrial heartland that a second wave of manufacturing apothecaries was established in the Midwest, mostly by enterprising physicians and pharmacists. The most important of these were Parke-Davis (Detroit, 1866); Eli Lilly (Indianapolis, 1876); Upjohn (Kalamazoo, Michigan, 1885); Abbott (Chicago, 1888); and G. D. Searle (Chicago, 1888).

The Importance of Marketing

The entrepreneurs who set up these companies began by producing the medicines they prescribed, if they were physicians, or dispensed, if they were druggists. As in Europe most medicines were natural products of plant origin, organic and inorganic salts formulated into syrups, pills, or powders. Because they were natural products or simple chemicals of known composition, they could not be patented, so technological competition was restricted to formulating more palatable forms and standardizing the content of the active principles.

Standardization became much easier after a Belgian surgeon, Adolphe Burggreave, had the idea of compressing weighed quantities of active principles into tiny granules, or pills, that patients could easily swallow. In the United States the automatic pill-compressing machine transformed the manufacturing apothecaries into regional and eventually national pharmaceutical companies by allowing them to increase their output exponentially and forcing them into expanding their markets beyond their towns and cities. This expansion intensified the competition among companies, which, because their drugs could not be protected by patents, had to turn to intensive marketing and sales efforts and to the development of distribution networks. Advertising was aimed initially at the consumer, but as the scientific principles underlying the composition and use of drugs were progressively understood, these efforts were directed at physicians, hospitals, and drugstores. Given the multitude and diversity of drugs available in the late nineteenth and early twentieth century, company name recognition became central to these advertising campaigns and to the commercial success of the advertised products. Companies employed traveling salesmen, published quasi-scientific journals in which their own products received prominent exposure, produced pamphlets describing individual drugs, and stressed their commitment to science and the medical profession.

Distribution was also an important factor as the frontier shifted westward and waves of settlers created new urban centers and thriving markets across the Alleghenies, in the industrial Midwest, and on the West Coast. Thus, of the two pillars on which the American pharmaceutical industry rests today—marketing and sales and research and development—marketing came first, followed by the adoption of a scientific base.

The First Research-Intensive American Companies

Apart from a considerable amount of quackery the American pharmaceutical companies up to the last decade of the nineteenth century were not far behind their European counterparts. The striking difference was their lack of involvement with academic researchers. In comparison with those in Britain, Germany, France, and Switzerland, American universities and teaching hospitals made few advancements in medicine, organic chemistry, and pharmacology in the 1800s.

The first science-based drug produced in the United States was the diphtheria antitoxin, which was manufactured in 1894 by state public health authorities. The Philadelphia Board of Health began production, but manufacturing apothecaries objected, arguing that the public sector was interfering in what should be a private, commercial endeavor. The first companies to produce the diphtheria antitoxin were H. K. Mulford and Company of Philadelphia and Parke-Davis of Detroit, which sought help in their endeavors from leading American universities. By establishing close ties with academics, these two companies also became the first American research-intensive pharmaceutical companies.

A machine for compressed pills that was developed and patented by Oberlin Smith and Abraham Morris and assigned to Harry K. Mulford was the basis of the manufacturing apothecary H. K. Mulford and Company, founded in 1891. The water-soluble pills they produced were a commercial success, and within two or three years Mulford had become a major producer with a wide range of products and a distribution center in Chicago. Milton Campbell, president and part owner of the company, saw the opportunity offered by the diphtheria antitoxin and hired Joseph McFarland to develop the company's own antitoxin. McFarland, who had studied bacteriology in Germany, was a professor at the Medico-Surgical College of Philadelphia and had led the diphtheria antitoxin project of the Philadelphia Board of Health. Within a short period McFarland produced the antitoxin, and Mulford built the required facilities for industrial production. McFarland also lured two colleagues from the University of Pennsylvania to Mulford: Leonard Pearson, a professor at the veterinary school who had studied at Robert Koch's Institute for Infectious Diseases in Berlin and, later, John Adams, professor of surgery at the veterinary school. The ties between H. K. Mulford and the University of Pennsylvania strengthened with time and helped transform the company from a manufacturer of medicinals to a developer of new drugs with sales offices in several foreign countries. In the 1920s Mulford merged with Sharp and Dohme, which in turn merged with Merck in 1952.

The development of Parke-Davis followed virtually the same pattern. S. P. Duffield, one of the three entrepreneurs who established the company in 1866, was a chemist who had studied with the renowned Justus von Liebig at the University of Giessen in Germany. While at Parke-Davis, Duffield pursued his academic interests, publishing a few papers in the 1860s in the *American Journal of Pharmacy.* About the same time the company also hired Albert Lyons, a physician from the University of Michigan who organized the company's quality-control laboratory.

To develop the diphtheria antitoxin, Parke-Davis hired E. M. Houghton in 1895. A professor of botany at the University of Michigan who had worked on the standardization of botanical drugs, Houghton was able to produce the serum that same year. Parke-Davis then proceeded to build close ties with the University of Michigan, supporting numerous graduates who worked on projects of common interest with the company and then hiring them when they completed their studies. In 1902 Parke-Davis became the first American pharmaceutical company to build its own research laboratory. That was also the year the company began to manufacture epinephrine (Adrenalin) by extraction from adrenal glands. Epinephrine was the first major contribution of the American industry to the pharmaceutical world.

Parke-Davis subsequently pursued work on hormones, a fashionable subject early in the century and of great medical and academic interest. Again the company's strong support of academic research paid off. Edward Kendall, a former employee working at the Mayo Clinic, isolated pure thyroxine, the thyroid hormone, in 1914, and Parke-Davis produced it commercially by extraction from thyroid glands. In 1928 Oliver Kamm of the University of Wisconsin isolated two hormones from the pituitary—vasopressin and oxytocin—which Parke-Davis produced commercially by extraction. The company also supported the work of Edward Doisy at the Washington University of St. Louis and of Russell Marker at Penn State. Doisy was eventually to receive a Nobel Prize for his work on sex hormones, while Marker was the founder of Syntex, the pharmaceutical firm that pioneered steroid research.

Factors Strengthening the American Industry

By the early 1900s the American pharmaceutical companies appeared destined to be imitators of the European innovators. With the exception of epinephrine all of the major contributions of the 1890s and 1900s came from Europe, particularly Germany. Advances made in the 1910s and 1920s both in Europe and in the United States were modest, as pharmaceutical companies pursued to their limits the innovations and research leads made at the turn of the century. The next wave of innovation would not begin until the 1930s, touched off by significant advances made in the 1920s and 1930s in organic chemistry and synthesis, in instrumentation, and in pharmacology. That American companies were able to make substantial contributions during this period was partly because of federal legislation.

In 1901 outbreaks of tetanus in Camden, New Jersey, and St. Louis, Missouri, were traced to contaminated smallpox vaccine. To prevent similar disasters, Congress passed the 1902 Licensing Act, which required manufacturers of vaccines, serums, and toxins to be licensed by the Secretary of the Treasury through the Laboratory of Hygiene. In 1906 Congress passed the Pure Food and Drug Act, which sought to ensure the safety of food and drugs, to regulate the labeling and marketing of drugs, and to bar manufacturers from making false claims about the efficacy of their drugs. Apart from weeding out many drugs whose claims were unsubstantiated and improving the purity of intermediates, these two federal laws forced companies to organize in-house quality-control laboratories and eventually to merge to attain the size and financial strength that allowed them to improve their scientific capabilities.

The federal government gave the American pharmaceutical industry an even bigger boost in 1917, when Congress passed the Trading with the Enemy Act, which permitted American companies to manufacture German patented drugs. But the most important factor in the growth of the industry was the great advances in research, teaching, and training made by the American universities. Indeed, without close cooperation with academic researchers, the companies could not have taken full advantage of the opportunities created by the abrogation of the German patents. And having benefited from these ties, several companies began to fund academic research that held promise for further innovation (J. P. Swann, 1990). As a result a few more companies joined H. K. Mulford and Parke-Davis in making a significant commitment to in-house research and innovation.

Companies Committed to In-House Research

The growth of the Dermatological Research Laboratory (DRL) is the most significant example of a company becoming research-intensive because of restricted trade during World War I. Jay Frank Schamberg, professor of dermatology at the University of Pennsylvania; John A. Kolmer, an immunologist; and George W. Raiziss, a biochemist, established DRL in 1912 in Philadelphia as a nonprofit institution. Funding came initially from a Philadelphia financier who suffered from psoriasis; hence the initial goal of DRL was to find a treatment for that condition. In 1915 Schamberg predicted that the British blockade would restrict the availability of German drugs in the United States and asked Raiziss to attempt to synthesize Salvarsan and Neosalvarsan, the antisyphilis drugs developed and patented by Hoechst. Raiziss obliged by completing them within the year. When the squeeze came and the price of Salvarsan skyrocketed, Schamberg persuaded Congress to pass legislation allowing American companies to manufacture German patented drugs, which it did after the United States had declared war against Germany in 1917. DRL was the first company to be licensed to produce a German drug whose patent was abrogated by the 1917 Trading with the Enemy Act.

The Dermatological Research Laboratory made substantial profits during the last two years of the war, and at the end of the hostilities the company was pressed by the federal government to clarify its nonprofit status. Schamberg used most of the profits to fund a newly established Dermatological Research Institute and sold DRL to Abbott Laboratories.

Abbott also became a research-intensive company as a result of the war situation. Since its establishment in 1888 the Chicago company had specialized in manufacturing high-quality alkaloid pills. By 1910 Abbott's managers had realized that the company's traditional line of products was no longer sufficient to support corporate growth or to meet the emerging science-based competition. Abbott's first commercial success at diversification was to manufacture a form of the disinfectant halazone (Aseptamide), originally a British discovery. But after the passage of the 1917 trading act, Abbott established a close relation with Roger Adams, a professor of organic chemistry at the University of Illinois. An illustrious American chemist, Adams within a few years synthesized, and Abbott Labs manufactured, several drugs formerly protected by German patents. These included the sedative barbital (Veronal), the topical anesthetic procaine (Novocain), the antigout agent cinchophen (Atophan), and—after the acquisition of DRL—arsphenamine (Salvarsan) and neoarsphenamine (Neosalvarsan). Abbott maintained its relation with the University of Illinois for many years, and many of the company's top executives came from the university's chemistry department. Ernest Vorwiler started as Adams's assistant, became head of research and development, and was later named chairman and chief executive. Elmer Vliet followed a similar path a few years later. Vorwiler was the champion of Abbott's work in anesthetics during the 1920s and 1930s, in which the company established a very strong and lasting CTT, manufacturing butacaine (Butyn [1920]), a topical anesthetic; buthetal (Neonal [1926]), a sedative; pentobarbital (Nembutal [1930]), an anesthetic, sedative, and a major commercial success; and thiopental (Pentothal [1936]), an intravenously administered, short-lasting anesthetic devoid of side effects.

"Colonel" Eli Lilly, who began as an apprentice in a drugstore, established the company bearing his name in 1876 in Indianapolis, Indiana, as a manufacturing apothecary producing pills, cough syrups, and plant extracts of medicinal value. He gradually improved the quality and standardization of his preparations, and by the end of the century, as his business expanded with the industrialization of the Midwest, Lilly had built a network of sales representatives across the United States.

The first pharmacist Lilly employed was his son, J. K. Lilly, who graduated in 1882 from the Philadelphia School of Pharmacy. Four years later another pharmacist, a botanist, and a pharmacologist were hired to form a research nucleus. They introduced the empty gelatin capsule in which powdered drugs were dispensed, which helped to standardize drug doses. The company sold the capsules to its competitors as well, and it was an important business for Lilly up to the 1940s. In 1911 the company built a laboratory dedicated to research and development and standardization.

Their first notable innovation was the introduction in the American market of the antiarrhythmic quinidine (Duraquin [1918]), a natural product found in quinine that remains an effective treatment of cardiac arrhythmias. Lilly's historic innovation, however, was a process for manufacturing pure insulin from cattle pancreas (1923). Insulin had been discovered in 1921 by Frederick Banting and Charles Best at the University of Toronto, and to make the product available to diabetics as quickly as possible, Lilly was given information necessary to develop a manufacturing process. George H. A. Clowes, Lilly's director of research and development, and George Walden, who discovered a method of isoelectric precipitation of insulin from the crude extract, were the major Lilly contributors to this pathbreaking project. This original and lucrative innovation gave rise to the first Lilly CTT, which continues to this day. In 1982 Lilly and Genentech jointly introduced a biotechnologically produced human insulin (Humulin).

During this period Lilly also became a producer of vaccines and developed an expertise in hypnotic sedatives. In 1923 the company introduced a commercially successful barbiturate-sedative—amobarbital (Amytal)—that could be taken orally or intravenously. Secobarbital (Seconal) followed in 1929. In 1926 Lilly introduced another major and lucrative innovation: a process for manufacturing ephedrine by extraction from the Chinese plant Ma-huang (*Ephedra equisetina*). A stimulant of the central nervous system, a cardiac stimulant, and a bronchodilator used extensively for treating asthma, ephedrine was discovered at Tokyo University in 1898 and rediscovered in the 1920s by K. K. Chen, whom Lilly hired to develop the process. The monopoly Lilly created for manufacturing ephedrine gave other companies the incentive to synthesize numerous similar products in the 1930s.

These were the most successful American pharmaceutical companies that adopted in-house research and development as the main source of new drugs. Some of their competitors had also realized the necessity for research and development, but their efforts did not bear fruit until the 1930s and 1940s. Upjohn, for example, introduced a series of lucrative but incremental innovations: the laxatives Caripeptic Liquid and Phenolax (1906); a standardized digitalis, Digitura (1919); and the antacid Citrocarbonate (1921). Its science-oriented management used some of the profits from these products to support academic research, notably in hormones, a strategy that led to Upjohn's preeminence in sex hormones and corticosteroids in the 1940s and 1950s.

Other companies innovated in specialty chemicals rather than pharmaceuticals. Among these Pfizer introduced in 1923 the fermentation process for manufacturing citric acid and created a CTT that led to processes for manufacturing vitamin C (ascorbic acid) and penicillin. Merck became a leader in the manufacture of very pure chemicals and pharmaceutical intermediates, including iodine, bismuth salts, and alkaloids. The company substantially lengthened its list of products after its merger with Powers-Weightman-Rosengarten in 1927.

Given the great technological advances made by the American chemical companies in the 1920s and their high profits and influence, it is not surprising that some of the best pharmaceutical companies concentrated their research on process, rather than product, development. What may appear strange and should not be overlooked is that the American chemical companies, some of which had become very large and profitable by the mid-1920s, did not enter the pharmaceutical industry as did the German and Swiss companies, Rhône Poulenc and ICI. Only American Cyanamid did so, largely by acquiring Lederle Laboratories in 1927.

Because of the abundance of oil, coal, and most sources of inorganic raw materials, the American chemical companies directed their R&D efforts to the manufacture of bulk chemicals and later to plastics and synthetic fibers. Because they did not challenge the European chemical companies in dyestuffs, they did not create the in-house skills in synthetic organic chemistry that were essential for innovation not only in dyestuffs but also in pharmaceuticals. Even DuPont was unsuccessful in its efforts to innovate in dyes. As a result the American chemical companies also missed the opportunities that were opened in the 1930s and 1940s with the emergence of the third generation of drugs, opportunities that were seized by those American pharmaceutical companies that were research intensive.

Wartime Technologies: The Watershed Years (1930–1960)

The 1930s, 1940s, and 1950s marked a watershed for the pharmaceutical industry. Not only did a third generation of drugs grow and mature, but the research and innovation methods and practices developed during the decades surrounding World War II, along with changes in corporate structures and the geography of the industry, would shape the pattern of technological advance, corporate structure, and the geography of the pharmaceutical industry for the second half of the twentieth century. This was the period when most pharmaceutical companies became committed to in-house research and when they radically changed the way they marketed many of their drugs. It was also the period when American companies became the acknowledged leaders of the industry.

As with so many other sectors of manufacturing, World War II was the defining event for the pharmaceutical industry during this period. Before and during the war the U.S. government's program to hasten the development of antibacterials, antimalarials, and anti-inflammatories encouraged several companies to enter the pharmaceutical field and others to step up their research capabilities. Similar measures taken by the British and German governments to meet the urgent needs of their armed forces were less effective and had less influence on their companies.

The third generation of drugs included five new technologies, all of them introduced between 1935 and 1942: vitamins, corticosteroid and sex hormones, sulfonamides, antihistamines, and antibiotics.

With few exceptions—Bayer, Hoechst, Rhône Poulenc, Burroughs Wellcome—pharmaceutical companies were not research intensive in the early 1930s, although they came to realize the merits, even the indispensability of in-house scientific research. They were also small in size and thus lacked the financial means to invest in research and development; growth of pharmaceutical markets was constrained by the limited therapeutic range of the available drugs. The third generation of drugs changed that situation radically: the versatility of the original innovations for each of the five technologies created ample opportunities for imitation by competing companies; government assistance to the pharmaceutical companies to meet the country's wartime needs enhanced their commitment to research and development and catalyzed the diffusion of these technologies; and the large markets opened by the drugs of the new technologies allowed existing pharmaceutical companies to grow and drew into the industry companies whose businesses were in other manufacturing sectors.

At the beginning of the period the American pharmaceutical industry was made up of three types of companies. First were companies like Merck and Pfizer, modeled after chemical companies, whose innovations were primarily in processes and who produced specialty chemicals in bulk for sale as intermediates. These companies shifted their strategies in the 1950s to market their own branded drugs directly to doctors and drugstores. The second type were the former manufacturing apothecaries, such as Lilly and Wyeth, who created R&D facilities to take advantage of the technologies made available through the federal wartime efforts. The third group was composed of diversified companies whose primary business was the manufacture of such proprietary goods as toiletries, cosmetics, and first-aid products and who entered the pharmaceutical field by acquiring or merging with pharmaceutical companies, again to take advantage of the opportunities offered by the government wartime programs and the expanding new pharmaceutical markets.

At the end of the war the government programs also ended, but by this time most of these companies had recognized the profitability of pharmaceuticals and also the value of intensive research. They consequently developed the corporate structures and the sales and marketing tools necessary to maintain a return on their investment in research and high profitability in a peacetime economy. Marketing strategies underwent a sea change, as companies sought to cope with the relatively large number of competitors and a wealth of new drugs. The number of diseases that could be effectively treated was growing, but so was the sheer number of drugs, their complexity, and the similarity of many of them to each other. Companies thus sought public recognition of their products through trademarks or brand names, and because most of these new drugs were available only by prescription, the companies shifted their marketing targets from consumers to doctors, drugstores, and others in the health professions.

The companies found a receptive audience—demand for new drugs to treat illnesses was high. The American industry benefited not only because domestic demand was strong but also because it had a near-monopoly of the world market; it was thus well situated to become the world leader, a position it still retains. The German industry, which had been the leader in pharmaceuticals, was devastated by the war and took several decades to recover. The industry in other European countries was not damaged quite as badly, but the domestic markets in those countries were comparatively small and weak. The great traditions of medicine and medical research in European universities and public research laboratories survived the war and, with the reconstruction of Western European economies in the 1950s and 1960s, helped the European pharmaceutical companies to rebound. By adopting the American model, several of them re-emerged as strong innovators and competitors, introducing in the early 1960s many of the technologies of the fourth generation of drugs.

THE TECHNOLOGIES OF THE THIRD GENERATION

The 1930s and 1940s were ripe for a new generation of drugs. Analgesics, antipyretics, hypnotics, anesthetics, antiprotozoal medicines, mercurial diuretics, serums, and vaccines introduced at the turn of the century had reached maturity. No matter how determined the efforts to continue to innovate in these areas—as were the antimalarial programs sponsored by the British and American governments during the war—these drugs could not yield radical innovations in numbers sufficient to sustain a vigorous industry. Moreover, scientific and medical advances in the first third of the century had prepared the ground for a new generation of drugs.

Five major technologies were developed in the 1930s and 1940s: vitamins; sulfonamides, sulfa-based drugs that provided the first effective treatment for a host of bacterial diseases; antibiotics, such as penicillin, streptomycin, and chloramphenicol, which soon largely displaced the sulfonamides in treating bacterial diseases; antihistamines, used primarily to treat allergic disorders; and hormones, including corticosteroids for use as anti-inflammatory drugs. Work on vitamins and hormones had begun in the 1910s and 1920s as researchers learned how to isolate some of these natural substances by extraction and purification. Some pharmaceutical companies had even begun to market a few pure hormone products, such as insulin to treat diabetes and thyroxine for thyroid disorders. Vitamins and hormones began their period of great growth, however, in the 1930s, when the development of analytical and spectroscopic methods requiring very small samples allowed researchers to elucidate the structure of these complex molecules. At the same time advances in synthetic organic chemistry allowed the development of processes to synthesize and manufacture vitamins and hormones in commercial quantities.

Although pharmaceutical companies manufacture most of the vitamins produced today, they are not considered drugs but food additives, sold in bulk to the food industry, or dietary supplements, sold to consumers through drugstores and supermarkets. Most people in the United States and Europe eat a mix of food that supplies the human body with the vitamins it needs. This was not the case, however, before the 1930s, when malnutrition led to vitamin deficiencies that caused crippling and even lethal diseases, especially in children.

Since the seventeenth century, physicians had identified foods whose presence in diets prevented illness and death: oranges, lemons, and vegetables protected people from scurvy, for example; unmilled rice prevented beri-beri (peripheral neuritis); and cod liver oil prevented rickets. It was logical, therefore, for medical researchers to conclude that these and other foodstuffs contained active principles essential to human health. Attempts to isolate and identify them began around 1900 in Britain, Holland, Germany, and the United States. These researches were not successful until the late 1920s, when the required analytical chemistry and spectroscopic methods became available. Thiamine (vitamin B_1), the deficiency of which causes beri-beri, was identified in 1926; ergocalciferol (vitamin D_2), the deficiency of which causes rickets, was identified in 1928. All other vitamins were discovered between 1931 and 1948.

Synthetic organic chemistry opened two other sectors in this period: sulfonamides and antihistamines. The discovery that the sulfonamides had antibacterial properties resulted from a well-planned and swiftly executed research project that Bayer initiated in 1927. Bacterial infections were the most common cause of disease and death in temperate climates and accordingly were an obvious target for a company with a great tradition in discovering "magic bullets" against pathogenic intruders in the human body. The project, which was aimed at finding a cure for lethal streptococcal infections, was conceived by Heinrich Hörlein, the director of pharmaceutical research of I.G. Farben; the leader of the research team was Gerhard Domagk, a physician who won a Nobel Prize for his discovery, in 1932, of the antibacterial properties of sulfamidochrysoidine (Prontosil). This drug, placed on the market in 1935, was the first to cure streptococcal infections.

The composition of histamine and some of its physiological properties (e.g., stimulation of the smooth muscle in the alimentary tract and the bronchioles, lowering of blood pressure) were discovered in 1910 by Henry Dale and his coworkers at Burroughs Wellcome. But histamine's role in the human body was difficult to establish because it is found in many tissues. Daniele Bovet, working with Ernest Fourneau at the Pasteur Institute, thought that if chemicals were found that acted as antagonists to histamine, they would help in identifying that role; the same method had been used previously for other physiological products. In 1939 Anne-Marie Staub, a Bovet student, accurately identified the structure of some of the chemicals that showed antihistamine properties. The first clinical experiments were conducted by M.-M. Mosnier with phenbenzamine (Antergan), which was found to be an antiallergic and was marketed by Rhône Poulenc in 1942.

Last but not least, the accidental discovery of penicillin (1928) and the development of processes for its commercial manufacture (1942) led to the discovery of many other antibiotics. Apart from the vitamins, whose properties could not be improved by molecular manipulation, the first innovations in these technologies served as models for imitation and acted as the engines of the dramatic expansion of the pharmaceutical industry in the 1940s and 1950s.

The key innovations of this third generation of drugs were spread fairly evenly among major pharmaceutical companies in the United States and Europe. As was the case with earlier drug technologies, these new technologies were products of a close cooperation between pharmaceutical companies and academic institutions prominent in the relevant academic disciplines: natural products and organic chemistry, pharmacology, physiological chemistry, biochemistry, biology, and physiology. Such institutions included the Pasteur Institute in France, the Medical Research Council Laboratory in Britain, the Mayo Clinic in the United States, the Kitasato Institute in Japan, the Zurich Polytechnic (Eidgenössische Technische Hochschule) in Switzerland, the Karolinska Institute in Sweden, and the University of Toronto in Canada. Numerous other academic laboratories in leading universities and teaching hospitals were also active participants in drug innovation and development during this period.

Government-Sponsored Projects

As Germany conquered one European country after another in 1939 and 1940, the U.S. government began to prepare for the eventuality of war. In 1940 it established the National Defense Advisory Committee, which in turn formed several other agencies, each dealing with a particular set of problems that war was likely to raise. One of these was the Office of Production Management, which was to ensure that the armed services had the materials they needed. After the Japanese bombing of Pearl Harbor in December 1941, the Office of Production Management was reconstituted as the War Production Board and given extensive powers to allocate resources and coordinate industrial production. To facilitate this effort, the Justice Department waived antitrust restrictions to permit companies to exchange information freely.

In 1941 the Office of Scientific Research and Development (OSRD) was established to ensure adequate research in scientific and medical problems related to national defense. The OSRD served as a center for coordinating, planning, and mobilizing scientific personnel and resources, aiding and supplementing where necessary the experimental and other research activities conducted by the armed services and other federal agencies. To this end it was authorized to enter into contracts and agreements with individuals, educational and scientific institutions, industrial organizations, and other agencies for studies, experimental investigations, and reports.The OSRD had six divisions, including the National Defense Research Committee, which directed the development of the first atomic bomb under the so-called Manhattan Project. Another division was the Committee for Medical Research (CMR), charged with research in military medicine.

One of the most successful of the CMR projects was penicillin research, clinical evaluation, and production. The project began in the summer of 1941, when Howard Florey and Norman Heatley, members of the Oxford University team that had established the drug's potential as an antibacterial, came to the United States to persuade American firms to undertake its development and commercial production. Florey and Heatley were put in touch with the U.S. Department of Agriculture's Northern Regional Research Laboratory in Peoria, Illinois, a laboratory with expertise in growing and characterizing molds. The CMR made funds available to the laboratory to carry out its investigations.

Within a few months the laboratory's researchers had made two key discoveries: They had isolated a far more productive strain of *Penicillium notatum* (Fleming's mold); and they dramatically increased the output of penicillin by using corn steep liquor, a by-product of the wet corn milling industry, in the fermentation broth.

To justify the expense entailed in moving from small-scale research to commercial production and to determine penicillin's spectrum of activity and its advantages over the sulfonamides, the OSRD funded clinical research projects on penicillin at fifty-eight academic and industrial laboratories, at a cost of $2.7 million. But because of the urgency imposed by the war, the OSRD also urged four companies—Lederle, Merck, Pfizer, and Squibb—to develop a commercial process. The four were chosen because of their expertise either in antibacterial drugs or in industrial fermentation. To increase the odds that the penicillin research would be successful, the CMR urged close cooperation among all the federal agencies involved as well as the research laboratories and the companies. Information and results were pooled at regularly scheduled conferences. Moreover, although the companies proceeded with their production research independently, they reported their progress to the CMR, which passed on to others information that would advance the program.

The great problem to be solved was how to increase the yields of penicillin. When the project began, the mold was grown on surface cultures, such as glass bottles or pans, and produced yields too low to meet demand. The only company to have experience with fermentation in submerged cultures was Pfizer, which used that method to produce citric acid (its main product) and fatty acids. Despite the risk of contaminating its own cultures, in 1943 Pfizer developed and built the first plant for manufacturing penicillin in a submerged culture. The cost was about $3 million; Pfizer itself was then a small company with annual sales of just $7 million.

Once the Pfizer process went on stream, the government selected 22 companies (out of 175 considered) to receive the construction materials and supplies necessary to produce enough penicillin to meet both civilian and military demands. It was estimated that some 200 billion units of penicillin were needed each month; production for all of 1943 amounted to only 21 billion units. In addition to Pfizer the companies receiving priority materials included Merck, which had developed its own submerged-culture process; Squibb; Winthrop; Abbott; Hoffmann–La Roche; Upjohn; Lederle; Parke-Davis; Lilly; Cheplin Laboratories; Cutter Laboratories; and Sharp and Dohme. In 1944, 1,663 billion units of penicillin were produced (about 138 billion a month); in 1945 production rose to 6,852 billion units (about 570 billion a month).

The federal government built six production facilities at a total cost of about $7.5 million. After the war these plants were sold to the companies that operated them for approximately $3.5 million. Sixteen other antibiotic plants were financed by private firms at a cost of $22.5 million. Of this amount $14.5 million was approved for accelerated amortization over a five-year period (compared with the usual ten to fifteen years).

Penicillin production was very profitable during the war, when the government bought all production at cost-plus prices, and in the late 1940s, when manufacturing capacity was insufficient to meet the explosive demand. By the end of the decade, however, the fast diffusion of the technology led to an oversupply, and prices tumbled. By 1956 only twelve of the original twenty-two firms were still manufacturing penicillin.

The financial cost to the U.S. government for the development of penicillin was modest compared with the therapeutic benefits for humankind and the technological benefits for the American pharmaceutical industry. Never before or since has a revolutionary technology diffused so quickly and so widely in the pharmaceutical industry. In the long run the expertise that American companies acquired in research, development, and production of natural antibiotics proved more significant than the profits they derived from penicillin. American companies monopolized the natural antibiotic market, which up to the late 1960s accounted for about 25 percent of the total drug market. Profitable antibiotics on the market in the 1950s and 1960s were chloramphenicol (Chloromycetin, Parke-Davis), erythromycin (Erythrocin, Abbott; Ilotycin, Lilly), chlortetracycline (Aureomycin, Lederle), tetracycline (Terramycin, Pfizer), streptomycin (Streptase, Merck), lincomycin (Lincocin, Upjohn), and gentamycin (Garamycin, Schering).

Another group of drugs—adrenocortical steroid hormones—was developed at roughly the same time as penicillin. The first steps for extracting, characterizing, and studying the physiological activity of adrenocortical hormones were taken in the early 1930s. The only known use for the hormones then was in the treatment of Addison's disease, an extremely rare and lethal disease caused by the destruction of the adrenal glands. A handful of academic and industrial laboratories in Switzerland and the United States had introduced and marketed either pure compounds or extracts for treating this rare disease. But the anti-inflammatory, anti-immunologic, and muscle-building properties of corticosteroids were not discovered until the late 1940s.

In 1941 an intelligence report that German pilots were fed with adrenal extracts to enable them to fly and fight at altitudes above 40,000 feet led the OSRD to initiate a research project on corticosteroids. Under the direction of Edward Kendall of the Mayo Clinic, the project involved cooperative agreements with Lederle, Merck, Parke-Davis, Schering, Squibb, and Upjohn. The report about the military uses of corticosteroids turned out to be false, and the program was discontinued. But not before Kendall and the participating academic and industrial organizations had identified many corticosteroids and developed important methods for their synthesis. Merck's Lewis Sarett, for example, developed a forty-two-step process for synthesizing cortisone. The companies also developed bioassays for evaluating the physiological properties of corticosteroids.

This expertise proved extremely valuable in 1948, when cortisone was found to be an effective treatment for rheumatoid arthritis. Because the technology had diffused among the participants of the OSRD project, all of them introduced corticosteroid drugs in the 1950s.

In contrast to the antibiotic and corticosteroid programs the federally sponsored project on antiprotozoal drugs had mediocre results because the technology had already matured. The Bayer drugs of the 1920s and 1930s were effective and were adopted as models for further innovation, a task that was made easier by the abrogation of Bayer's patents. Although the war created a need for the manufacture of antimalarial drugs, the technology hardly diffused outside Sterling, which represented Bayer's interests in the United States, while 50 percent of its equity capital belonged to the chemical company General Aniline and Film (GAF), I.G. Farben's American subsidiary company.

Wartime Appropriations and Reparations

The exigencies of war also led to the diffusion of German technologies by noncommercial means. Schering, a subsidiary of Schering AG, was managed during the war by the Alien Property Custodian. Since the 1930s Sterling-Winthrop had produced under license for the American market the drugs of Bayer and

Hoechst while 50 percent of its equity was owned by GAF. At the end of the war Schering became an independent American company, and Sterling bought back the German interest. As German patents and technology were included among the war reparations, the German government was required to pay the Allies, both companies—and some others—produced German drugs under their own trademarks after the end of the war. These measures were not as effective as the 1917 Trading with the Enemy Act because most of the technologies of the German companies were mature by the end of World War II.

Only the sulfonamides among the modern technologies of the 1930s and 1940s were of German origin, and they had already diffused to the rest of Europe and the United States. The analgesics meperidine (Demerol) and methadone (Dolophine), developed by Hoechst, and the bronchodilator isoproterenol, developed by Boehringer, were produced by a score of European and American companies and served as models for the synthesis of some cardiovascular and central nervous system drugs and analgesics.

ASCENT OF THE AMERICAN PHARMACEUTICAL INDUSTRY

To a great extent the U.S. government's wartime policies led to the emergence of the American pharmaceutical industry as the undisputed worldwide leader. The American companies had made considerable advances in the 1930s and had come to recognize the importance of in-house research. But with the exception of Merck none of them was on a par with the German and Swiss companies, Rhône-Poulenc in France, or Burroughs Wellcome in Great Britain. The federal war effort encouraged corporate research and development, widened and deepened the companies' cooperation with academic institutions, and catalyzed the diffusion of the new technologies across the industry. By the end of World War II fifteen American companies could fairly be described as research intensive.

Merck: A Research-Intensive Model

One company uniquely situated to take full advantage of the government wartime programs was Merck & Company. In the late 1920s Merck's management had made a conscious decision to transform the company into a research-intensive organization of the first order, and that decision soon bore fruit. Merck's contributions to pharmaceutical innovation in the 1930s and 1940s were outstanding: The company led in the development of manufacturing processes for vitamins and in the development and clinical use of penicillin (1941), and introduced streptomycin (1946) and cortisone (1948). Merck was thus instrumental in launching three of the five technologies of the period (vitamins, antibiotics, and hormones).

Established in New York in 1891 as a sales branch of the German company E. Merck of Darmstadt, Merck & Company sold fine chemicals, some of which were pharmaceutical intermediates. They included photographic chemicals, chloral hydrate, iodides, bismuth salts, acetanilide, salicylates, disinfectants, and alkaloids, all specialties of the German parent. At the turn of the century the American branch began to produce some of these chemicals, and by 1910 it produced more than it imported. But it continued to depend on its German parent for technology. During World War I, Merck had to sever its links with E. Merck and buy back the stock owned by the German company, which had been appropriated by the federal government. The ties between Merck & Company and E. Merck were partially restored in 1932 by an agreement to exchange technical information and to use the family name as a trademark.

In 1926 George W. Merck Jr. succeeded his father as head of the company and began to transform it into the research-intensive organization that it is today. In 1927 Merck merged with the Philadelphia firm of Powers-Weightman-Rosengarten, a manufacturer of fine chemicals and pharmaceuticals ranging from iodides and bismuth to quinine and the antisyphilis drug neoarsphenamine. This merger more than doubled Merck's size (Powers-Weightman was bigger than Merck at the time of the merger) and enabled the company to shift to a research-intensive approach.

Merck, however, differed from other American companies in the way it became research intensive. Companies did not usually make substantial investments in research until they had a major commercial success. Such a success gives a company confidence, helps management to recognize the merits of in-house R&D, and leads to the investment of some of the profits in further R&D. Merck, in contrast, made the investment in R&D before it had any commercial success. In 1933 the company established the Central Research Laboratory and the Merck Institute for Therapeutic Research, using funds that came from its traditional business. George Merck was not a scientist, but he had the foresight to recognize the dependence of the emerging pharmaceutical industry on scientific advance. He was convinced that original research would lead to profitable products in the long run and that only creative scientists could succeed in this task. The aim of the two research facilities, Merck once said, was "to provide an environment where genuinely creative minds could be so protected that their mental powers of thought, study, and imagination can concentrate on problems of great difficulty."

Randolph Major, Merck's first director of research and development, decided to focus the company's research efforts on vitamins. Both the emphasis on synthetic organic chemistry and on process innovation, which became the strongest traditions of Merck research, originated in this work on vitamins. Found mostly in traces in natural products, vitamins had to be synthesized if they were to be produced industrially; and given the complexity of the structures of many of them, low-yielding and multistep laboratory processes had to evolve into simpler, high-yielding industrial processes so that the final product could be sold at affordable prices. In the 1930s pharmaceutical companies in the United States and Europe were working hard to elucidate the chemical structure of vitamins, but the development of manufacturing processes had been overlooked. At the time only natural extracts relatively rich in vitamins were in use.

Envisioning that synthetic vitamins could be produced in bulk and sold at prices that would allow for their use as food additives, Major saw an opportunity for Merck to become a protagonist in this development. The great specialists in vitamins—Frederick Gowland Hopkins, Richard Kuhn, Paul Karrer, Albert Szent-Georgyi, and Adolf Windaus—were all Europeans, and most of them were tied to European pharmaceutical companies. Therefore Merck was forced to develop its own expertise.

To that end Major took several important steps. He hired many promising young scientists from the best universities, including Max Tishler, who developed both a new synthesis of and an industrial process for manufacturing riboflavin (vitamin B_2), and Karl Folkers, who led Merck's effort to isolate and elucidate the molecular structure of the vitamins in general and of vitamins B_6 and B_{12} in particular.

Major hired some of the most renowned American professors in organic and natural products chemistry as consultants, including Louis Fieser and Homer Adkins. He set up a physical chemistry department equipped with all the analytical tools that might prove useful, including infrared and nuclear magnetic resonance spectroscopy, vapor-phase chromatography, and X-ray crystallography. Merck was one of the first institutions in the United States, corporate or academic, to use these new research methods. The company also encouraged its scientists to publish the results of their research in scientific and professional journals, even exerting pressure on the patent department to speed up its procedures to avoid delaying publication.

Moreover, the corporate research agenda was based to a considerable extent on suggestions from the research staff, and promising researchers were encouraged to return to universities to obtain higher degrees. Thus, Selman Waksman, professor of soil microbiology at Rutgers University and the discoverer of streptomycin, commented in 1942 that the Merck Research Laboratories compared favorably with the best academic departments in facilities, research policies, and creativity.

In 1934 Robert R. Williams, a chemist at Bell Laboratories and an adjunct professor at Columbia University, asked Merck for support in developing an industrial process for manufacturing vitamin B_1 (thiamine), which he had isolated and characterized. The process was successfully completed in 1935 and was licensed to Merck in 1937. Thiamine soon accounted for about 10 percent of corporate sales, creating the confidence the company's scientific and business managers needed to pursue this line of research and initiating Merck's first CTT.

In 1937 Merck also introduced processes for manufacturing vitamin B_2, nicotinic acid, and vitamin C (ascorbic acid); processes for vitamin B_6 and calcium pantothenate followed in 1940. In 1949 Merck crowned its work in this area by isolating and manufacturing cyanocobalamin (vitamin B_{12}) by extraction. Thus within a short span of time, Merck had become the leading American manufacturer of vitamins, and its research staff had built the necessary self-confidence for tackling other original projects.

One of those was the synthesis of new sulfonamides, a technology that fitted with the company's expertise in synthetic organic chemistry. Merck's contribution turned out to be modest, but serendipity introduced a welcome twist: One of the first sulfonamides synthesized by Max Tishler, sulfaquinoxaline, was found to be highly effective in preventing coccidiosis, a widespread poultry disease. In 1942, after five years of testing, Merck began to produce the drug commercially. The ability to control coccidiosis enabled poultry breeders to develop the mass production methods that made the modern broiler industry possible. Sulfaquinoxaline was for Merck the radical innovation that created a new technology and a new market and was the origin of their CTT in veterinary drugs and feed additives.

Accelerators of Technology Diffusion

In addition to being a unique model of a research-intensive company, Merck was one of three companies in this period whose business operation helped accelerate the diffusion of technology through the American industry. Merck, Pfizer, and Syntex all operated like chemical companies rather than pharmaceutical companies, selling high value-added specialty products in bulk to other pharmaceutical companies, which then formulated, packaged, marketed, and sold the products under their own trade names. For Merck and Pfizer process innovation rather than product innovation was their business up to the early 1940s. They manufactured well-known products using new processes that cut production costs and increased their profits.

That was the goal of the two companies when they undertook the development of manufacturing processes for vitamins; the difference was that vitamins were known in academic laboratories but were new products for the pharmaceutical industry. Merck and Pfizer stuck to their traditional way of doing business, selling vitamins in bulk to other pharmaceutical companies, which then formulated, packaged, and sold them as proprietary products. Thus Parke-Davis, Lilly, Smith Kline and French, and Upjohn, companies that made only minor contributions to vitamin research and development, sold numerous trademarked dietary and nutrition products. Merck and Pfizer adopted a similar policy when they produced penicillin and streptomycin, selling the drugs under their own brand names while also manufacturing them in bulk for formulation, packaging, and sale by other drug companies.

Syntex, during its Mexican period (1943–58), followed the same policy with sex hormones, progestins, and corticosteroids. By selling progesterone and estradiol to other manufacturers at one-hundredth and one-tenth of their market prices, respectively, Syntex gave to a host of companies—particularly those that had participated in the government corticosteroid project—the opportunity to use the drugs as intermediates for further molecular manipulation and to produce steroid drugs with improved therapeutic effects. Syntex also accelerated the process of diffusion by licensing its processes to third parties.

Pfizer and Company

In 1849, in New York, Charles Pfizer and Charles Erhart established Pfizer and Company as a manufacturer of such medicinal chemicals as santonin, a plant extract used against parasitic worms; iodoform; borax; boric acid; potassium iodide; and refined camphor. In 1880 the company began to produce citric acid from imported crude citrate of lime. Citric acid would remain Pfizer's major product for the next seventy years, although at the turn of the century they were also producing bromides, strychnine, tannin, benzoates, and salicylates.

The maritime embargo during World War I interrupted Pfizer's supply of raw materials for making citric acid, which by then had reached 420 tons a year. In 1917 James Currie of the U.S. Department of Agriculture discovered a way to make citric acid by fermenting sugar in the presence of the black bread mold *Aspergillus niger* and licensed the process to Pfizer. Six years later Pfizer had built an industrial plant that was making fifty tons of citric acid a year using this new method. The company made incremental improvements to the process over the next few years and by the late 1930s was producing 12,000 tons a year. The company had also applied its fermentation methods to other organic acids, such as tartaric, fumaric, and gluconic. Altogether 70 percent of Pfizer's sales were of fermentation products; nearly half were of citric acid. The company had established a strong CTT in industrial fermentation, which reached its peak in 1943 with the deep-tank fermentation process for manufacturing penicillin.

In the 1930s Pfizer also began to manufacture vitamins, both by licensing processes from third parties (vitamin C [1937] and vitamin A [1940]) and by introducing its own processes (vitamin B_2 [1938] and vitamin C [1940]). By 1940 vitamins accounted for 11 percent of its sales.

The similarities between Pfizer and Merck in the prewar period are obvious. The significant difference between the two companies was Merck's immense effort in the 1930s to become a research-intensive pharmaceutical company specializing in organic chemical synthesis, while Pfizer remained under the strong influence of its CTT, making specialty chemicals by fermentation.

The two companies cooperated during the war in the development of penicillin: In step with their CTTs, Merck produced the first pure sample of penicillin to be used clinically in the United States, and Pfizer developed the deep-tank fermentation process that made the commercial production of penicillin possible. By the late 1940s Pfizer was the biggest penicillin producer in the world.

Syntex

Syntex, which made its mark in hormones and particularly contraceptives, was established in 1943 in Mexico City by Russell Marker, a professor of chemistry at Pennsylvania State University, and two European chemists, Emeric Somlo and Federico Lehman, who owned a small company that made hormones by extraction.

In the 1930s and early 1940s the main hurdle slowing the development of sex and corticosteroid hormones was their scarcity. Minute quantities were produced by extraction from large volumes of physiological products or animal organs. By the mid-1940s Ciba and Merck had developed synthetic processes using bile acids as starting material, but these processes were cumbersome and low yielding.

Marker discovered a partial synthesis of progesterone from a plant-derived natural product, diosgenin, and found that the roots of the Mexican yam tree (*cabeza de negro*) yielded significant quantities of that substance. Less than a year after this process was commercialized, the price of progesterone fell from $80 a gram to $15; by 1950 the price had dropped to just $3 a gram, so that many pharmaceutical companies started to use progesterone as the basic intermediate for synthesizing corticosteroid drugs.

In 1946 Marker sold his interest in the company, which then hired George Rosenkrantz as research director. Rosenkranz had studied under Leopold Ruzicka, the Nobel laureate in natural products chemistry. At Syntex, Rosenkrantz developed commercial processes for manufacturing testosterone and deoxycorticosterone from diosgenin. He also hired outstanding scientists to pursue work on corticosteroids, among them Carl Djerassi, who developed synthetic routes for manufacturing estrogens and a variety of steroids and corticosteroids, including norethindrone, the active component of most contraceptive pills. By the early 1950s the price of estrogens had fallen from hundreds of dollars a gram to about $10. By the mid-1960s half of the steroids worldwide were made from diosgenin.

Operating in Mexico with a small, highly trained staff and without a marketing department, Syntex specialized in process technologies, which were widely licensed, and corticosteroid intermediates, which were sold in bulk to pharmaceutical companies. Only late in the 1950s, when the commercial potential of the oral contraceptive pill became evident, did Syntex move to the United States and become a true pharmaceutical company, selling as well as manufacturing their own trademarked products.

DIFFUSION OF TECHNOLOGY BY ACQUISITION

The fast diffusion of the technologies of the 1930s and 1940s and the profitability of the pharmaceutical companies drew into the industry several companies that manufactured diverse proprietary products. The most important of these were Bristol-Myers, American Home Products, Warner-Lambert, and

Johnson and Johnson. The traditional strengths of these companies were their marketing and sales expertise and their business acumen in expanding through acquisitions and mergers, both of which they applied to their pharmaceutical businesses. Indeed they had to follow the acquisition route because their expertise in research and development was modest at best. Entry by in-house innovation or even licensing was out of their reach. Entry into a research-intensive sector by acquisition is relatively easy if the acquiring company has the cash, but to remain competitive for a long period of time, as all four companies did, and even to emerge among the leaders, as American Home Products and Bristol-Myers did, is another matter.

These four companies succeeded largely because they gave considerable independence to the managements of the companies they acquired, which allowed their pharmaceutical subsidiaries or divisions to operate as research-intensive companies even when they generated only a small fraction of the companies' sales and profits, as they did in the 1950s and 1960s.

As the diversified companies deepened and widened their pharmaceutical business and acquired companies in other research-intensive sectors, such as scientific instruments and optical products, the importance and influence of R&D and innovation grew for the companies as a whole. This process accelerated in the 1980s, when the diversified companies sold off many of their traditional proprietary product businesses; as a result their sales of ethical pharmaceuticals—drugs that can be dispensed only with a physician's prescription—exceeded 50 percent of the value of their total sales (turnover), and their R&D expenditures approached those of the "traditional" pharmaceutical companies.

American Home Products

American Home Products was established in 1926 when several independent companies that manufactured household products were consolidated under a single management. In 1932 the company bought the Wyeth Corporation from Harvard University. (John Wyeth, the last member of the Wyeth family, had bequeathed the assets of his company to the university.) Wyeth was one of the biggest manufacturing apothecaries, selling hundreds of "traditional" medicines. Its business relied on innovative machines for making tablets and on strong marketing, but not on product or process innovation. Wyeth became the backbone of American Home Products' pharmaceutical business.

In the 1940s Wyeth participated in the federal penicillin project, became a producer of penicillin and streptomycin, and even introduced some patentable forms (Bucillin and Amphocillin). Wyeth also manufactured sulfonamides and vitamins and continued to sell a wide range of antacids, biologicals (serums and vaccines), fungicides, cardiotonics, and laxatives.

Recognizing the promise of pharmaceuticals, American Home Products embarked on a series of acquisitions in the 1940s that broadened the kinds of pharmaceuticals the company sold as well as the markets for them. The companies acquired were Whitehall-Pharmacal, a manufacturer of proprietaries, including the analgesic Anacin and the dentifrice Kolynos; the International Vitamin Corporation (1941); Reichel Labs (1942), which manufactured biologicals, plasma, and vaccines; the Ayerst, McKenna and Harrison Company Ltd. (1943), a large Canadian business that produced vitamins; biologicals; and sex, pituitary, and corticosteroid hormones, including the combination of estrogens sold under the trade name Premarin (which is still earning $500 million a year); and the Fort Dodge Serum Company (1945), a manufacturer of veterinary products. Thus, by the end of the 1940s, American Home Products had penetrated through acquisition all the pharmaceutical markets that Merck had entered through innovation. Once in American Home Products' fold these companies were encouraged to deepen their commitment to in-house R&D. Wyeth, for example, established the Wyeth Institute of Applied Biochemistry and introduced several innovations in the late 1940s and 1950s in antibiotics, analgesics, sulfonamides, and cardiovascular and central nervous system drugs.

In the 1950s American Home Products further concentrated its pharmaceutical activities by divesting many of its businesses in household products; by acquiring Ives Pharmaceuticals, which specialized in cardiovascular drugs; and by licensing drugs from third parties, particularly from European companies. This latter program was extensive and lucrative. Among the pharmaceuticals American Home Products licensed were the cardiovasculars pentolinium (Ansolysen from May and Baker) and cyclandelate (Cyclospasmol from Brocades), the antihistamine promethazine (Phenergan from Rhône-Poulenc), the central nervous system drugs primidone (Mysoline from ICI) and meprobamate (Miltown from Wallace; Equanil from Wyeth-Ayerst), and the anesthetic halothane (Fluothane from ICI). In the 1950s American Home Products was among the first American companies to launch first-generation cardiovascular and central nervous system drugs, albeit through licensing.

By the end of the decade American Home Products was still a diversified company, but its main business was pharmaceuticals. Ethical drug sales accounted for 45 percent of turnover; proprietary drugs, 24 percent; household products, 17 percent; and foods, 14 percent.

Bristol-Myers

Bristol-Myers was established in 1887 in Clinton, New York, as a manufacturing apothecary making standard medicinal preparations. By the turn of the century the company had moved to New York City, where the commercial success of its first proprietary products—Sal Hepatica, a laxative, and Ipana, a medical toothpaste—provided the small company's bread and butter. In the 1920s the company gave up the manufacture of its medicinal preparations to concentrate on toiletries. Market research, aggressive marketing, and advertising became the strengths of the business, which in the 1930s was extended to tooth, shaving, and paint brushes.

Although it had no experience in ethical drugs or in fermentation processes, Bristol-Myers recognized the potential for future growth offered by the federal penicillin project. The company therefore acquired the Cheplin Biological Research Laboratories of Syracuse, New York, which had qualified to participate in the project. Cheplin manufactured injectable drugs and had experience in industrial fermentation as a producer of acidophilus milk—cow's milk fermented with bacteria.

The financial and technical support provided by the federal government was essential to the transfer of fermentation technologies from Pfizer, Merck, and the government laboratories to Bristol-Myers. In 1945 Bristol-Myers bought the government-owned penicillin plant it had operated during the war and became a producer in its own right. The company also hired the best talent it could find to pursue research on antibiotics. The first successes came with the tetracyclines, largely through cooperation with Pfizer, but the company's greatest achievement was its work in semisynthetic penicillins. As a result Bristol-Myers established a very strong CTT in antibiotics that continues to this day. The company is a striking example of the influence that government technological policy can have on the technical and business orientation of a private firm.

In the late 1950s Bristol-Myers merged with Mead and Johnson, a research-intensive pharmaceutical company, and Westwood Pharmaceuticals, a manufacturer of dermatological drugs. The merger with Mead and Johnson led to another specialty—antineoplastic drugs. Despite these acquisitions and the considerable strengthening of in-house research Bristol-Myers remained a highly diversified company, and ethical drugs never accounted for more than a third of sales until its merger with Squibb in 1989.

Warner-Lambert

In contrast to American Home Products and Bristol-Myers, which by extensive acquisitions and in-house R&D took advantage of the opportunities offered by the technologies of the 1930s and 1940s, Warner-Lambert and Johnson and Johnson became strong competitors in the pharmaceutical industry only after a second stage of acquisitions and mergers took place in the late 1950s.

Warner-Lambert had its beginnings in the 1940s, when Warner, a manufacturer of toiletries, merged with Gustavus Pfeiffer and Company of St. Louis, a manufacturing apothecary. Ethical pharmaceuticals made up only 23 percent of the small company's sales. That changed when Warner undertook a series of mergers and acquisitions, the most important of which were Chilcot Labs (1952), which manufactured ethical pharmaceuticals, and Lambert (1956), which made proprietary pharmaceuticals. By the end of the 1950s Warner-Lambert was the fourth largest of the pharmaceutical companies, after Merck, Lilly, and Pfizer. Proprietaries accounted for 47 percent of sales, ethical pharmaceuticals for 24 percent, plastic and glass packaging for 14 percent, and cosmetics for 15 percent. Research expenditures were less than 3 percent of sales, however, and the innovations Warner-Lambert introduced were all incremental. The company sought to redress that situation in 1959 by establishing a central laboratory, the Warner-Lambert Research Institute, in Morris Plains, New Jersey. Until its merger with Parke-Davis in 1969, however, Warner-Lambert was a successful seller of pharmaceuticals, but it was not structured to be either research intensive or innovative.

Johnson and Johnson

Johnson and Johnson was established in 1874 as a manufacturer of surgical dressings, antiseptics, and disinfectants, most of which were discovered by Joseph Lister in the 1860s. It diversified into first-aid products and later baby foods. Since the turn of the century Johnson and Johnson has expanded its business through acquisitions and mergers. In 1940 it acquired its first pharmaceutical company, Ortho Pharmaceuticals, which specialized in gynecological products.

Johnson and Johnson's contribution to the technologies of the 1940s was modest, and its ethical pharmaceuticals business accounted for less than 10 percent of turnover. This figure doubled with the acquisitions of McNeil Laboratories (1958), the Swiss company Cilag Chemie AG (1959), and the Belgian company Janssen (1961). Janssen has since been responsible for Johnson and Johnson's most important innovations in the pharmaceutical field.

Research Intensity in the American Companies

The structure and competencies of companies determine to a great extent their business strategies, but radical technological changes create new opportunities that force companies to adapt to the exigencies of the new technologies and to change their traditional structures and markets. The technologies of the 1930s and 1940s had just such a profound effect on the pharmaceutical industry.

The American pharmaceutical companies that became the leaders of the industry in the second half of the century were a mixed lot. They included

- Research-conscious, process-innovating companies operating on the model of the chemical industry (Merck, Pfizer, Syntex);
- Research-intensive pharmaceutical companies that had contributed to the development of the turn-of-the-century technologies (Abbott, Lilly, H. K. Mulford, Parke-Davis);
- Pharmaceutical companies that had responded to the challenges offered by the technologies of the 1930s and 1940s but that still had made only a modest commitment to in-house R&D (Ayerst, Lederle, Searle, Smith Kline and French, Squibb, Upjohn, Wyeth);
- Horizontally diversified companies selling proprietary products that had joined the industry by merging with or acquiring pharmaceutical compa-

nies (American Home Products, Bristol-Myers, Johnson and Johnson, Warner-Lambert); and

- The subsidiaries of the Swiss companies, which were research intensive, had their own laboratories and manufacturing plants in the United States, and operated like American companies (Hoffmann–La Roche, Ciba-Geigy, Sandoz).

To remain competitive in this profitable business after World War II, many of these companies had to change their traditional methods of business. No longer could they rely solely on making and selling hundreds of "low-tech" proprietary drugs, other proprietary consumer products, or bulk specialty chemicals. Instead they had to adopt the practices of research-intensive companies that were developing, manufacturing, and selling "high-tech" ethical drugs. To this end they had to establish large R&D departments, staff them with highly trained scientists and engineers organized in multidisciplinary teams, create cooperative networks with academic institutions, organize patent departments, and establish large marketing and sales departments using sophisticated methods addressed to the medical profession. They also realized that if they were to compete successfully in that industry, they had to spend at least as much on R&D as their competitors.

This extensive restructuring took many years to complete. The reorganization of marketing, for example, took place in the 1950s (see following section). Some companies took longer to restructure than others, and some were unable to complete the restructuring at all and either merged with or were acquired by competitors. Others resorted to strategic mergers to acquire expertise they were unable to create internally. The speed with which each company achieved this transformation often depended on management's resistance to changing established business practices, on the methods chosen to bring about the changes, and on the company's technological and business performance during the years of change.

The companies that had invested in R&D in the 1920s and 1930s—Abbott, Lilly, Merck, Mulford (then in the fold of Sharp and Dohme), and Parke-Davis—did not need to invest much more to take advantage of the government-sponsored projects and the technological opportunities they opened. By the end of the 1940s these companies were spending about 2.5 percent of their annual sales on R&D. Such companies as Lederle, Pfizer, Schering, Searle, Squibb, and Upjohn invested more heavily—5 to 10 percent of annual sales—to create research departments able to respond to these opportunities. The horizontally diversified companies that entered the industry by acquisition and whose pharmaceutical sales were about 10 to 30 percent of their turnover, spent only about 2 percent on R&D. Thus, by the end of the 1940s, about fifteen American pharmaceutical companies were on their way to becoming research intensive but had by no means reached that goal.

THE POSTWAR DECADE

The second stage of the transformation of the American pharmaceutical industry was completed in the 1950s when the companies developed corporate structures and practices for marketing and sales in a peacetime economy that ensured both the return of their investments in R&D and a high level of profitability. This was facilitated by strong demand and opportunities for incremental innovation associated with the stages of technological maturity and decline.

Maturity of the Third-Generation Technology

In the 1950s natural antibiotics and corticosteroids went through their stages of growth and maturity; sulfonamides, antihistamines, and vitamins matured and declined. Thus there was a wealth of opportunities for corporate innovation, profits, and growth, particularly in the case of natural antibiotics. Many of these drugs were effective against a wide spectrum of disease-carrying bacteria, including some that had developed resistance to penicillin. Furthermore, unlike penicillin, which was not patented, and streptomycin, the patents for which were given by Merck to Rutgers University to allow its manufacture by more than one company, the innovating companies were able to patent the new antibiotics; their manufacturers thus made substantial profits for ten years or so until the semisynthetic penicillins and cephalosporins took over the market.

The demand for natural antibiotics led pharmaceutical companies and research laboratories to mount major programs to screen soil samples collected from around the world. Thousands of bactericidal chemicals were isolated; fifty were actually commercialized. J. D. Bernal, the biochemist and science historian, described that search in a rather pejorative way as a "gold rush" and a "feverish search among large groups of organisms for anything that will work" rather than a "properly conducted scientific prospecting operation." The search for new natural antibiotics was, however, typical of the sequence of events following the introduction of a commercially successful radical innovation, the scientific origins of which are not fully understood at the time. Companies try to acquire a share of the market through incremental innovation, which leads to further scientific and technological advances until the commercial potential of the technology has been exhausted. In short, science and technology advance along parallel lines. What may have caused Bernal's reaction was the great number of companies that undertook such searches.

The other group of drugs that was further developed in the 1940s but that did not peak until the 1950s was the corticosteroid and sex hormones. Because of the patent-sharing agreements among the participants of the wartime corticosteroid hormone projects and the availability in the 1950s of steroid intermediates at low prices, many innovations of the 1950s were introduced nearly simultaneously by a score of companies. As a result the prices of many corticosteroids, including prednisone and prednisolone, which were introduced in

1954, dropped so quickly that by the end of the 1950s they were hardly profitable. Moreover, the seriousness of the side effects associated with corticosteroid therapy placed their therapeutic and commercial potential under a cloud. For those reasons the companies that had established CTTs in corticosteroids (Syntex, Searle, Schering, and to some extent Upjohn) did not grow as much as did their competitors who had CTTs in antibiotics (Pfizer, Lilly, and Bristol).

The other technologies of the 1930s and 1940s still offered some small opportunities for innovation and profit but not to the extent of the antibiotics and corticosteroids. With the discovery of vitamin B_{12} in 1948 the search for new vitamins drew to a close. The technology reached the stage of decline because chemical manipulation could not improve further on the properties of vitamins, and their commercial potential was fully exploited by those process-innovating companies that sold vitamins in bulk.

The sulfonamides also matured and began their decline, both because the technology had diffused so quickly and because they were often replaced by antibiotics with superior properties and fewer side effects. Sharp and Dohme, Lederle, and Squibb continued their work with sulfonamides, which led to the discovery of the thiazide and carbonic anhydrase inhibitor diuretics. European companies pursued research on sulfonamides more vigorously. ICI, May and Baker, and Geigy all owed their entry into the pharmaceutical business to sulfonamides, while Ciba made commercially successful radical innovations in the field with sulfathiazole (Cibazol [1940]). In the 1950s some German companies introduced sulfonamides for treating diseases other than those carried by bacteria, such as the hypoglycemic (antidiabetic) tolbutamide (Rastinon from Hoechst) and the anticonvulsant sulthiame (Ospolot from Bayer).

Like the sulfonamides, antihistamines were a European discovery. The first antihistamine, phenbenzamine (Antergan), was developed and introduced by Rhône Poulenc (1942). The first antihistamine made in the United States was introduced by Ciba (tripelennamine, or Pyribenzamine) in 1946, but the race for imitation was launched by the introduction of three great commercial successes that soon followed: diphenhydramine (Benadryl from Parke-Davis [1946]), chlorpheniramine (Chlor-Trimeton from Schering [1948]), and dimenhydrinate (Dramamine from Searle [1949]). Despite the versatility of the main models for antihistamines and the many companies that were able to introduce incremental innovations in the 1950s, the technology matured and declined quickly because the incremental innovations did not improve on the therapeutic effects of the original drugs. Nor did they eliminate the side effects, notably that of drowsiness; this was not achieved until the 1980s.

The opportunities offered by antibacterial and anti-inflammatory drugs had an orienting effect on the research and innovation of the American companies in the 1950s. Consequently, most of the first effective cardiovascular and central nervous system drugs were introduced by European companies.

American companies had an important advantage in the 1950s: Their home market was by far the strongest because the European and Japanese economies were still struggling to recover from the war. With the exception of the Swiss, European companies had no access to the U.S. market except through licensing their innovations to the American companies (as did the Japanese companies in the 1980s and 1990s). Early access to new technologies gave licensees the opportunity to catch up with the innovators and to become competitors within a short period, sometimes by introducing more effective drugs than the originals and by appropriating markets that might otherwise have been beyond their reach. For example, Rhône-Poulenc developed the first tranquilizer, chlorpromazine (Largactil [1952]) and licensed it to Smith Kline and French (Thorazine [1953]). The profits helped turn the American company into a prosperous and research-intensive company, which subsequently introduced two innovations of its own, prochlorperazine (Compazine [1956]) and trifluoperazine (Stelazine [1958]), both of which were commercially successful. Hoechst licensed the first synthetic hypoglycemic tolbutamide (Rastinon [1956]) to Upjohn, which sold it profitably as Orinase and followed it with two commercially successful innovations: tolazamide (Tolinase [1966]) and glyburide (Micronase [1984]). The company that profited most from such licensing was American Home Products, which marketed ICI's products in the United States. These included the anesthetic halothane (Fluothane [1958]); the anticonvulsant primidone (Mysoline [1952]); and the first beta-blocker, propranolol (Inderal [1964]). Inderal continued to have annual sales of $200 million in the 1990s.

Research and Development Expenditures

The research and development expenditures for fourteen companies, both in millions of dollars and as a percentage of sales, show that most of them were not research intensive in 1950 but that—with a few exceptions—they were well on their way to becoming so by the end of the decade. That the innovative activity of this decade, as reflected in the numbers of new drugs and patents, was achieved with such a low level of R&D expenditures can only be attributed to the coincident growth and maturity of the technologies of the 1930s and 1940s (Table 7).

A company's level of R&D investment is largely determined by the profits it makes from in-house innovations. Because most American companies introduced profitable innovations in the 1940s and the 1950s, they were able to make healthy investments in R&D. The exceptions were three of the diversified companies (American Home Products, Johnson and Johnson, and Warner-Lambert), which had depended on the technologies of the pharmaceutical companies they acquired or on licensing for their products. Sterling, which also did not make substantial investments in R&D during the 1950s, depended instead on Bayer's technology. Smith Kline and French had not contributed much to the technologies of the 1930s and 1940s, but its commercialization of chlorpromazine in the United States made it one of the fastest growing and

Table 7. Research and Development Expenditures—Comparison by Company in the 1950s*

Company	1950		1959		1950–1959		
	In millions of dollars	Percentage of sales	In millions of dollars	Percentage of sales	In millions of dollars	Percentage of sales	Number of patents
Abbott	10.0	2.0	40.0	6.0	230	4.0	189
Bristol-Myers	7.5†	2.2	35.0	4.5	220	4.5	118
Eli Lilly	27.5	3.3	105.0	10.0	580	6.3	202
Johnson & Johnson	NA	NA	NA	NA	NA	NA	88
Merck	25.0	2.5	95.0	7.5	650	6.0	830
Parke-Davis	18.5	2.5	55.0	5.0	340	4.0	316
Pfizer	20.0	6.0	77.5	5.3	430	4.5	335
Schering-Plough	10.0	8.5	45.0	10.0	200	7.8	114
Searle	10.0	10.0	20.0	10.0	160	10.0	446
Smith Kline and French	10.0	5.0	60.0	9.0	330	6.5	63
Sterling	NA	NA	NA	NA	200†	2.0	300
Upjohn	27.5	6.0	85.0	9.5	530	8.4	866
Warner-Lambert	NA	2.0†	30.0	2.7	140†	2.2†	43

Data from company annual reports and *Chemical Abstracts*.
* In constant 1995 U.S. dollars. NA = not available.
† Estimated.

more profitable companies in the second half of the 1950s, which, in turn, led to considerable investments in R&D.

The Marketing of Pharmaceuticals

The 1950s witnessed not only the growing research intensity of the industry but also dramatic changes in the marketing and sales of pharmaceuticals. Until the 1950s pharmaceutical companies followed one of two paths for marketing their products. The few companies that were research intensive adopted the marketing practices of the chemical industry, patenting their manufacturing processes and synthetic products and selling their products in bulk as intermediates in the production process to industrial customers.

Most of the pharmaceutical companies, however, although sensitive to the necessity of R&D, were primarily manufacturing apothecaries. Most of their drugs—laxatives, cough syrups, analgesics, antiseptics, and the like—could not be patented, did not require medical prescription, and were sold directly to the consumer as over-the-counter (OTC) proprietaries. Because of the infinite number of possible combinations and formulations, trademarks and intense promotional marketing were indispensable. The diversified companies followed similar practices, not only for their drugs but for their other proprietary products, such as toiletries, cosmetics, house cleaning products, and confectioneries.

The technologies of the 1930s and 1940s and their fast diffusion forced companies in both categories to change their marketing strategies for two main reasons. First, the innovations of the period widened the markets for drugs by lengthening the list of diseases that could be cured or treated. The rush of many companies to participate in these new and lucrative markets flooded them with many formulations of the same active ingredient or with many drugs whose active ingredients differed only slightly from one another, but enough to allow their manufacturers to make diverse claims about their efficacy. (Penicillin, streptomycin, vitamins, and cortisone are examples of the former; antihistamines and sulfonamides of the latter.) With the exception of some vitamins all these products were ethical pharmaceuticals. It therefore became essential for companies to inform, and in many instances, to educate the physicians about their products' properties and uses, as most of these drugs were unknown when the physicians were attending medical school. The target of marketing thus shifted from consumers to professionals and acquired a scientific dimension.

Second, the fast pace of diffusion and the increase in the number of pharmaceutical companies led to a dramatic increase in the number of drugs available to treat any particular disease. Patents, which protected the active ingredients and their generic names, were not always adequate to create monopoly or oligopoly markets, so to defend their innovations, companies began to use company-specific trademarks. As this practice spread across the industry, companies promoted and physicians prescribed drugs by their trade name—sometimes adding the name of the manufacturer—rather than by the generic name. For example, the generic erythromycin was marketed as Ilotycin by Lilly and Erythrocin by Abbott. Drugs have been sold since as proprietary products by a generic name (usually protected by patents) and a trade name (protected by a trademark).

Merck and Pfizer, which had experienced vanishing profits as the technology for penicillin, streptomycin, and vitamins diffused throughout the industry, decided to market their drugs exclusively under their trademarks. Both companies realized that in the short run their decision would cost them the good will of their traditional industrial customers, many of whom, in any event, were by then competitors in antibiotics and vitamins. To survive, Merck and Pfizer had to appeal directly to physicians, hospitals, and drugstores, all of whom were confronted with the increasing complexity of the drugs themselves, with new modes of therapeutic action, and with competition for their attention from a greatly increased number of producers and products. They therefore created large sales departments that used sophisticated methods of marketing and advertising. Pfizer not only increased its sales force, which grew from twenty-five employees in 1949 to a thousand in 1953, but also upgraded its marketing efforts by acquiring, in 1952, Roerig, a company experienced in marketing ethical and proprietary drugs. Merck followed a similar pattern, merging in 1952 with Sharp and Dohme, a company that not only had an illustrious tradition in pharmaceutical research but was also strong in marketing.

Syntex, in contrast—and to some extent because of its foreign base—continued with its traditional way of doing business in the 1950s, extending it only by joint ventures with American companies (Lilly, Pfizer, and Parke-Davis, among others). Thus the diffusion of corticosteroid technology accelerated, and only in the late 1950s and early 1960s, after Syntex had moved from Mexico to the United States, did the company establish its own marketing department. This comparatively late switch in its marketing strategy explains in part why Syntex, which was acquired by Hoffmann–La Roche in 1992, remained a medium-sized company.

Traditional pharmaceutical companies, such as Abbott, Lilly, and Upjohn, had marketing and sales departments in place as the 1950s began. Indeed, for many of them, marketing was their strength in the first half of the century when product differentiation rather than technological innovation was the name of the game. These companies thus had only to adapt their methods to the changing commercial environment.

The horizontally diversified companies were traditionally manufacturers of proprietary products and, hence, far stronger in marketing than in research. Their need, apart from stressing the scientific dimension of marketing, was to enlarge their pharmaceutical business, which they did by acquiring or merging with traditional research-intensive pharmaceutical companies.

Thus the American pharmaceutical industry consolidated the advantages it had gained during World War II by investing heavily in in-house R&D; developing large, specialized marketing departments; and building the corporate structures and practices necessary to operate profitably in a peacetime economy. These steps not only ensured that the American industry would continue to dominate the worldwide pharmaceutical market, but also positioned the industry to respond to the scientific, technological, and commercial opportunities of the second half of the century.

Growth and Profitability

The economic picture for the pharmaceutical industry was bleak during the first half of the 1950s but robust in the second half. From 1950 to 1954 demand was flat and profitability disastrous; from 1954 to 1959 both were vigorous. The U.S. recession during the Korean War was responsible for the poor performance in the early years of the decade. High taxation sharply reduced profits, which for the companies that dealt predominantly in pharmaceuticals fell from 13 to 17 percent to about 8 to 9 percent. Stock value for these companies remained flat (Figure 4). These negative effects were reinforced by a set of industry-related factors. Overproduction of penicillin, streptomycin, and vitamins caused their prices to collapse, affecting all producers of those products.

At the same time newer antibiotics and corticosteroids were still in their early stages of growth, and market demand for them was just beginning to grow. As a result most American companies, which had been gearing up to take advantage of strong domestic demand and a near-monopoly of overseas markets, cut back their planned investments in new factories and in some cases in R&D.

The end of the Korean War signaled the beginning of a period of worldwide economic growth that—with minor business cycle interferences—lasted until the mid-1970s. The end of the 1950–53 recession coincided with the growth stages of antibiotics and corticosteroids and the introduction of the first effective antihypertensives and central nervous system drugs.

Between 1954 and 1959 the companies dealing mostly in pharmaceuticals regained their profitability, which reached 14 to 18 percent by the end of the decade. Stock values for these companies more than tripled (see Figure 4). Profits for the less innovative pharmaceutical companies and for the diversified companies were in the 9 to 10 percent range.

The pharmaceutical companies also grew in size in the 1950s. In 1950 American Home Products, Johnson and Johnson, and Sterling were the largest, each with employment around 11,000 and sales ranging from $180 million to $250 million. Ethical pharmaceuticals accounted for less than half of their turnover. The next-largest group included four innovative companies of the 1920s and 1930s—Abbott, Lilly, Merck, and Parke-Davis. These companies employed between 6,000 and 8,500 people and had sales ranging from $100 million to $160 million. The remaining companies were small, employing 1,000 to 3,000 and with sales ranging from $22 million to $90 million (Table 8).

By 1959 only two companies Searle and Schering, could still be considered small. Smith Kline and French employed only 4,100 people, but it had sales of $154 million. All the others employed from 6,000 to 17,500 people and had sales of $150 million to $600 million.

For these companies it became clear during the 1950s that manufacturing pharmaceuticals was more profitable than making other proprietary products—which is still true. Profitability in the 1950s was directly proportional to the percentage of total sales accounted for by pharmaceuticals. The profitability of the strongly diversified companies (where the ratio of pharmaceutical sales to total sales was 0.1 to 0.45) ranged between 5.0 and 8.2 percent, while that of the predominantly pharmaceutical companies (whose ratio of pharmaceutical sales to total sales averaged 0.6 to 1.0) ranged between 11.0 and 21.0 percent. Accordingly, almost all industry acquisitions and mergers of the 1950s involved other pharmaceutical companies, as both predominantly pharmaceutical and diversified companies sought to take advantage of the opportunities for innovation offered by the growing and mature technologies and to strengthen their positions in the by-now-peaking markets.

Although high profitability is essential for R&D–intensive companies and is amply rewarded by the stock markets, corporate growth is equally important.

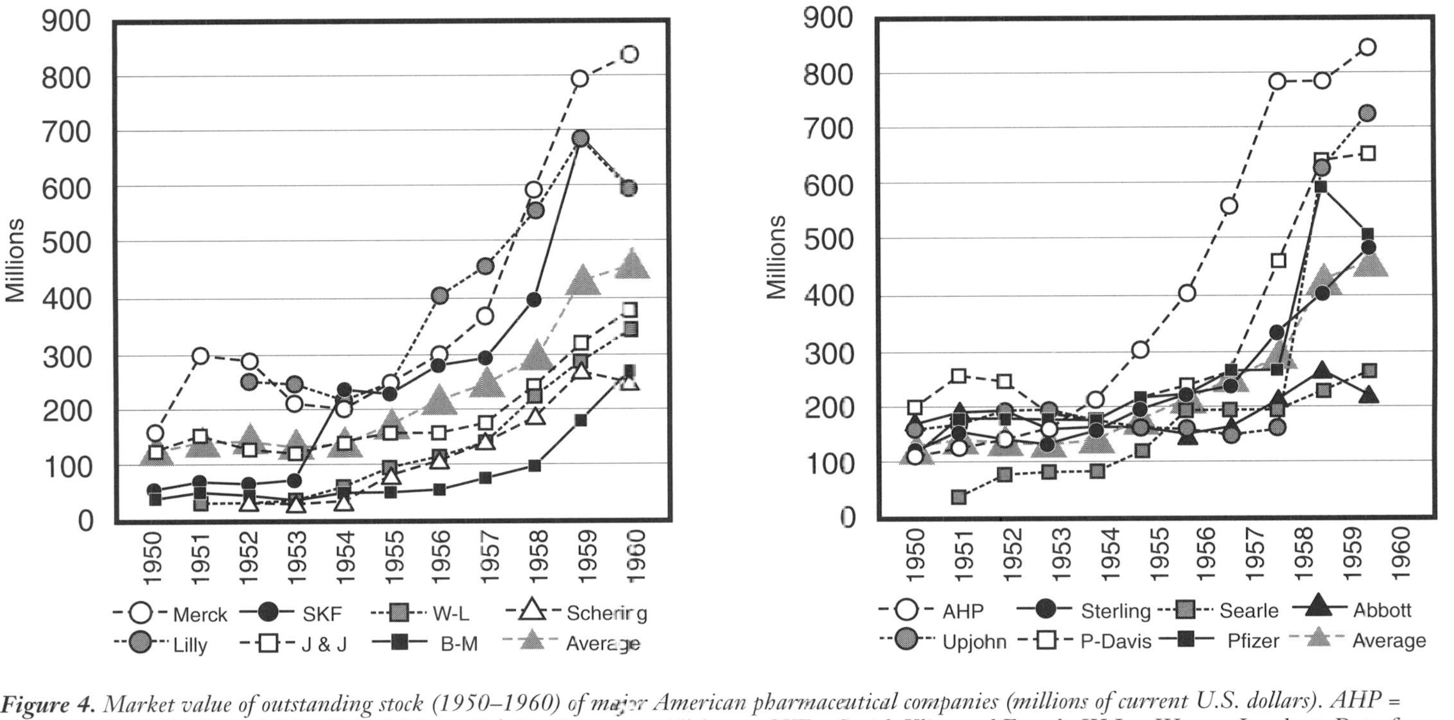

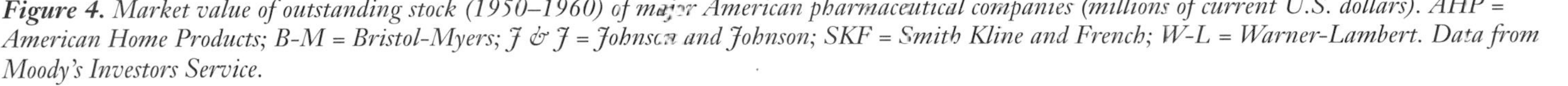

Figure 4. *Market value of outstanding stock (1950–1960) of major American pharmaceutical companies (millions of current U.S. dollars). AHP = American Home Products; B-M = Bristol-Myers; J & J = Johnson and Johnson; SKF = Smith Kline and French; W-L = Warner-Lambert. Data from Moody's Investors Service.*

Table 8. Growth of Total Sales, Ethical Pharmaceutical Sales, and Employment by Company in the 1950s (in millions of constant 1995 U.S. dollars)

Company	1950				1959				1950–1959			
	No. employees (in thousands)	Total Sales	Ethical Pharm. Sales	Percentage	No. employees (in thousands)	Total Sales	Ethical Pharm. Sales	Percentage	Total Sales	Ethical Pharm. Sales	Percentage	Profit-ability*
Abbott	6.0	505	450†	90†	8.7	700	450	64	5,880	4,175†	71†	11.3
American Home Products	11.0	1,235	520	42	17.2	2,570	1,140	44	17,070	7,625	44	8.6
Bristol-Myers	2.0	360	90	25	6.0	750	160	21	4,900	1,110	22	6.4
Johnson & Johnson	11.2	1,125	112†	10†	13.7	1,735	173†	10†	15,275	1,527†	10†	4.5
Lilly	7.7	835	710†	85†	9.7	1,065	905	85	9,190	7,810†	85†	13.3
Merck	6.0	835	625†	75†	11.6	1,235	925	75	10,830	7,800†	75†	10.8
Parke-Davis	8.6	735	685	93	10.6	1,095	1,018	93	8,430	7,840†	93†	13.8
Pfizer	3.4	425	245†	58†	17.5	1,450	825†	57†	9,510	5,420†	57†	10.7
Schering	1.1	110	95	86	3.7	455	350	77	2,580	2,115	82	15.3
Searle	1.0	115	115	100	1.2	195	195	100	1,600	1,600	100	21.0
Smith Kline and French	1.6	270	256	95†	4.1	770	732†	95†	4,965	4,717†	95†	15.5
Sterling	11.0	900	405†	45†	14.2	1,305	444	34	10,775	4,300†	40†	7.9
Upjohn	3.0	445	400†	90†	5.7	895	800	90	6,290	5,660†	90†	12.9
Warner-Lambert	3.0	325	70	22	10.0	1,090	260	24	6,545	1,705	26	7.4

Data from company annual reports and estimates from them.
*Profitability: After-tax profits as percentage of total sales for the decade.
†Estimated.

Growth—even at lower rates of profit—provides financial stability and confidence in the future, both necessary for a company to embark upon risky, long-term research projects. Three companies (Lilly, Merck, and Upjohn) that had introduced important and commercially successful innovations in the 1940s and 1950s made a strong commitment to internally generated growth by spending more than 6 percent of their turnover on R&D and another 6 percent on capital investment. (All three have continued that commitment into the present.) Pfizer came a strong fourth by spending nearly 7 percent of turnover on new plants and 4.5 percent on research. A notable case is that of American Home Products, whose low aggregate percentages in both research and capital investment camouflage relatively large sums.

On the negative side Abbott, Parke-Davis, and Sterling all underinvested in R&D and new facilities in the 1950s and were to suffer the consequences in the 1960s and 1970s. Similar fates would befall Schering and Searle, not because they did not commit themselves to research and innovation, but because their relatively small size did not allow them to invest on the same scale as their larger competitors. Last, Bristol-Myers, Warner-Lambert, and Johnson and Johnson invested much less in research and new plants, depending instead on mergers and acquisitions for their growth in the 1950s.

THE EUROPEAN PHARMACEUTICAL INDUSTRY

In the 1930s the European research-intensive companies had an advantage over their American counterparts because they had helped to launch several technologies and open their accompanying markets at the turn of the century. These included antiprotozoal drugs, anesthetics, analgesics, hypnotics, serums, and vaccines. In addition European companies launched two of the technologies of the third generation—sulfonamides and antihistamines—and were leaders in developing sex and corticosteroid hormones, with Schering AG, Organon, and Ciba among the protagonists. But in vitamins—with the exception of Roche—and antibiotics the role of the European companies was less important than the contributions of the European academic institutions.

World War II was a period of slowdown, even standstill, in innovation for all but a handful of the European companies. The war effort in Britain and Germany and the occupation of France and other European countries interrupted both academic and industrial research other than that directly related to defense. At the end of the war companies had to overcome physical destruction and economic hardship to pick up the threads from the 1930s.

The speed with which countries and companies recovered their research and innovation capabilities after the war differed markedly. The Swiss industry and companies were the least affected by the war and went through a "golden" period. The German industry and companies were affected the most and never regained the leadership position they had held before the war. The other European pharmaceutical industries and companies fell somewhere between these two extremes, and their achievements in the 1950s were primarily scientific and technological rather than commercial. Their recovery might have taken longer had it not been for the outstanding advances of European academic institutions. Their work in the chemical transmission of nervous impulses opened for the pharmaceutical industry the major sectors of cardiovascular and central nervous system drugs. The full impact of these technologies was felt in the 1960s and 1970s, but the first drugs introduced in the 1950s were harbingers of the European industry's resurgence.

The Swiss Industry

Until the late 1920s the Swiss chemical companies had made only modest contributions to pharmaceuticals (see "The Swiss Industries" above). Competition and low growth in the dyestuffs industry forced these companies to look for opportunities to diversify. For the Swiss companies, pharmaceuticals and the emerging sector of pesticides fit well with their expertise in organic synthetic chemistry and the marketing of high value-added products, while staying within the limits imposed on them by geography.

As is commonly the case in the history of research-intensive companies, the main thrust in these early stages of development came from two areas. First was the close cooperation with national academic institutions, which were among the leaders in scientific disciplines and technologies relevant to emerging or newly established technologies. Second was the introduction of commercially successful radical innovations, which served as models for further corporate R&D and led to strong, long-lasting, and lucrative CTTs.

The Department of Organic Chemistry of the Eidgenössische Technische Hochschule in Zurich, the leading academic laboratory in natural products chemistry, which was under the leadership of Leopold Ruzicka, played a significant role in the emergence of the Swiss pharmaceutical companies because many of its professors were also industrial consultants. Work on steroids by Ruzicka and Tadeus Reichstein, later the director of the School of Pharmaceutical Chemistry in Basel, led to Ciba's predominance in sex hormones and corticosteroids. Paul Karrer and Richard Kuhn's work was instrumental in Hoffmann–La Roche becoming the leading European producer in vitamins.

The Swiss companies also received expertise from an unexpected source in the 1930s, when the country provided refuge to Jewish scientists from Germany and other central European countries, which were soon to be occupied by the Nazis. This wave of immigrants greatly benefited not only Swiss medicinal research but other sectors as well. Among the immigrants was Leo Sternbach, who discovered the benzodiazepines chlordiazepoxide (Librium) and diazepam (Valium) while working in Roche's research laboratories in Nutley, New Jersey. The companies had greatly expanded their business in the United States in the prewar years, both to provide insurance against a German takeover of Switzerland and to take advantage of the American market.

Geigy was a latecomer to the pharmaceutical industry, attracted in the mid-

1930s by the introduction of the sulfonamides and the opportunities they appeared to offer. Its first product was the antibacterial sulfadicramide (Irgamide [1938]), while Ciba introduced one of the most successful sulfas, sulfathiazole (Cibazol [1940]). Shortly after antihistamines were discovered, Geigy and Ciba produced their own versions: Ciba made tripelennamine (Pyribenzamine) and antazoline (Antistine), and Geigy, halopyramine (Synpen), all in the 1940s. Thus, with the exception of antibiotics, the Swiss companies contributed to the development of all the pharmaceutical technologies of the 1930s and 1940s.

An important first was Ciba's early involvement in antihypertensive drugs. After closely watching the British research on peripheral dilators, Ciba introduced the vasodilators tolazoline (Priscol [1940]) and, most important, hydralazine (Apresoline [1949]), an orally active antihypertensive vasodilator that was still in use in the late 1990s.

After the war, two of the Swiss companies became leaders in central nervous system drugs: Geigy with imipramine (Tofranil [1959]) and Roche with iproniazid (Marsilid [1958]). Both during and after the war Sandoz pursued its work on ergot alkaloids, introducing the oral uterine stimulant ergometrine (1934) and synthesizing LSD, or lysergic acid diethylamide (1938–47), which before its hallucinogenic properties were discovered was considered to be better than ergometrine. Sandoz's strong CTT in this area led it to the introduction in the 1980s of synthetic ergot medicines for the treatment of Parkinson's disease.

The British Industry

Up to the early 1930s the only British research-intensive pharmaceutical company was Burroughs Wellcome, which did not play a significant role in any one of the technologies of the third generation of drugs. Indeed, Burroughs Wellcome turned down an opportunity to develop penicillin and made only modest contributions in sulfonamides. Yet the emphasis Burroughs Wellcome placed on basic research led the company to develop very original drugs in therapeutic sectors that were to become important in the 1960s and beyond. Cooperation with the U.K. Medical Research Council Laboratory, headed at the time by Henry Dale, led to the early antihypertensive bretylium (Darenthin [1959]), which was still being used as an antiarrhythmic in the 1990s and which opened a new sector for the company. Cooperation with the Chester Beatty Institute led Burroughs Wellcome to the antineoplastic drugs biosulphan (Myleran [1950]) and chlorambucil (Leukeran [1952]).

Burroughs Wellcome's greatest successes, however, came from its American research laboratory. In 1940 Donald Woods, working at Oxford University, showed that the antibacterial properties of sulfanilamide arose from the drug's interference with the process of rapid reproduction, which is essential for the survival of parasites, be they bacteria, protozoa, or viruses. While searching for compounds with similar properties, George Hitchings and Gertrude Elion, working out of the company's Tuckahoe, New York, laboratory, initiated a research program in 1942 to synthesize purine and pyrimidine analogs. This long-term, highly speculative research project was immensely successful; it led to highly innovative and commercially lucrative drugs in a score of therapeutic markets and created a CTT for the company. Among these drugs were the antibacterial drug pyrimethamine (Daraprim [1949]), the antineoplastic drugs 6-mercaptopurine (Purinethol [1953]) and azathioprine (Imuran [1968]), the antigout agent allopurinol (Zyloprim [1966]), the combination of two antibacterials, co-trimoxazole and sulfamethoxazole (Septrim [1970]), and the antiviral agent acyclovir (Zovirax [1985]) (see Chapter 6).

The great British discovery of the third generation of drugs was penicillin. Neither Alexander Fleming, who serendipitously discovered it in 1928, nor his team of researchers had the skills in chemistry required to isolate penicillin in pure form and characterize it. That had to wait until 1939, when Howard Florey and his coworkers at Oxford University undertook the task. Within a year they had isolated penicillin and conducted pharmacological tests on animals and clinical trials that proved its antibacterial properties and low toxicity. Florey then approached British companies about undertaking the industrial development of penicillin. Boots and Burroughs Wellcome declined. Glaxo and ICI showed some interest, but Glaxo had little expertise in chemistry and ICI lacked expertise in industrial fermentation. Furthermore, wartime pressures convinced Florey that success would not be achieved in Britain, so he decided to cross the Atlantic to try to persuade American companies of penicillin's merits.

Subsequent reports from clinical trials in both Britain and the United States led the British government in 1942 to form the Therapeutic Research Corporation of Great Britain, Ltd., to pool the results on specified projects, the most important of which was the manufacture of penicillin in quantities sufficient to meet the needs of the armed forces. Britain's major pharmaceutical firms participated in the Therapeutic Research Corporation and produced penicillin, albeit by surface cultures. Glaxo built four factories, which produced 7.5 billion units in 1944, or about 80 percent of the British output. In 1945 Merck transferred its deep fermentation process technology to Great Britain through licensing agreements with Glaxo and the Distillers Company, and its subsidiary Commercial Solvents. By the end of the war, British production had reached 100 to 120 billion units a month.

The lessons drawn by the British government from the experience with penicillin led it to establish the National Research and Development Corporation in 1948 "to secure when the public interest so requires the development or exploitation of inventions resulting from public research and any other invention not being sufficiently developed or exploited." Ten years later Florey's Oxford research team and the National Corporation played a key role in the development of both cephalosporins by Glaxo and semisynthetic penicillins by Beecham.

The introduction of the sulfonamides in the mid-1930s gave a strong incentive to several companies to undertake research in pharmaceuticals. May and

Baker, which was acquired by Rhône-Poulenc in 1927, became one of the first innovators by introducing sulfapyridine (MB 693) in 1938. Imperial Chemical Industries, which was established in 1925 by the merger of four chemical companies to compete with I.G. Farben, formed a pharmaceutical research group within its famous Dyestuffs Research Laboratory. That group championed sulfonamide research and introduced related antimalarial drugs during the war, such as chlorguanide (Paludrine [1944]).

In general, British contributions to the third generation of pharmaceuticals were modest, especially when compared with those of the 1960s and beyond. The strength of the industry lay with its academic institutions and the Medical Research Council Laboratory, which participated in nearly all of that period's innovations. But the potential of British academic research was recognized and appreciated by many American companies, which established research laboratories in Britain in the 1950s and 1960s to take advantage of it.

The French Industry

France's major pharmaceutical company, Rhône-Poulenc, experienced its golden age during the 1930s, with major innovations in sulfonamides and the discovery of antihistamines. The latter came about in 1942, when the company, in cooperation with the Pasteur Institute, introduced the first antihistamine, phenbenzamine (Antergan). Rhône-Poulenc also began to manufacture vitamins, amino acids, and eventually antibiotics. In 1952 the company made another major innovation with its introduction of chlorpromazine (Largactil), the first neuroleptic agent. This discovery, which was directly related to Rhône-Poulenc's work on antihistamines, not only launched the development of drugs for treating disorders of the central nervous system but also revolutionized the therapy of psychiatric patients. Within a few years the number of patients in psychiatric hospitals worldwide was reduced by half.

In the postwar years Rhône-Poulenc became one of the largest chemical and pharmaceutical companies in the world. Between 1956 and 1966 the company acquired Theraplix and a majority stake in Laboratoires Roger Bellon and Laboratoires Merrieux. The latter had been established in Lyon at the turn of the century by a student of Pasteur and produced vaccines and veterinary drugs. Once in Rhône-Poulenc's fold it became one of the world's top producers of vaccines for humans; after its merger with Connaught Laboratories of Canada (1994), it became the world's largest producer of vaccines and serums.

Despite its size and its strong ties with the Pasteur Institute, Rhône-Poulenc's contributions to pharmaceutical innovation since the 1960s has not matched its work of the 1930s and 1940s.

The Decline of the German Industry

Just as World War II was a catalytic event for the American pharmaceutical industry, it was also a crucial factor in the decline of the German pharmaceutical industry. The physical destruction, the abrogation of patents, the interruption of industrial activities were all severe blows to the functioning of German research-intensive companies.

An even more critical factor, however, was the German industry's inability to remain innovative in the face of new competition from other European countries and the United States. The merger of Hoechst and Bayer in the fold of I.G. Farben reduced competition between the two, costing them the technological edge they had held for so many decades. With the exception of the sulfonamides, which Domagk had pioneered at Bayer, and Schering's work on sex hormones, German companies played only a modest role in the introduction of the technologies of the 1930s and 1940s. Synthetic organic chemistry had led them to drugs that relieved only the symptoms of diseases rather than their causes—analgesics and antipyretics, for example—and to drugs that fought such disease-causing intruders as protozoa and bacteria. But, as Heinrich Hörlein, director of pharmaceutical research at I.G. Farben recognized as early as 1932, the German companies could not match the advances that British and American academic institutions had made in physiological chemistry, which was essential for research on hormones; vitamins; and cardiovascular, anti-inflammatory, and central nervous system drugs—drugs that could directly counter the physiological consequences of disease.

In short, German pharmaceutical companies were never able to recover their leading position—not only because of the war but also because their weakness was predominantly technological. A comparison of the postwar performances of the German chemical and pharmaceutical industries illustrates the point. Although Germany's chemical companies suffered more wartime loss, both in physical destruction and intellectual property, Bayer, Hoechst, and BASF re-emerged as the three world leaders of the chemical industry in the early 1970s largely because their technologies and R&D of the 1930s and 1940s were among the most advanced in the world and led to a substantial number of key innovations in the immediate postwar period. For example, although DuPont, Dow, and ICI, among others, introduced many new polymers—for example, polyethylene, nylon, polyesters, and polyacrylics—the German companies responded with polyurethanes, low-pressure polyethylene, and Nylon 6 and closely followed the Allied wartime leaders in synthetic fibers, organic chemicals, dyestuffs, and pesticides, eventually recovering their leadership. That Bayer and Hoechst, the two leading chemical companies, did not succeed in doing the same with their pharmaceutical businesses further supports the hypothesis that technology, and not only the war, was at the root of the German pharmaceutical industry's decline.

By necessity and under the influence of their tradition in organic synthetic chemistry, after World War II, the German companies pursued their prewar leads in sulfonamides, hypnotics, antiprotozoals, and analgesics and extended their work in antihistamines. But this period was generally characterized by isolated innovations, such as Hoechst's antidiabetic drug tolbutamide (Rastinon [1956]) and the coronary dilator prenylamine (Segontin [1958]) and Boehringer's antihypertensive agent etilefrin (Effortil [1950]). The recovery of the German industry had to await the advent of the fourth generation of pharmaceuticals.

The Maturing of an Industry (1960–1980)

The fourth generation of drugs, which encompassed the 1960s and 1970s, was one of particularly promising innovation based on advances in the life sciences. It was also a period in which the American pharmaceutical industry experienced great uncertainty before embarking on the corporate strategies that would place it among the most prosperous manufacturing sectors in the economy.

Unlike earlier generations of drugs, which worked primarily by countering disease carrying organisms, the fourth generation was aimed largely at correcting illnesses caused by heredity, diet, and environmental factors. Among the radical innovations were drugs to treat hypertension and other heart diseases, anxiety, and depression. Breakthough developments also occurred in antibacterial and nonsteroidal anti-inflammatory drugs (NSAIDs). And the first effective contraceptive pills came on the market, triggering profound social and cultural changes throughout the world. These new technologies, which reached their peak in the 1970s, opened new lucrative markets and extended old ones, giving the pharmaceutical industry ample opportunities for expansion and profit.

These innovative drugs were launched during a period of immense economic growth in the United States, Western Europe, and Japan. The industrialized world enjoyed very low unemployment, and consumers could afford a wide variety of household goods and products, many of which had been considered luxuries before and immediately after the war. Demand for pharmaceuticals was also high. Not only were drugs available to treat hundreds of diseases and conditions, but most industrial countries introduced public health insurance programs that paid for at least part of the costs of prescription medications.

Yet, in the midst of this widespread economic growth, the managers of many American drug companies in the early 1960s began to doubt whether research, innovation, manufacturing, and marketing of drugs, profitable as it was, would continue to be so in the future and whether the pharmaceutical industry could support rates of growth comparable to those of other manufacturing sectors. Several factors combined to cause these doubts. First, the competitive environment for pharmaceuticals in the 1960s was not that of the 1940s. The innovating companies of the 1960s protected their products by numerous patents that covered the model, a considerable number of related structures, and the manufacturing process, thus strongly restricting the possibilities for swift imitation. Government-sponsored programs were not aimed at the development of specific drugs, as they had been in the 1940s, but instead supported long-term academic research on cancer, cardiovascular illnesses, and other widespread and complicated diseases and therefore did not offer opportunities for drastic diffusion of technology. Companies had to depend on their own laboratories and expertise and on the research carried out by academic institutions, particularly those whose research they supported. Furthermore, new federal regulations governing the approval of new drugs prolonged the time it took for those drugs to reach their market and begin to produce revenue for their manufacturers.

Thus, relatively few companies made profits in the 1960s from the models of the fourth generation. Companies that were not among the innovators did not break into the new technologies until the 1970s, while many others, unable to replicate the technologies of the innovative companies, had to wait even longer for the expiration of the basic patents. The only opportunity for accelerated diffusion of technology that persisted in the 1960s was the licensing by the European companies of their innovations to American companies.

Although many commercially successful radical innovations with great potential for imitation were introduced in the 1960s, their effect on the pharmaceutical markets was not fully felt until the 1970s. Together with strategic steps the companies took to strengthen their economic performance, the drugs of the fourth generation were responsible for the considerable growth and the return of confidence that most American pharmaceutical companies enjoyed in that decade.

TECHNOLOGIES OF THE FOURTH GENERATION

By the early 1960s the technologies of the 1930s and 1940s had matured, and many were in decline. Patents on the early antibiotics, corticosteroids, antihistamines, sulfonamides, and vitamins had expired, and many companies, including newcomers, were producing them as "generics"—that is, selling them under their generic name without trademark—or introducing incremental innovations with little if any improvement in therapeutic properties over drugs already on the market. As a result a great many products crowded the markets, sharpening competition and squeezing profits.

The commercially successful radical innovations that launched the fourth cluster of technologies were introduced between 1959 and 1964. They included, among antihypertensive drugs, the thiazide diuretic chlorothiazide (Diuril from Merck [1959]) and the beta-blocker propranolol (Inderal from ICI [1964]); among central nervous system drugs, the tranquilizer haloperidol (Haldol from Janssen [1959]), the antidepressant imipramine (Tofranil from Ciba-Geigy [1959]), and the anxiolytic chlordiazepoxide (Librium from Roche [1960]); among antibacterial agents, the semisynthetic penicillin phenethicillin (Broxil from Beecham [1959]) and the cephalosporin cephaloridine (Ceporin from Glaxo [1964]); the NSAIDs indomethacin (Indocin from Merck [1963]) and ibuprofen (Brufen from Boots [1964]); and the contraceptive pill (Enovid from

Searle [1961]). These models offered a great scope for further innovation by imitation and molecular manipulation.

Advances in Life Sciences

The new drugs of the 1960s and 1970s were made possible by significant theoretical and experimental advances in the life sciences. The elucidation of the mechanisms of diverse physiological processes at the cellular level were necessary for designing and synthesizing effective drugs and for interpreting the action of serendipitously discovered drugs.

This knowledge was indispensable in the case of the chemical transmission of nervous impulses in the peripheral autonomic and central nervous systems. The research leading to the discovery that nervous impulses trigger physiological processes in multiple and diverse organs by means of chemical transmitters can be traced back to the turn of the century, when epinephrine and acetylcholine were isolated, characterized, and recognized as physiological products. But the identification of their role as chemical transmitters proved a long and controversial research project because of the experimental and theoretical difficulties involved and because most academic researchers early in the century were convinced that electrical rather than chemical mediation was responsible for the transmission of nervous impulses. Not until the mid-1930s was sufficient evidence accumulated to persuade medical researchers of the critical role of chemical transmitters. Industrial interest in the transmitters of nervous impulses and their physiological effects was kept alive by the discovery of the bronchodilator and antiasthmatic properties of the sympathomimetic amines (ephedrine in 1926 and isoproterenol in 1940).

The first antihypertensive drugs that were found to interfere with the release of natural chemical transmitters appeared in the early 1950s, but most of these drugs showed many unwanted side effects. Just as natural chemical transmitters trigger several physiological processes, so did the initial drugs modeled after them. The decisive breakthrough came in 1948 when Raymond P. Ahlquist attributed the diversity of the effects of epinephrine to the presence in tissues of two types of adrenergic receptors: α-adrenergic receptors, associated with epinephrine's most excitatory effects (such as peripheral vasoconstriction) and one inhibitory effect (intestinal relaxation); and β-adrenergic receptors, associated with most inhibitory functions (such as vasodilation and bronchodilation) and one excitatory function (myocardial stimulation). This extremely important hypothesis, which was to revolutionize medicinal research, suggested that chemicals could be synthesized to selectively block one type of receptor, allowing the transmission of epinephrine-mediated impulses only by the other. This seminal theory led Sir James Black to the discovery of the β-adrenergic blockers (propranolol [Inderal from ICI, 1964]), which lower blood pressure by blocking the adrenergic stimulation of the heart during nervous excitation or physical effort; this action slows cardiac rhythm and the need for oxygen without blocking peripheral vasoconstriction, which is triggered through the α-adrenergic receptor.

In contrast to the antihypertensive drugs the radical innovations that opened the sector of CNS drugs in the late 1950s were serendipitous discoveries. Apart from the commercial success of these drugs, which led to the vigorous expansion of the sector in the 1960s and 1970s, they served as tools in elucidating nervous transmission in the central nervous system. This work led to the identification of other chemical transmitters, such as serotonin, dopamine, and γ-aminobutyric acid (GABA); of their receptors; and of their role in mental illnesses, such as schizophrenia, depression, anxiety, and Parkinson's disease.

Advances in the life sciences also made possible radical departures in three other technologies of the 1960s and 1970s, all of which were related to technologies of the 1930s and 1940s. Thiazide diuretics, commercialized by Merck in 1958, were related to the sulfonamides but were discovered after a lengthy study of the functions of the kidney in the 1940s and 1950s by Sharp and Dohme and Merck.

Semisynthetic penicillins and cephalosporins were related to natural antibiotics but were developed only after the central and active parts of the molecules of penicillin and cephalosporin were isolated, characterized, and produced commercially. The study of the relations between the structure of the antibiotics and their medicinal properties led to the synthesis of numerous products in the 1960s and 1970s. Some of these were effective against a wide spectrum of bacteria, and others killed bacteria that had developed resistance to penicillin and other antibiotics.

The first oral contraceptive pill was not a new chemical entity but a combination of an estrogen and a progestin that were synthesized in the 1940s. That formulation was made possible, however, only after the study of the hormonal and other chemical changes that accompany the female fertility cycle. That study allowed researchers to determine the quantities of the active principles that provided effective birth control while minimizing, if not eliminating, side effects.

The NSAIDs were the only technology of the fourth generation to be discovered by the classical method of synthesizing and screening chemicals structurally related to existing drugs to improve their therapeutic properties and eliminate their side effects. The physiological mechanisms by which anti-inflammatories function were not clarified until the late 1970s.

The Bandwagon of the 1970s

The maturing of the technologies of the 1960s, combined with the geographical spread of the innovating companies and strong consumer demand, created the strongest bandwagon effect in the history of the pharmaceutical industry. While the NSAIDs were still in their growth stage, contraceptive pills, antihypertensive agents, semisynthetic penicillins and cephalosporins, and CNS drugs (imipramine, chlorpromazine, and benzodiazepines) all came to maturity during the 1970s, making innovation by imitation possible for a great many companies, including many that had not previously been innovators. Japanese, Italian, and French companies all joined the bandwagon, as did a few Swedish, Danish, and Belgian companies. For example, thirty new analgesics and anti-inflammatories, mainly NSAIDs, were introduced in the 1970s, including in-house innovations from twelve companies that were entering the sector for the first time. Fifty-five new CNS drugs were introduced; twenty-one companies entered that sector for the first time. And thirty-eight new antibacterial drugs were introduced, with nine companies innovating for the first time. The innovative activity in sectors that had reached the stage of decline was much weaker. For example, there were seventeen antiprotozoal drugs (mainly for veterinary use), with only four companies entering the sector; seven antihistamines, with only two companies entering; and twenty-seven corticosteroids (twelve of which were for topical use), with only five new entrants.

Numerous other radical innovations introduced in the 1960s were lucrative for the innovating companies but did not offer the same potential for imitation, either because of their complex chemical structure or because of the unique way in which they functioned. Among these drugs were the synthetic antibacterial nalidixic acid (NegGram from Sterling [1963]), the antihypertensive methyldopa (Aldomet from Merck [1962]), the antibiotics gentamycin (Garamycin from Schering [1966]), and clindamycin (Cleocin from Upjohn [1964]), and the diuretic triamterene (Dyrenium from Smith Kline and French [1964]).

Beginning in the mid-1970s some of the radical innovations that would lead to the technologies of the fifth generation were introduced. Most were based on further specialization in blocking or enhancing the action of chemical transmitters of the autonomous nervous system or on interference with the action of enzymes. Drugs were also introduced to aid in the treatment of cancer, a disease for which few effective drugs have been found.

Prazosin (Minipress from Pfizer [1976]), an antihypertensive α_1-blocker; cimetidine (Tagamet from Smith Kline French [1977]), an antipeptic ulcer histamine H_2-inhibitor; and nifedipine (Adalat from Bayer [1974]), an antianginal and antihypertensive calcium channel blocker, were the model drugs that helped launch the fifth generation of pharmaceuticals (discussed in more detail in the following sections). Meanwhile the antineoplastic agents doxorubicin (Adriamycin from Farmitalia [1974]), tamoxifen (Nolvadex from ICI [1977]), and cisplatin (Platinol from Bristol [1978]) provided effective treatment for some forms of cancer.

Of particular importance was the discovery of recombinant DNA processes (1975–78). These processes insert segments of one organism's DNA—nucleic acids that form the molecular base for heredity—into the DNA of another organism, causing the latter to produce proteins identical to those of the former. The potential of these methods for diagnosing and treating disease and for studying biological processes at the molecular level was unparalleled and would lead to the formation of the biotechnology industry. Moreover, the use of computers in imaging such physiological structures as receptors and enzymes helped in the design of molecular structures that could block or enhance the action of chemical transmitters or enzymes. The revolutionary advances in biological sciences were thus offering many new opportunities for radical innovations that would lead to the fifth generation of drugs.

These advances certainly did not mean the abandonment of the classical methods for identifying molecules with therapeutic properties, synthesizing them, and formulating them into appropriate medicinal forms. Even screening of soil samples to identify new bactericidal metabolites was continued with success. But as is so often the case, the new methods were developed in academic laboratories, where they were recognized as potentially powerful tools for pharmaceutical R&D, and companies began to take notice and to apply them.

TIGHTER GOVERNMENT REGULATION

The revolutionary scientific and technological developments of the 1940s and 1950s and the vast opportunities they created dramatically changed the structure and business practices of the pharmaceutical industry. Nowhere was this more apparent than in the United States, where high rates of innovation and fast diffusion of technology brought dozens of new drugs to the market every year. Between 1953 and 1962 no fewer than 564 new drugs were released, 462 of which were considered to be new chemical entities by the Food and Drug Administration (FDA). Many of these, however, were in fact imitative drugs without significant therapeutic advantage over drugs already on the market.

To some extent this proliferation of new drugs was caused by a lack of strict federal regulations. Existing FDA requirements for approval of new drugs, which were adopted in 1938, were concerned primarily with toxicity and did not require the companies to submit evidence on the therapeutic efficacy of a new product or on its differences from existing drugs. Nor were companies required to inform the FDA of their intention to initiate clinical trials or to submit progress reports on the trials themselves. Furthermore, the FDA was required to act on a new drug application within sixty days of its submission unless the agency deemed the information provided to be incomplete.

The Kefauver Committee

Government regulatory systems respond to revolutionary technological changes only when their consequences begin to affect the established economic and social environment. By the late 1950s the high profitability of the pharmaceutical industry led to charges of price fixing, and in 1958 Congress established a committee under the chairmanship of Senator Estes Kefauver to investigate industry practices. The committee found that direct manufacturing costs for new drugs accounted for only 32 percent of sales income and that after-tax profits for the industry were 21 to 22 percent, twice the average profits for all manufacturing. Attributing the high profitability to the introduction of drugs that were more expensive but not necessarily more effective than the drugs they replaced, the committee began hearings on the pharmaceutical business that were to last three years.

The pharmaceutical industry did not deny its high profitability but pointed out that the committee evaluations did not take into account R&D or marketing expenditures or the numerous candidate drugs that never reached the marketplace. Industry spokesmen said that the numerous new drugs that were introduced resulted from the fast rate of postwar scientific and technological advance, which frequently made even recently discovered drugs obsolete. Under those conditions high sales volume and hence profits from any one drug could not be expected to last for long, and thus the survival of the companies depended on continuous successful innovation, which entailed both technological and financial risks. Consequently, the industry argued, price controls on new drugs would stifle innovation to the detriment of public health.

The Thalidomide Tragedy

During the course of the hearings the most tragic incident in the history of the pharmaceutical industry took place. Gruenenthal Chemie, a small German company, discovered a sedative called thalidomide (Pantosediv). It was sold without prescription in Germany from 1957 to 1961 and was also produced under license in France, the United Kingdom, and other European countries. (It was also licensed for manufacture in the United States, but the FDA found the information provided inadequate and withheld approval.) Clinical tests in Germany had not included pregnant women, so thalidomide's teratogenic effects on developing fetuses were discovered only after thousands of babies were born with horribly malformed limbs. The tragedy led to demands around the world for stricter controls on the development and approval of new drugs. In the United Kingdom, for example, Parliament established the Dunlop Committee to investigate the practices of the British pharmaceutical industry.

The thalidomide incident turned the Kefauver Committee's attention to FDA regulatory practices and legislation that would prevent similar incidents from happening. Congress enacted the Kefauver-Harris Drug Amendments in October 1962, and the FDA's new regulations came into effect in February 1963. Proposals dealing with antitrust and monopoly control, including pricing of new drugs, length of patent life, and obligatory licensing to competitors, fell by the wayside in the rush to protect the American public from tragic accidents.

The new FDA regulations required companies to submit information on preclinical toxicity tests (carcinogenicity, teratogenicity, and mutagenicity tests were added later), on the proposed clinical trials, and on the qualifications of the investigators. The companies also had to show that the clinical subjects had consented to participate after being informed of the potential risks involved. Detailed progress reports on the clinical trials had to be submitted regularly. To win approval for a new drug, a manufacturer had to provide the FDA with "substantial evidence" through well-controlled investigations that the drug was effective as well as safe for its proposed indications.

The new regulations had several significant effects on the drug industry:

- The delay between submission of a new drug application and its approval by the FDA increased substantially from seven months to up to eight years. Today it has decreased to an average of approximately two years.
- The nominal effective patent life of seventeen years was shortened by an average of four years because of the prolongation of preclinical and clinical trials and the longer time required to win FDA approval.
- The number of new chemical entities reaching the clinical-trial stage fell by an estimated 60 percent or more.
- The introduction of new drugs in the United States fell about 70 percent, from 564 in the decade before 1962 to 166 in the decade after 1962. Drug introductions in the United States lagged behind France, Germany, and the United Kingdom by an average of 1.0, 1.6, and 2.1 years, respectively.

The pharmaceutical industry blamed the decline in drug introductions on the FDA regulations and said they were stifling innovation. Although the additional time involved in the approval process delayed the arrival of new drugs on the market in the 1960s, the industry's charges were overstated. The companies did not take into account that the technologies of the fourth generation of drugs were just beginning to diffuse, while the technologies of the third generation were declining, providing few opportunities for innovation. Furthermore, the effectiveness of defensive patenting by the innovating companies limited the opportunities for swift imitation by their competitors. That the decline in the number of new drugs reaching the market was a temporary phenomenon can be seen in Figure 5, which shows the numbers rising again in the mid-1970s and the 1980s.

The number of new drugs first introduced in the American market also declined after the new regulations went into effect. In 1960–62 about two-thirds of all new drugs were introduced in the United States before being offered in other countries. By 1972–74 only about 30 percent of all new drugs were offered first in the United States, even though many of the others were discovered by U.S. companies. Once again the pharmaceutical industry blamed the FDA regulations for the decline, and once again the claims were exaggerated. The American companies did not acknowledge the resurgence of innovation among European companies, who by this time had overcome the devastation of World War II and rebuilt their research laboratories. In the 1960s and 1970s they were benefiting from strong national economies and the establishment of the European Common Market. European companies made significant con-

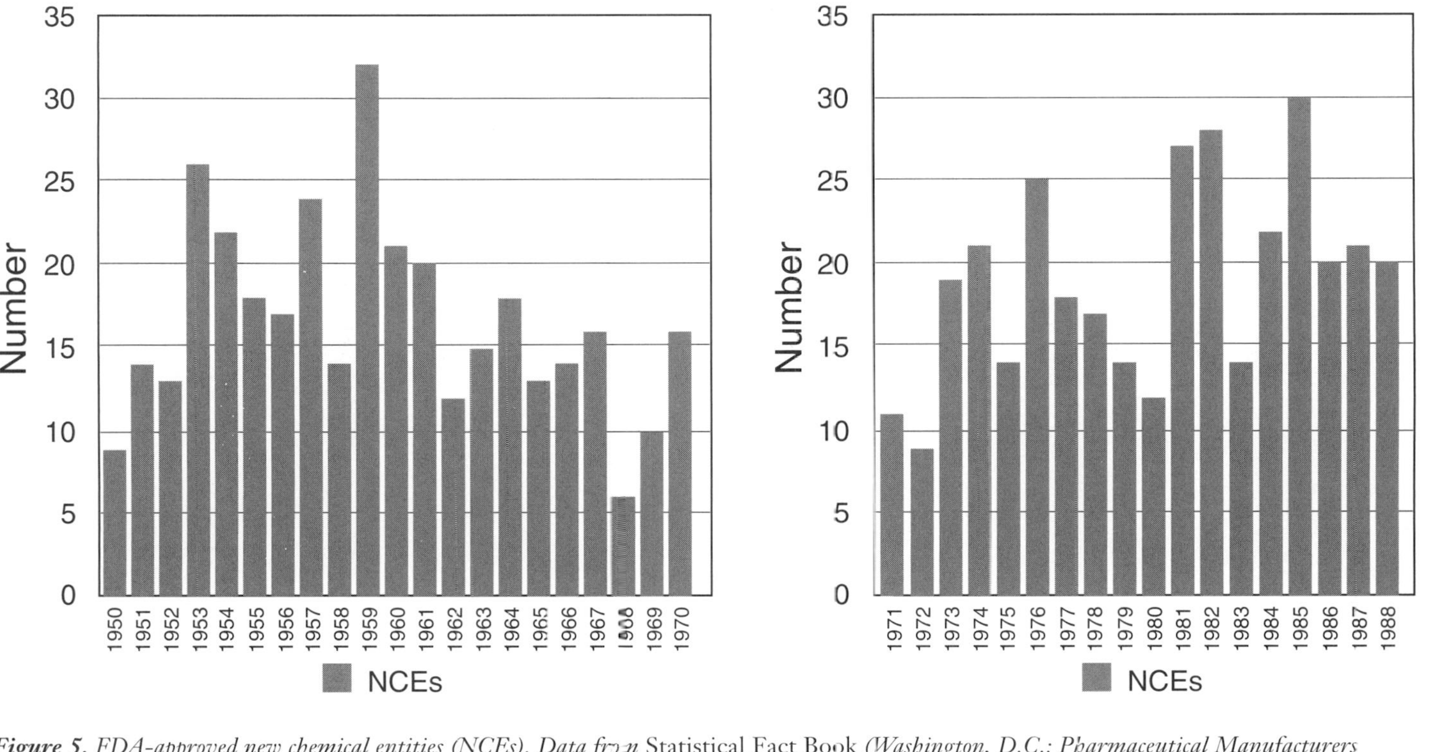

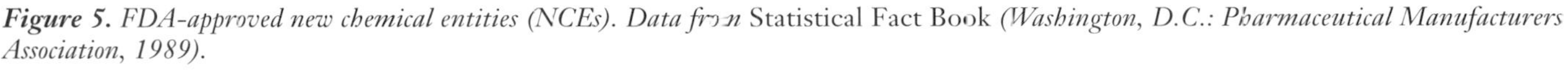

Figure 5. *FDA-approved new chemical entities (NCEs). Data from* Statistical Fact Book *(Washington, D.C.: Pharmaceutical Manufacturers Association, 1989).*

tributions to the technologies of the fourth generation of drugs: Beecham in semisynthetic penicillins; Glaxo in cephalosporins; ICI in antihypertensive beta-blockers; Boots and Ciba in NSAIDs; and Janssen, Ciba, Hoffmann–La Roche, and Rhône-Poulenc in CNS drugs.

The American companies also did not acknowledge that they launched many of their new drugs through their subsidiaries in Europe to circumvent the FDA regulations and to take advantage of clinical testing in Europe, which was generally less cumbersome and less expensive. This tactic also eliminated any competitive advantage European companies could have enjoyed from the disparity in regulations. In the mid-1970s American companies launched 60 percent of their new drugs in Europe before they were approved by the FDA. That number declined to just 20 to 25 percent in the mid-1980s after European governments instituted regulations similar to those in the United States.

The most positive reaction of the pharmaceutical industry to the new regulations was the development of far more efficient laboratory methods for testing new chemical entities in vitro, on isolated tissues and organs, and on animals. These new laboratory methods curtailed the number of candidate drugs that were submitted to clinical trial for any given research project and went a long way toward demonstrating therapeutic activity.

The 1962 Drug Amendments generated a great deal of controversy between the industry and the government that lasted for more than a decade. Industry staff and academics wrote tens of books and hundreds, if not thousands, of papers giving often-conflicting evaluations of the effect of the new regulations on drug innovation and on the economics of the pharmaceutical industry. It is not known whether the regulations themselves were beneficial or whether the industry was flexible and skilled enough to circumvent the restrictions that would have been detrimental to its business. In any event, thirty-five years later, the amendments form the backbone of regulations governing the industry worldwide, the industry is still one of the most innovative sectors of manufacturing, the drugs introduced since 1962 have greatly benefited society, and the thalidomide incident has not been repeated.

MARKET DEMAND AND CORPORATE POLICIES IN THE 1960S: SEARCHING FOR SECURITY

Although the innovative side of the American pharmaceutical industry may have been thriving as the fourth generation of drugs began, the business side was not growing at the same rate. Despite the growth of the drug companies in the 1950s, most of them were still only medium sized in the early 1960s; their opportunities for further growth appeared modest compared with those of other manufacturing sectors or of such basic industries as iron and steel, aluminum, coal, oil refining, and power generation. Nor did pharmaceuticals seem to offer the same opportunity for growth as some sectors of the chemical industry, such as petrochemicals, plastics, synthetic fibers, and fertilizers.

Larger families and increasing personal incomes had created an unprecedented demand for housing, furniture, home appliances, and automobiles. Demand was also strong for consumer products that before World War II had been considered luxuries—cosmetics; toiletries; better, cheaper, and more easily prepared foodstuffs; and comfortable and fashionable clothing. These sectors were not as profitable as pharmaceuticals, and most of them required much higher investments in manufacturing plants; but they grew much faster and were creating gigantic companies. The pharmaceutical companies appeared unable to follow that trend.

To deal with this problem of growth, the American pharmaceutical companies adopted two policies: They expanded overseas sales, particularly in the thriving Western European markets, and they diversified horizontally by merger and acquisition into markets of proprietary products.

International Expansion

In the 1940s and 1950s most American pharmaceutical companies had established marketing and sales operations and sometimes manufacturing plants in Latin America, Britain, and British Commonwealth countries, particularly Canada and Australia, while selling their products through local representatives in European, Asian, and African countries. In the 1950s these companies realized that some of their representatives had created local monopolistic markets and operated with profit margins superior to those in the home market. The formation of the European Economic Community in 1958 raised the possibility that the American companies might soon encounter trade barriers in Europe, which were to be erected to protect European companies rebuilding after the war. Several of these companies had already provided evidence of their resurgence in innovation and might soon pose a new source of competition.

To ensure their position in the European market, the American companies began to establish their own outlets in Western Europe during the early 1960s. In addition to large marketing and sales organizations in most Western European countries, this expansion included manufacturing facilities in strategically selected countries and even some R&D laboratories. Most of these labs were located in Britain because of its outstanding academic system and expertise in disciplines related to medicinal research, the much lower salaries of highly trained researchers, and the common language and heritage. A few companies established or acquired laboratories in Belgium, France, Germany, Italy, and Switzerland.

The dramatic effects of this European expansion are illustrated in Table 9. For many American pharmaceutical companies overseas sales increased by about 10 percent of total sales between 1960 and 1969. Apart from the strategic considerations involved, this growth was beneficial for the American companies because it did not require the discovery of new drugs and thus coincided perfectly with the slowdown in the diffusion of innovation. This trend continued in the 1970s, and by 1979 overseas sales accounted for about 40 to 50 percent of the markets of the major American pharmaceutical companies. The establishment of overseas subsidiaries also helped the American companies launch their innovations in European countries, thus both avoiding the delays imposed by the new FDA regulations and maintaining their competitive advantage over European companies.

Horizontal Diversification

During the 1940s and 1950s the high profitability of pharmaceuticals attracted several companies that sold diverse proprietary products into entering the industry by acquiring or merging with pharmaceutical companies. At the same time several pharmaceutical companies sought to strengthen their technical and market positions by acquiring or merging with other drug makers. Corporate growth and profitability in this period depended primarily on the sale of pharmaceuticals and specifically on the ability of companies to innovate. The stock market clearly approved of this strategy: Most stocks of research-intensive companies scored the highest gains in the 1950s, while most of the horizontally diversified companies scored the lowest (see Figure 4).

In the 1960s this trend was turned upside down. Ethical drugs, despite their much higher profitability, could not generate by themselves the high rates of growth necessary for the survival of companies as independent entities, Merck being the exception to this rule. Why was this true when higher profitability benefits shareholders in terms of distributed income and capital gains and allows managements to increase R&D expenditures and capital investment? It seems that investing in less profitable sectors of proprietary products provided stability to corporations by balancing the risks inherent in the dynamics of technological innovation with those of the business cycle. Cash flows from well-established trademarked proprietary products were more secure than those of many ethical drugs whose commercial lives were frequently shortened by the introduction of ever-more-effective drugs. The clustering of technologies created an uneven distribution of opportunities for innovation, the very element that the pharmaceutical companies depended upon for their survival. A steady flow of new drugs—and profits—is impossible because the nature of innovation itself is erratic, making the introduction of commercially successful radical innovations and the initiation of a corporate technological tradition easier at some stages than at others. Furthermore, while corporate technological traditions strongly enhance a company's innovative capabilities in some areas, they restrict its ability to take advantage of the opportunities offered in other areas. A degree of normality could be gained by diversifying into less research-intensive and innovation-dependent sectors, a fact that was appreciated by both managers and shareholders.

Proprietary products did not require large investments in R&D, but mainly intensive marketing and sales efforts commensurate with the expertise of both

Table 9. Overseas Sales as Percentage of Total Sales by Company (1960–1980)*

Company	1960 (%)	1969 (%)	1979 (%)
Abbott	27	28	40†
American Home Products	17	22	33
Bristol-Myers	21	16	NA
Johnson & Johnson	26	39	46
Lilly	16†	25	40
Merck	28	39	47
Pfizer	38	48	52†
Schering-Plough	22	45	47
Searle	17	28	40
Smith Kline and French	14	19	35
Sterling	38	38	46
Upjohn	13	30	40
Warner-Lambert	27	22	44†

Data from company annual reports.
* NA = not available.
† Estimated.

pharmaceutical and horizontally diversified companies. Large, diversified companies became the landmark of the manufacturing industry in the 1960s and 1970s, particularly in the chemical industry, where the introduction of numerous polymers and synthetic fibers was rapidly opening new markets. Big companies with diverse markets appealed to the investor of the 1960s and 1970s as being less dependent on the vagaries of any one market and allowing managements to shift resources from profitable sectors to others going through adverse phases in technology or demand.

Moreover, many of the sectors into which companies diversified were closely related to pharmaceuticals either because of their technology or their markets: Veterinary products were related to medicinal research because some drugs, such as the anthrax vaccine and many antiprotozoal agents, were equally effective for humans and animals. Sulfas and antibiotics, growth hormones, and vitamins made animal farming and the broiler industry possible, and even some CNS drugs were found useful in treating house pets. Such health care products as disposable sutures, bandages, injection needles, intravenous solutions, nutrients, and disinfectants were sold to hospitals, one of the most important market outlets for pharmaceuticals; and cosmetics and toiletries—the most common sector for diversification after veterinary products—required both chemical and medicinal expertise and were distributed largely through drugstores, another prime market outlet for pharmaceuticals. Another important area of diversification for pharmaceutical companies was eye care products, such as simple sight-correcting lenses, contact lenses, diverse ocular preparations, and operating equipment. Here, too, drugstores, physicians, and hospitals were the companies' main customers.

Table 10. Horizontal Diversification of American Pharmaceutical Companies in the 1960s and 1970s (major acquisitions only)

Company	Acquired Companies	Business	Year
Abbott	M & R Dietetic Labs	Nutritionals, baby formula	1963
	Many companies	Medical electronics, clinical laboratories	
	Faultless Rubber Company	Hospital care services	1960s
American Home Products	Ecco Products	Housewares	1965 (divested 1982)
	E. J. Brach	Confectioneries	1965 (divested 1986)
	A.R. Lite	Cookware	1968 (divested 1982)
	The Prestige Group	Housewares	1970 (divested 1982)
	Corometric Medical	Medical instruments	1973
Bristol	Clairol	Toiletries, cosmetics	1959
	Drackett	Household products	1965 (divested 1992)
	Mead & Johnson	Pharmaceuticals, baby foods	1967
	Westwood	Dermatological medicines	1968
	Zimmer	Orthopedic prosthetics	1971
	Unitek	Orthodontic, dental materials	1979 (divested 1987)
Johnson & Johnson	Janssen	Pharmaceuticals	1961
	Godman-Scritzef	Surgical instruments	1971
	Dr. Karl Hahn Inc.	Tampons	1974
Lilly	Distillers	Pharmaceuticals	1963
	Elizabeth Arden	Cosmetics	1980 (divested 1980s)
	Ivac	Electronic medical instruments	1978 (divested 1990s)
	Cardiac Pacemakers	Pacemakers	1978 (divested 1990s)
Merck	Calgon	Water purification; environment	1968 (divested 1980s)
	Quinton	Proprietaries	1967 (divested 1980s)
	Baltimore Coil	Industrial refrigeration	1969 (divested 1980s))
	Kelco	Chemicals from kelp	1972
	Hubbard Farms	Poultry genetics	1972

Pfizer	C. K. Williams	Metals, pigments,magnetic iron oxide	1962 (divested 1990s)
	Leeming & Paquin	Proprietary health products	1962
	Coty	Cosmetics	1963 (divested 1990s)
	Quigley	Industrial refractories	1968 (divested 1990s)
	Howmedica	Hospital, orthopedic, dental supplies	1972
Schering	Plough	Consumer proprietaries, household products, Maybelline cosmetics	1970 (partially divested 1980s)
	Dr. Scholl	Foot health products	1979 (divested 1980s)
Searle	Nuclear Chicago	Scientific, industrial instruments	1960s (divested 1981)
	Fermo Labs	Scientific instruments	1960s (divested 1981)
	Buchler Instruments	Scientific instruments	1960s (divested 1983)
	(many)	Diagnostics	1970s
Smith Kline and French	Norden Labs	Veterinary products	1960
	Sea and Ski	Sports equipment, sunglasses	1965 (divested 1983)
	Branson Instruments	Ultrasonics	1965 (divested 1983)
	Avocet	Health foods	1966
	Clinical Labs	Medical services	1970
	Hydron	Contact lenses	1979
Sterling	Hilton-Davis	Pigments, dyes, optical brighteners flushed colors for inks	1962 (divested 1980s)
	Lehr & Fink	Toiletries, cosmetics (including Dorothy Gray), disinfectants	1966
Upjohn	Carwin	Polyurethanes	1962 (divested 1984)
	CPR International	Polyurethanes	1963 (divested 1984)
Warner-Lambert	American Chicklet	Confectioneries	1962
	American Optical	Optical instruments, lasers, corrective lenses	1966 (divested 1982)
	Eversharp Schick	Shaving blades, toiletries	1969
	Parke-Davis	Pharmaceuticals	1969
	Entenmann's	Packaged foods	1969 (divested 1982)

A few pharmaceutical companies ventured into such unrelated sectors of proprietaries as household products, foods, sports equipment, and chemicals. Table 10 presents the most important acquisitions of the major pharmaceutical companies in the 1960s and 1970s. (It also shows that this trend was reversed in the 1980s.)

By 1969 the top stock market performers included the most diversified companies: American Home Products, Bristol-Myers, Warner-Lambert, Johnson and Johnson, and Pfizer. Merck and Lilly were the only research-intensive companies among the top performers (Figure 6). Table 11 shows the growth of fourteen American pharmaceutical companies in the 1960s by listing total sales, pharmaceutical sales, and profitability. Table 12 presents R&D expenditures for the same companies and gives the number of patents each was granted during the decade. Several observations can be made using these data:

- Most of the companies grew vigorously as a result of these new mergers and acquisitions. For the first time there were three very large companies—American Home Products, Pfizer, and Warner-Lambert, each with more than thirty thousand employees—and another four—Bristol-Myers, Johnson and Johnson, Merck, and Sterling, with more than 20,000 employees. American Home Products passed the $5 billion mark in sales; Bristol-Myers passed the $4 billion mark; Pfizer and Werner-Lambert the $3 billion mark; Merck, Lilly, Sterling, and Johnson and Johnson the $2 billion mark; and only Searle had sales of less than $1 billion.
- As a result of diversification most traditional pharmaceutical companies saw pharmaceuticals decrease as a percentage of total sales during the decade. Lilly's percentage decreased not because of acquisitions but because it developed in-house a very profitable business in herbicides. Lilly was also unique in that it was the only pharmaceutical company during the decade to acquire another pharmaceutical company (the pharmaceutical business of the U.K. Distillers Company, Ltd). The other three companies that acquired or merged with pharmaceutical companies were diversified companies striving to strengthen their pharmaceutical business: Bristol-Myers acquired Mead Johnson and Westwood; Warner-Lambert acquired Parke-Davis; and Johnson and Johnson acquired Janssen. It is not then surprising that pharmaceuticals accounted for a higher percentage of these companies' sales by the end of the decade.
- That company managers were skeptical about the effectiveness of in-house R&D is reflected in the numbers. Although the dollars spent on R&D increased, expenditures as a percentage of total sales stayed about the same or even decreased during the 1960s. The diversification into less research-intensive sectors exacerbated this trend.
- Stock value trends in the 1960s indicate that the business cycle had only a minor effect on performance and that the effect was more pronounced in the shallow recession of 1960–62 than in the deeper recession of 1969–72.

Table 11. Growth of Total Sales, Ethical Pharmaceutical Sales, and Employment by Company in the 1960s (in millions of constant 1995 U.S. dollars)

Company	1960				1969				1960–1969			
	Number of Employees (in 1,000s)	Total Sales	Ethical Pharm. Sales	Percentage	Number of Employees (in 1,000s)	Total Sales	Ethical Pharm. Sales	Percentage	Total Sales	Ethical Pharm. Sales	Percentage	Profitability*
Abbott	9.0	890	500	56	17.3	1,890	600	33	12,910	6,035	47	9.6
American Home Products	18.7	2,710	1,160	43	40.4	5,785	2,085	36	40,925	15,635	38	10.2
Bristol-Myers	6.0	825	180	22	26.0	4,220	905	21	21,875	4,305	19	8.0
Johnson & Johnson	17.0	1,750	263	15	21.4	2,950	530	18	22,775	3,735	16	7.0
Lilly	10.1	1,005	900	90	18.4	2,450	1,700	69	16,400	11,780	72	12.6
Merck	11.3	1,230	800	65	21.1	2,940	2,055	70	19,200	13,970	73	15.5
Parke-Davis	11.8	1,130	900	80	14.8	1,240	1,060	85	11,350	9,980	88	11.4
Pfizer	18.0	1,520	900	59	30.0	3,810	1,900	50	26,780	12,840	48	10.0
Schering†	3.6	475	390	82	8.0	1,010	889	88	6,575	5,650	86	11.9
Searle	1.5	210	210	100	7.9	745	580	78	4,785	3,600	75	21.5
Smith Kline and French	4.6	725	625	86	9.4	1,575	950	60	11,700	8,530	73	16.1
Sterling	14.0	1,375	465	34	23.4	2,705	935	35	19,450	6,770	35	9.0
Upjohn	6.0	900	810	90	11.6	1,690	1,255	74	12,230	10,050	82	13.1
Warner-Lambert	10.0	1,120	305	27	38.0	3,675	2,000	54	21,170	7,700	35	9.0

Data from company annual reports and estimates from them.
*Profitability: after-tax profits as percentage of total sales for the decade.
†Figures include all proprietary pharmaceutical products.

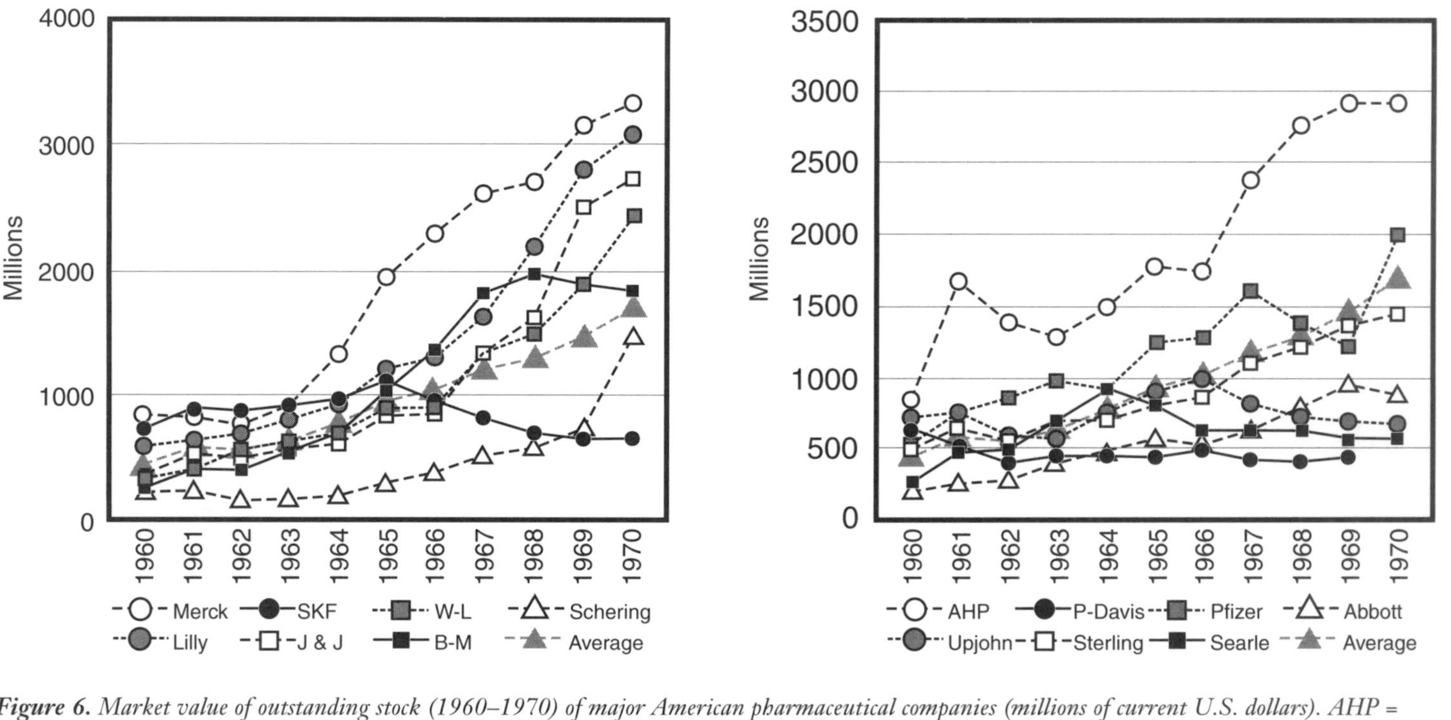

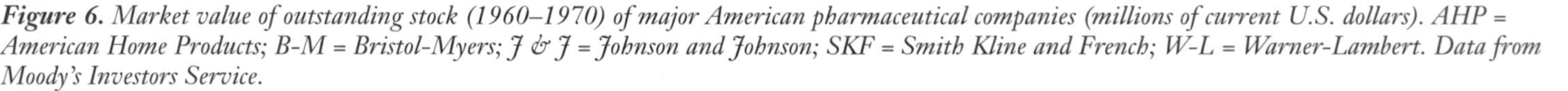
Figure 6. *Market value of outstanding stock (1960–1970) of major American pharmaceutical companies (millions of current U.S. dollars). AHP = American Home Products; B-M = Bristol-Myers; J & J = Johnson and Johnson; SKF = Smith Kline and French; W-L = Warner-Lambert. Data from Moody's Investors Service.*

By the end of the 1960s the values of individual stocks had diverged sharply: In 1960 they ranged between $200 million and $800 million; in 1969 the range was from $400 million to $3.5 billion (Figure 6). The companies with the highest stock values at the end of the decade were those that had grown the most and, to a lesser extent, those like Merck and Lilly, that had introduced fourth-generation drugs. Research intensity and profitability did not reflect stock value performance: Among the most research-intensive companies only Merck and Lilly did well in the stock market. The stock value of Searle, Smith Kline and French, and Upjohn declined despite their high commitment to R&D.

MARKET DEMAND AND CORPORATE POLICIES IN THE 1970S: ENSURING GROWTH

The corporate policies of the 1970s mimicked those of the 1960s with one important difference: The skepticism about ethical pharmaceuticals had evaporated. The intention in the 1970s was not to run away from pharmaceuticals but to ensure growth.

Demand for pharmaceuticals grew substantially in the 1970s despite the economic recession that began in 1973 and the beginning of the decline of manufacturing in the Western world. By the end of the decade the pharmaceutical industry had emerged as one of the few manufacturing sectors that not only survived the crisis but prospered.

This expansion occurred because the technologies of the fourth generation had reached maturity, by which time many key patents had expired and companies had become familiar enough with the technologies to circumvent the patents that were still in force. As a result the industry introduced numerous diuretics, antihypertensives, cholesterol reducers, tranquilizers, anxiolytics, antidepressants, semisynthetic penicillins and cephalosporins, analgesics, and NSAIDs. Because of the wide therapeutic scope of these drugs the relatively modest markets of the 1950s expanded both qualitatively and quantitatively. By the late 1970s physicians could choose the drugs they prescribed from far longer lists than at any earlier period, which led pharmaceutical companies to intensify dramatically their marketing and sales efforts.

The established drug companies, particularly those in Britain and the United States, grew significantly. Many employed twenty thousand to thirty thousand people and had annual sales of $5 billion to $10 billion, making them comparable to the large chemical companies. Among them, only Merck and Upjohn remained predominantly pharmaceutical companies, with sales of ethical drugs accounting for between 65 and 70 percent of total sales. Ethical drug sales amounted to 50 to 55 percent of turnover for most of the other traditional pharmaceutical firms, such as Squibb, Searle, and Pfizer, and to between 18 and 41 percent for the horizontally diversified companies (Table 13). For Abbott, which experienced a renaissance in the 1970s by diversifying into hospital care

Table 12. Research and Development Expenditures and Patents by Company in the 1960s (in constant 1995 U.S. dollars)*

Company	1960		1969		1960–1969		
	U.S. Dollars (in millions)	Percentage on Sales	U.S. Dollars (in millions)	Percentage on Sales	U.S. Dollars (in millions)	Percentage on Sales	Patents
Abbott	50.0	5.8	110.0	6.0	825	6.4	398
American Home Products	NA	NA	NA	NA	810†	2.0	NA
Bristol-Myers	36.5	4.4	155.0	3.7	825	3.8	251
Johnson & Johnson	NA	NA	165.0	5.7	1,000†	4.4	311
Lilly	113.0	11.0	250.0	10.2	1,660	10.1	294
Merck	120.0	9.6	275.0	9.3	1,790	9.3	1,533
Parke-Davis	70.0	6.2	85.0	6.9	755	6.7	363
Pfizer	75.0	5.0	125.0	3.2	985	3.7	743
Schering-Plough	47.5	10.0	80.0	7.9	565	8.6	174
Searle	25.0	11.0	80.0	10.7	475	9.9	635
Smith Kline and French	77.0	9.4	132.0	9.2	1,115	9.5	488
Sterling	40.0†	3.0†	75.0†	3.0†	585†	3.0†	400
Upjohn	90.0	10.0	170.0	10.0	1,265	10.3	1,207
Warner-Lambert	35.0	3.0	135.0	3.7	755	3.4	268

Data from company annual reports and *Chemical Abstracts*.
* NA = not available.
† Estimated.

products, instruments, and devices, this figure dropped to just 24 percent. Sales of ethical drugs accounted for 63 percent of total sales for Smith Kline and French in 1979, but that was almost entirely attributable to the outstanding performance of Tagamet, which came on the market in 1977. Otherwise, Smith Kline and French remained a highly diversified pharmaceutical company.

This distribution of sales resulted from the adoption of two opposing policies. Most of the predominantly pharmaceutical companies that had not diversified in the 1960s did so in the 1970s. Lilly, Schering, Smith Kline and French, and Squibb all adopted the merger-acquisition route; Abbott and Upjohn adopted the in-house innovation route. In contrast, the strongly diversified companies strengthened their commitment to ethical drugs by both in-house innovation and acquisitions of or mergers with pharmaceutical companies that had failed to grow in the 1960s. Bristol-Myers, American Home Products, and Pfizer, which had diversified extensively in the 1960s, all increased ethical drug sales as a percentage of total sales. Johnson and Johnson's ethical drug sales held steady, while Warner-Lambert's fell (see Table 13 for sales data, Table 10 for major acquisitions in the 1970s.)

This drive for diversification caused a slowdown in R&D expenditures in the 1970s (Table 14). In the long run, however, diversification had a beneficial effect on pharmaceutical innovation because it generated the growth and increased cash flows that helped sustain the companies' commitment to research during periods of modest innovation. With the exception of Merck most large American pharmaceutical companies today are those that diversified in the 1960s and 1970s or even before, while many of the purely pharmaceutical companies of the 1960s and 1970s were later taken over by or merged with some of their competitors.

Largely because of their diversification the value of pharmaceutical stocks reflected the general trends of the economy rather than their individual performance in innovation in pharmaceuticals (Figure 7). The stocks of nearly all the pharmaceutical companies moved sharply upward until 1973, following the trend established in the late 1960s, and then declined sharply, reaching bottom in 1976. Although pharmaceutical sales were not affected by the economic recession, their acquired businesses were. Moreover pharmaceutical sales were also vulnerable to the significant drop in the parity of the U.S. dollar because nearly half of their sales were in Western Europe and Japan. Starting in 1977, the stocks began to recover lost ground, a trend that would accelerate in the 1980s. A notable exception was Smith Kline and French; its introduction of the antipeptic ulcer drug Tagamet in 1977 raised the value of its outstanding stock from $1 billion to $4 billion in just two years. That experience provided a foretaste of what was to become almost a commonplace in the 1980s, when many companies showed similar performances on the strength of a single commercially successful product.

Table 13. Growth of Total Sales, Ethical Pharmaceutical Sales, and Employment by Company in the 1970s (in millions of constant 1995 U.S. dollars)

Company	1970				1979				1970–1979			
	Number of Employees (in 1,000s)	Total Sales	Ethical Pharm. Sales	Percentage	Number of Employees (in 1,000s)	Total Sales	Ethical Pharm. Sales	Percentage	Total Sales	Ethical Pharm. Sales	Percentage	Profitability
Abbott	18.3	1,970	670	34	29.2	4,180	1,015	24	28,000	8.025	29	8.2
American Home Products	41.8	5,940	2,190	37	50.2	7,785	3,210	41	72,905	28,070	39	10.5
Bristol Myers	24.0	4,260	1,000	24	34.0	6,330	2,155	34	54,280	16,500	30	7.8
Johnson & Johnson	38.2	4,310	820*	19	71.8	9,990	1,865*	18	67,925	13,050*	30	8.5
Lilly	26.4	2,545	1.825	72	37.2	5,250	2,620	50	37,600	21,895	58	15.2
Merck	22.3	3,210	2,395	73	30.8	5,100	3,885	69	43,710	30,990	71	15.7
Pfizer	35.0	4,010	2,040	51	41.0	6,315	3,290	52	51.215	25,795	50	8.9
Schering-Plough	11.6	1,730	1,120	65	26.0	3,565	1,855	52	24,040	14,750	61	14.4
Searle	8.2	865	570	66	17.5	2,215	1,210	55	17,520	8,875	51	9.6
Smith Kline and French	10.2	1,500	840	56	25.3	4,135	2,185	63	20,230	10,790	53	12.9
Squibb†	35.0	3,370	1,460	43	27.0	3,395	2,000	59	34,165	17,710	52	8.5
Sterling	23.5	2,770	965	35	27.0	3,485	915	26	31,835	9,500	30	8.0
Upjohn	28.2	1,710	1,195	70	32.4	3,610	2,335	65	26,485	17.770	67	9.0
Warner-Lambert†	55.5	5,405	2,000	37	58.0	7,130	2,450	34	65,120	23,430	36	7.2

Data from company annual reports and estimates from them.
* Profitability: After-tax profits as percentage of total sales for the decade.
† Includes sales of proprietary pharmaceutical products.

Table 14. Research and Development Expenditures and Patents by Company in the 1970s (in constant 1995 U.S. dollars)

Company	1970		1979		1970–1979		
	Dollars (in millions)	Percentage of Sales	Dollars (in millions)	Percentage of Sales	Dollars (in millions)	Percentage of Sales	Patents
Abbott	115	6.0	200	5.1	1,515	5.3	458
American Home Products	120	2.0	205	2.6	1,710	2.3	NA
Bristol-Myers	165	3.9	235	3.7	1,950	3.6	492
Johnson & Johnson	175	4.0	440	4.5	3,010	4.4	432
Lilly	260	10.3	410	7.9	3,250	8.6	1.030
Merck	295	9.2	430	8.5	3,620	8.3	1,623
Pfizer	135	3.3	315	5.0	2,290	4.5	918
Schering-Plough	90	5.2	175	5.3	1,395	5.8	75
Searle	95	11.0	135	6.5	1,450	8.2	439
Smith Kline and French	135	9.0	250	7.4	1,655	8.2	610
Squibb*	150	4.5	160	4.7	1,500	4.4	967
Sterling	75	2.7	110	3.2	965	3.0	458
Upjohn	180	10.5	295	8.6	2,400	9.0	1,401
Warner-Lambert†	230	4.3	215	2.9	2,295	3.5	700

Data from company annual reports and *Chemical Abstracts*.
* Squibb figures do not appear before 1970 because it was a part of Olin Mathieson.
† Warner-Lambert figures also include Parke-Davis, with which they merged on 1969.

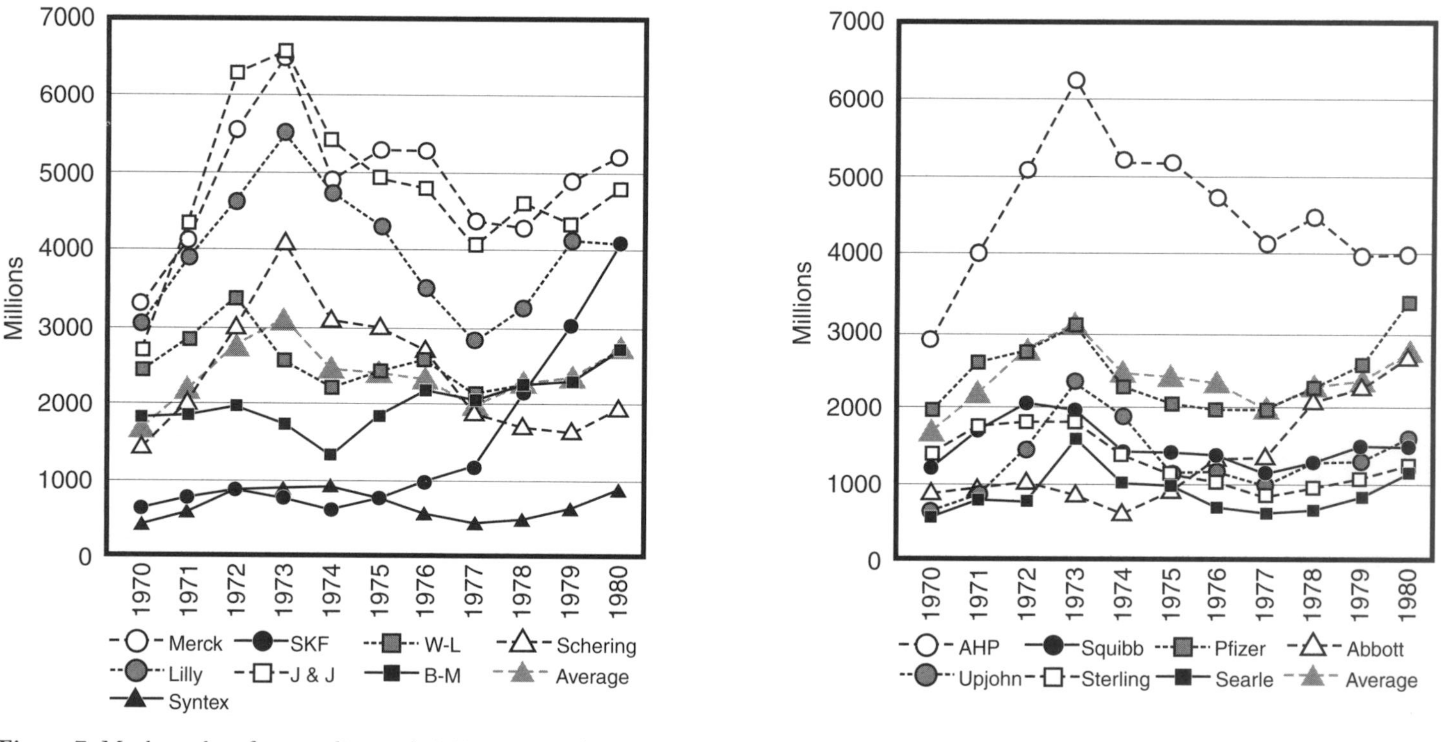

***Figure* 7.** *Market value of outstanding stock (1970–1980) of major American pharmaceutical companies (millions of current U.S. dollars). AHP = American Home Products; B-M = Bristol-Myers; J & J = Johnson and Johnson; SKF = Smith Kline and French; W-L = Warner-Lambert. Data from Moody's Investors Service.*

Blockbusters, Biotechnology, and the Globalization of Companies (1980 and Beyond)

The 1980s and early 1990s were almost a reverse of the previous two decades, with the 1980s being an outstanding period for the pharmaceutical industry both technologically and commercially and the 1990s being a period of self-doubt and restructuring. The phenomenon of the "blockbuster" drugs—a handful of drugs that earned their innovating companies $500 million or more in a single year—appeared in the 1980s. Technologically the decade represented the completion of the industry's shift in focus from organic chemistry to the life sciences and biotechnology, advances that promised to solve many of the remaining mysteries of human health and disease and to offer unprecedented commercial opportunities.

Some of the euphoria of the 1980s waned in the early 1990s. Managers of pharmaceutical companies described this period of revolutionary change as one that offered far more challenges than opportunity for growth and profits. Yet, when viewed through the lens of history, the early 1990s appear similar to the late 1950s and early 1960s. Both periods were characterized by promising technologies that had not yet reached their full potential, by a marketplace crowded with products of older technologies and competing companies, by strong government intervention and criticism of some of the industry's business practices, and by a changing pattern of demand that included important geographical market shifts.

Such periods of change create stresses and strains for the companies, which must adopt new structures and business practices to ensure continued growth and profitability. Some inevitably lose their competitive edge, while others succeed and emerge stronger. Although some markets are already well served by available drugs and are unlikely to offer much opportunity for new growth, the potential for advances in molecular biology, biotechnology, and the other life sciences indicate that the pharmaceutical industry as a whole will survive as a research-intensive and profitable arm of manufacturing.

TECHNOLOGIES OF THE FIFTH GENERATION

In the mid-1970s and early 1980s a new generation of drugs was introduced that initiated the fifth and latest cluster of technologies in pharmaceuticals. The major characteristic of this generation of drugs was the industry's shift away from organic chemistry, pharmacology, and systematic screening of molecules with promising structural features to the life sciences—physiology, pharmacology, biology, biochemistry, biophysics, enzymology, and molecular biology. This shift led to most of the commercially successful radical innovations that launched the fifth cluster of technologies.

This new generation of drugs originated largely from advances made in the understanding of the diversity, structure, function, and specificity of the sites in tissues and organs that receive chemical transmissions from the autonomous and central nervous systems, triggering diverse physiological and pathological processes. Companies sought to build on the introduction of the beta-blockers and the CNS drugs of the 1960s, using both traditional research methods and newly developed laboratory and computer-assisted techniques in the life sciences. These new techniques allowed scientists to identify receptor sites, their biological structures, and their distribution among diverse tissues and organs and to deepen their understanding of the biological mechanisms by which synthetic drugs block or enhance the action of physiological messengers.

These advances led in turn to the study of enzymes—the catalysts responsible for most biochemical processes in living organisms. Scientists began to identify their active sites, to determine their structure and the mechanisms by which they function, and to apply that knowledge to designing chemicals that block or enhance the biochemical processes the enzymes catalyze. Many of the systematically or serendipitously discovered drugs of previous decades served as research tools for studying physiological and pathological processes in tissue, animal, and clinical tests, and the advances made in the understanding of these processes led to further research and innovation.

A major step in this direction was the discovery and application of biotechnology methods (recombinant DNA and monoclonal antibodies) to produce biological proteins and use them either to diagnose or to treat many diseases. Biotechnology, defined as the application of biological organisms, systems, and processes to agriculture and manufacturing, is revolutionizing research in the life sciences and the fight against disease. This is particularly the case for cancer and viral diseases, for which relatively little progress has been achieved.

The shift to the new research techniques has been gradual, in part because of the natural pace of scientific advance and in part because of institutional resistance to change. Many of the research projects that led to the introduction of fifth-generation drugs applied either classic or modern technologies, or both.

Many of the innovations of the 1980s were in cardiovascular (particularly antihypertensive) and CNS drugs. The frequency in industrialized societies of chronic cardiovascular and CNS diseases that require long-term medication heightened demand for effective and safe drugs, and the prices that new drugs fetched in the 1980s made both markets attractive to the industry. For both sectors, progress beyond the achievements of the 1960s called for a much deeper understanding of the variety of receptors and the mechanisms by which physiological transmitters and medicinal chemicals act. This need opened a vast and challenging field for academic and industrial researchers.

Cardiovascular Innovations

Among the numerous cardiovascular drugs introduced in this period, revolutionary treatments for hypertension, thrombosis, and high cholesterol levels became outstanding therapeutic and commercial successes. At least three new technologies were initiated and by the end of the 1980s had reached their stages of growth and even of maturity: calcium channel blockers, used to treat angina pectoris and hypertension; ACE inhibitors for treating hypertension and heart failure; and hypolipidemics, which control cholesterol levels.

Examples of radical innovations are the α_1-adrenergic blocker prazosin (Minipress from Pfizer [1973]); the calcium channel blocker nifedipine (Adalat from Bayer [1974]); the ACE inhibitor capoten (Captopril from Squibb [1977]); the hypolipidemic lovastatin (Mevacor from Merck [1987]); and the antithrombotics urokinase (Abbokinase from Abbott [1977]) and streptokinase (Streptase from Hoechst [1978]), and the tissue plasminogen activator Activase from Genentech (1987).

Innovations in CNS Drugs

In the 1980s numerous transmitters and receptors of the central nervous system were identified and the functions of agonists and antagonists elucidated. This research helped explain both the effectiveness and the side effects of CNS drugs already on the market and provided leads for designing new, improved products.

This process is illustrated by the benzodiazepine anxiolytics, one of the most successful categories of CNS drugs. When they were introduced in the early 1960s, the mechanisms by which they exerted their anxiolytic and hypnotic effects were unknown. Starting in 1977 Roche researchers located a benzodiazepine receptor by using radioactive-labeled diazepam (Valium) and identified γ-aminobutyric acid (GABA) as a chemical transmitter of the CNS. This knowledge led to the development of benzodiazepine anxiolytics with moderate sedative effects, selective anticonvulsants, antidotes for benzodiazepine overdose (such as flumazenil [Romazicon from Roche, 1987]), and a diazepine antidepressant, alprazolam (Xanax from Upjohn [1980]).

Similarly, research on agonists and antagonists of dopamine, a known transmitter of nervous impulses in the brain, using the tranquilizer chlorpromazine and the antiparkinsonian drug levodopa, led to the elucidation of the role of dopamine in schizophrenia and to the introduction of numerous antiparkinsonian drugs, such as bromocriptine (Parlodel from Sandoz [1984]) and pergolide (Permax from Lilly [1989]).

The most important advance made in CNS drugs in the 1980s resulted from research on serotonin, a neurotransmitter of the CNS and the alimentary canal. This research showed that all the actions of serotonin could be interpreted by assuming that it had at least eleven types of receptors. Selective agonists and antagonists were synthesized, some through computerized molecular design, to define serotonin's role in diverse diseases and their symptoms, such as migraine, headaches, nausea, anxiety, and depression. Serotonin-related research led to the introduction in the 1980s and 1990s of a score of effective, commercially successful drugs, including the tranquilizer buspirone (Buspar from Bristol-Myers [1984]) and the antidepressant fluoxetine (Prozac from Lilly [1986]). Other serotonin-related drugs that were developed during this period opened completely new therapeutic markets; these included sumatriptan

(Imitrex from Glaxo [1990]) for treating migraine and ondansetron (Zofran from Glaxo [1990]), an antinausea drug for patients undergoing chemotherapy for cancer. The promise of these new markets led competing companies to begin to imitate these drugs in the 1990s.

Antipeptic Ulcer Drugs

Another major innovation of the fifth generation of drugs was cimetidine (Tagamet from Smith, Kline and French), the first effective drug for treating peptic ulcers and one that created a new technology. Until Tagamet came on the market in 1977, people who suffered from peptic ulcers were not adequately treated and often required surgery. Drug researchers knew that histamine triggered the secretion of gastric acids, but none of the known antihistamines blocked histamine's action on the secretion of gastric acid. In the late 1960s Sir James Black, then a consultant to Smith Kline and French, suggested that there might be two types of histamine receptors, with only one of them located in the stomach. It took Smith Kline and French seven years to come up with a chemical that could block that receptor.

Tagamet was the first of the "blockbuster" drugs, with annual sales exceeding $1 billion. It was followed by a score of antiulcer imitations; the most important of these was ranitidine (Zantac from Glaxo), which became a blockbuster in its own right, reaching annual sales of $2 billion. In 1987 Astra, a Swedish company, introduced a new antipeptic ulcer drug, omeprazole (Losec or Prilosec), which functions by inhibiting the formation of gastric acid. Prilosec later formed the basis of the Astra-Merck joint venture. By 1998 its annual sales had risen to the level of $2 billion.

Antineoplastic Drugs

In the 1980s and 1990s academic and industrial laboratories undertook a major effort to develop drugs for treating cancer. By the late 1990s researchers had made great advances in understanding the physiological processes by which neoplasms form and spread and had discovered some drugs that could control and sometimes eliminate them. But the great variety of the compositions and structures of anticancer drugs indicated that researchers still had not found technologies that could lead to better therapy through molecular manipulation. Nonetheless about twenty new antineoplastic drugs were launched between the late 1970s and the late 1990s, including tamoxifen (Nolvadex from ICI [1977]), cisplatin (Platinol from Bristol-Myers [1978]), etoposide (VePesid from Bristol-Myers Squibb [1993]), paclitaxel (Taxol from Bristol-Myers Squibb [1992]), and flutamide (Eulexin from Schering [1982]).

Other Important Drugs

Several other innovations of the 1980s and 1990s were aimed at illnesses seldom dealt with in the past, namely, viral and age-debilitating diseases. These drugs include the antiviral agents acyclovir (Zovirax from Burroughs Wellcome [1983]), used to treat herpes zoster, and zidovudine (Retrovir from Burroughs Wellcome [1985]), used in the treatment of AIDS; terfenadine (Seldane from Merrell [1987]), the first antihistamine with no sedative side effects, which became a great commercial success; the immunosuppressive agent cyclosporine (Sandimmune from Sandoz [1983]), which made organ transplants possible; and a set of drugs for age-debilitating diseases, such as tacrine (Cognex from Warner-Lambert [1993]) for treating Alzheimer's disease, alendronate (Fosamax from Merck [1994]) for controlling osteoporosis, and finasteride (Proscar from Merck [1992]) for controlling benign prostate enlargement. Many of these drugs have potentially great therapeutic and commercial value and have been widely imitated by competing companies.

New Drug Delivery Systems

A major breakthrough in the fifth generation of drugs was the development of improved methods for delivering medicines orally and transdermally. These techniques have dramatically increased the potential of many drugs—both conventional and biotechnologic—because of their ability to control the concentration levels of a drug in the bloodstream, which improves efficacy, reduces side effects, facilitates dispensing, and expands indications for individual drugs. Alza's Oros oral system, adopted by Pfizer for its hypertensive drugs Minipress XL and Procardia XL, was one of the most successful of these technologies. Implantable or injectable systems, which were the subject of intense research efforts, may have fewer technical problems than oral delivery systems and are used to deliver numerous drugs, including contraceptives and nitroglycerin for patients with angina pectoris.

Biotechnology

Perhaps the most promising set of innovations introduced in the 1980s and 1990s are those based on biotechnological techniques and processes. The first process was that of recombinant DNA, discovered in 1973–75 by Herbert Boyer and Stanley Cohen of the University of California, San Francisco. They introduced foreign genes into bacterial DNA, causing them to produce corresponding proteins. This historical discovery created the basis for use of bacteria as "factories" for producing proteins of medicinal importance and steered research toward understanding the relation between the structure of individual genes and the proteins they produced.

Another major discovery, made by British researchers in 1985, was the cell fusion technique by which monoclonal antibodies are produced. These antibodies are used to locate specific targets—such as tumors—and to carry specific chemicals to them; the antibodies thus acted as "magic bullets." The realization that both cancers and viral diseases were linked to DNA intensified interest in molecular biology among researchers seeking treatments for these

Table 15. Marketed Biotechnology Drugs as of 1993 (excluding diagnostics)

Generic Name	Trade Name	Use	Company	Country	Year of Commercialization
Human insulin	Humulin	Diabetes	Genentech/Lilly	United States	1983
Human growth hormone	Protropin	Delayed growth	Genentech/Schering	United States	1985
Human growth hormone	Humatrope	Delayed growth	Lilly	United States	1987
Interferon α-2a	Roferon-A	Cancer, hepatitis C, Kaposi sarcoma, leukemia	Genentech/Roche	United States/ Switzerland	1986
	Intron A		Biogen/Schering	United States	1986
Interferon α-n3	Alferon N	Genital warts	Interferon Sciences	United States	1989
Muromonab–CD3	Orthoclone OKT3	Transplant rejection	Ortho-Biotech	United States	1986
Hepatitis B vaccine	Recombivax HB	Hepatitis B	Merck	United States	1986
	Engerix-B	Hepatitis B	SmithKline-Beecham	United States/ United Kingdom	1989
Alteplase (TPA)	Activase	Thrombosis	Genentech	United States	1987
Erythropoietin	Epogen	Anemia, renal failure	Amgen	United States	1989
	Procrit		Ortho-Biotech	United States	1990
Interferon γ-lb	Actimmune	Chronic granulomatous disease	Genentech	United States	1990
Sargramostim (GM-CSF)	Leukine	Bone marrow transplants	Immunex	United States	1991
Filgrastim (G-CSF)	Neupogen	Neutropenia; adjunct to cancer chemotherapy	Amgen	United States	1991
Aldesleukin, interleukin-2	Proleukin	Renal cell carcinoma	Cetus/Chiron	United States	1992
Antihemophilic factor (rAHF)	Recombinate	Hemophilia A	Baxter	United States	1992
	Kogenate		Bayer/Miles	Germany	1993
Interferon β-1b	Betaseron	Multiple sclerosis	Berlex	Germany	1993

diseases. Study of the life sciences flourished in academic institutions, and hundreds of small, entrepreneurial companies, based around university research teams and funded by venture capital, were formed to use biotechnology to produce new drugs. Boyer, for example, formed Genentech, and Berg, Amgen.

The application of biotechnology was the first time in the history of the pharmaceutical industry that the earliest medicines based on revolutionary academic advances were developed outside the established research-intensive companies. Busy with developing and marketing the blockbuster drugs of the 1980s, the big companies were convinced that the small biotechnology companies would eventually have to come to them for help in developing and marketing their discoveries. And that is exactly what happened when Genentech licensed the first bioengineered human hormone insulin (Humulin [1983]) to Lilly and the antineoplastic interferon A (Roferon-A [1986]) to Roche.

By the mid-1980s many of the biotechnology companies still had not produced anything of commercial value, and venture capitalists and other investors began to question the wisdom of their investments. Alarmed by this skepticism, several of the biotech companies turned to the development of diagnostic tests based on recombinant DNA and monoclonal antibodies. Ironically, perhaps the most successful—and certainly one of the most needed—diagnostic test was developed by a pharmaceutical company; Abbott Labs in 1985 introduced a test that could detect the AIDS virus in human blood collected for transfusions, which ensured a safe blood supply.

In 1985 biotechnology companies also began to introduce their own important and commercially successful bioengineered products; these included human growth hormone (Genentech [1985]); the antithrombotic tissue plasminogen activator (tPA) (Genentech [1987]); and the red corpuscle producer erythropoietin (Amgen [1989]), all of which soon passed the $100 million mark in annual sales (Table 15). These successes caught the attention of the established pharmaceutical companies, which realized not only that bioengineered drugs could be as lucrative as drugs produced using more traditional means but also that the most successful biotechnology companies could become strong competitors. The pharmaceutical companies thus forged numerous bilateral agreements and joint ventures with biotechnology firms to develop specific promising medicines and processes, and many of the established companies acquired biotechnology companies to assimilate their expertise into their own research activities.

According to a survey by the Boston Consulting Group, by 1993, one-third of the research projects mounted by the major pharmaceutical companies were based on biotechnology; in 1980 only 2 percent of all research projects had focused on biotechnology. Biotechnological research continued to accelerate in the 1990s, while the number of new drugs and diagnostics produced using biotechnology grew at a slower pace (Table 16). This pattern was typical of the growth period of a new technology. Most of the biotech research projects were aimed at discovering treatments for cancer and viral diseases, while others focused on vaccines, growth factors, and hematological agents. Taken together the trends both in research and in business strategy leave little doubt of the importance of biotechnology for the future technological and commercial performance of the pharmaceutical industry.

PERFORMANCE IN A SLOWING ECONOMY

The slowdown in economic growth in the industrialized world, which began in the 1970s, had two successive—and opposing—effects on the pharmaceutical industry. The industry at first appeared to be immune to the slowdown. Indeed the outstanding business and commercial performance of pharmaceutical companies in the 1980s—American, European, and Japanese—resulted primarily from the decline of the low value-added sectors of the manufacturing industry in industrialized countries and their relocation to developing countries with abundant and cheap labor, particularly in the Far East. The contraction in the United States and Europe of such basic industries as coal, steel, and petrochemicals, even the auto industry in the United States and the United Kingdom, led to high unemployment and the shift of the work force from high-paying manufacturing jobs to low-paying service sector jobs. These factors, in turn, dampened the demand for housing, furniture, home appliances, cars, and a wide range of consumer goods—all sectors that drove economic growth in the 1960s and 1970s. Most of these technologies had matured and diffused widely, so only products based on technologies with fast rates of advance and high value-added proprietary products could sustain high returns on invested capital: computers, equipment and machinery incorporating advanced electronics, aircraft, luxury products, defense industries, and pharmaceuticals. These industries benefited from the flow of capital away from the traditional manufacturing sectors.

The pharmaceutical industry was perhaps the greatest beneficiary because, apart from the advances made in the 1980s in the sciences and technologies on which it depended, its markets were not cyclical and were partially protected by public and private health insurance that softened the adverse effects of unemployment and reduced family income. Thus the stocks of pharmaceutical companies skyrocketed in the 1980s, particularly those of Merck, Bristol-Myers, and American Home Products (Figure 8), and venture capitalists were eager to back the hundreds of small biotechnology companies that mushroomed in that decade.

By the 1990s, however, the effects of the slowdown were taking their toll on the pharmaceutical industry, which saw its stock value begin to slide, with decreases becoming particularly noticeable in 1992 (Figure 9). The weak economic growth of the late 1980s eroded the financial basis for both private and public health insurance systems at the same time that health care costs were rising at least three times faster than overall consumer prices. This situation

Table 16. Early Development of Biotechnology Drugs

Category	1989	1990	1991	1993
Approved drugs	9	11	14	19
Drugs and vaccines in development	80	104	132	143
Companies developing drugs or vaccines	45	51	58	63
Products in development by type and use				
Blood-clotting factors (hemophilia)	2	2	4	1
Colony stimulating factors (CSFs) (immune system stimulants)	7	9	8	6
Dismutases (transplants)	3	3	1	1
Erythropoietins (red blood cell production)	4	5	4	1
Growth factors (cell proliferation regulators)	2	7	11	9
Human growth hormones (stimulants for growth of long bones)	3	5	7	4
Interferons (cancer, viral diseases)	12	13	16	11
Interleukins (immune system stimulants)	12	14	13	10
Monoclonal antibodies (cancer, viral diseases)	25	41	58	50
Recombinant soluble CD4s (AIDS)	2	4	2	2
Tissue plasminogen activators (TPAs) (thrombosis)	—	—	4	1
Tumor necrosis factors (cancer)	4	4	2	3
Vaccines	13	15	18	20
Others	6	4	10	23
Total	**80**	**104**	**132**	**143**

Data from a survey by the U.S. Pharmaceutical Manufacturers Association (1994).

was exacerbated by the demand for public health care from the growing number of unemployed and aging in most industrial countries. Under those conditions government and private health insurance systems had to reduce their costs. Although pharmaceuticals are a relatively small part of the health insurance package (10 percent in the United States and about 20 percent in Europe), drug companies had to absorb their share of the reductions.

Used to working in a rapidly changing technical and commercial environment, the large pharmaceutical companies moved quickly to cope with the slower growth and to restore the confidence of their investors. Nearly all companies reorganized their operations to reduce their work forces. Most also took steps to strengthen their competitive position in existing markets, to enter the growing generic and over-the-counter drug markets, and to expand into underserved areas of the world. With these steps the major companies sought to position themselves to take full advantage of the potential offered by biotechnology and advances in the life sciences.

The "Blockbuster" 1980s

The stellar performance of the pharmaceutical companies in the 1980s had little to do with the diffusion of biotechnology, whose commercial effects were not felt until the 1990s. The industry instead relied on radical innovations based

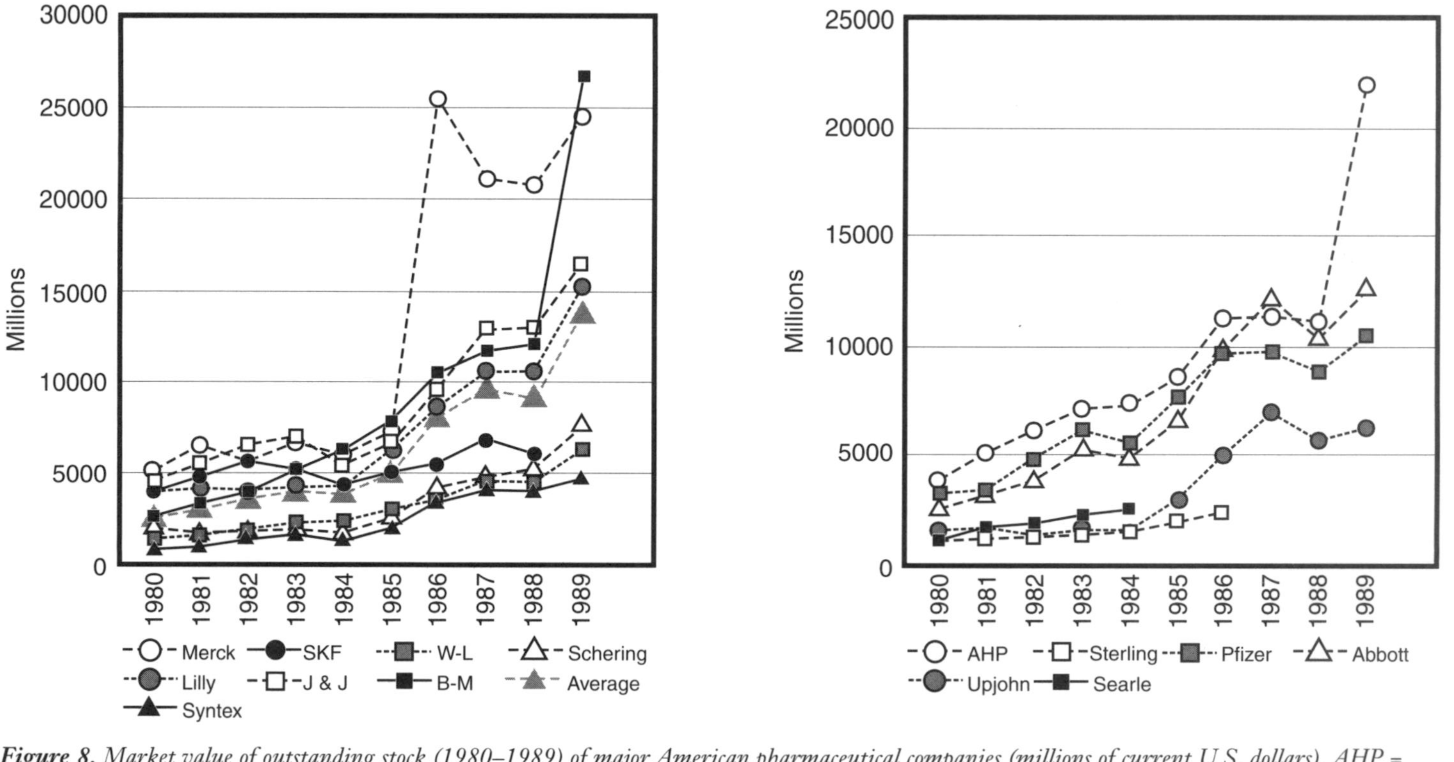

Figure 8. *Market value of outstanding stock (1980–1989) of major American pharmaceutical companies (millions of current U.S. dollars). AHP = American Home Products; B-M = Bristol-Myers; J & J = Johnson and Johnson; SKF = Smith Kline and French; W-L = Warner-Lambert. Data from Moody's Investors Service.*

on more traditional and new approaches to drug design. But the main cause of the upswing was the decline of the traditional manufacturing sectors, which did not exhibit the strengths of the pharmaceutical industry as a research-intensive sector with a robust and promising scientific and technological basis.

The prices that new products fetched in the 1980s inflated the dollar value of the markets, particularly the large markets for antibacterial, cardiovascular, antineoplastic, CNS, antiulcer, and analgesic products. Not only were highly original drugs with substantial improvement in therapeutic value profitable, so too were those drugs that had only minor differences from existing drugs and that were able to carve out a market niche. Some drugs with great therapeutic advantages were promoted worldwide according to long-planned marketing strategies to capture very large chunks of the market. These drugs became the "blockbuster" products of the 1980s, reaching annual sales of $1 billion and more. Smith Kline and French's antiulcer drug Tagamet was the first blockbuster, quickly followed by Squibb's ACE inhibitor Capoten and Pfizer's NSAID Feldene (1979). Many others followed in the 1980s (Table 17). For the first time in the history of the pharmaceutical industry, some companies derived up to 50 percent of their profits from only one product, while companies that failed to produce a blockbuster lost their competitive edge.

Despite the great profits made from these drugs, excessive dependence on them introduced an element of vulnerability in the long-term prospects of companies. The strength of the markets created by these drugs provoked increased efforts for imitation and improvement of their properties by competitors. Some of these imitations—for example, Glaxo's antiulcer drug Zantac, Lilly's antibacterial agent Ceclor (1979), and Upjohn's anxiolytic-hypnotic drug Halcion (1982), were immensely successful and won considerable market share.

Furthermore, the inevitable approach of the date of expiration of the key patents portended the crowding of the market with generics and the erosion of the original drug's profitability. Major drug companies defended their products against the growing number of generic drugs produced both in the United States and overseas by stressing the trade name of the original drug and capitalizing on the goodwill of the medical profession. Thus the proprietary character of drugs became as important as the research and development that went into producing them in the first place.

The result was a proportional increase in R&D expenditures and even higher increases in spending on marketing and sales. Companies used these resources to widen the scope both of their technologies and their markets.

In addition to the significant research costs and risks entailed, the application of new technologies and the opening of new therapeutic markets required innovating companies to break up some of their powerful CTTs. For CTTs that had reached the stage of diminishing returns, such as antihypertensive beta-blockers, semisynthetic penicillins, and chlorpromazine-type antipsychotics, this effort was not difficult. But where a company's CTTs were still strong, attempts at diversifying in-house innovation into new sectors had to

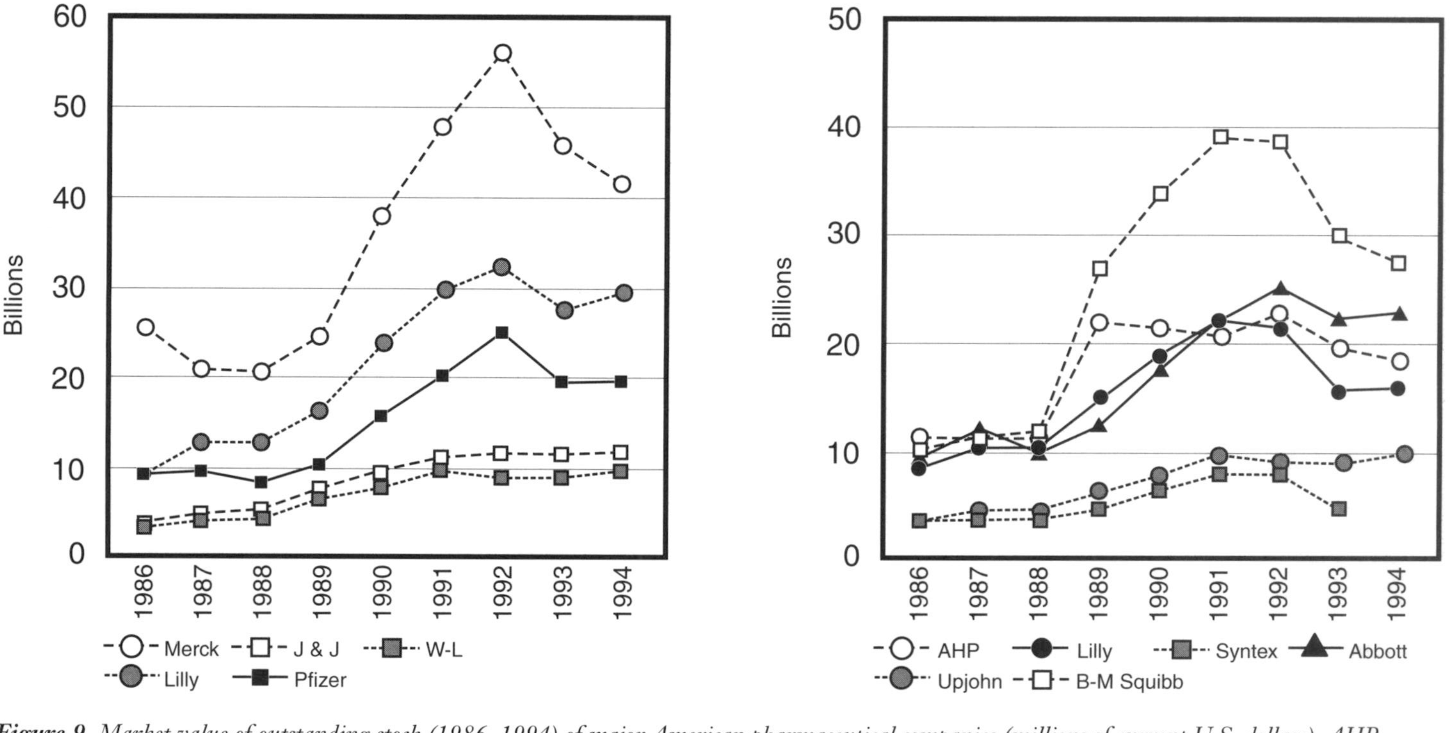

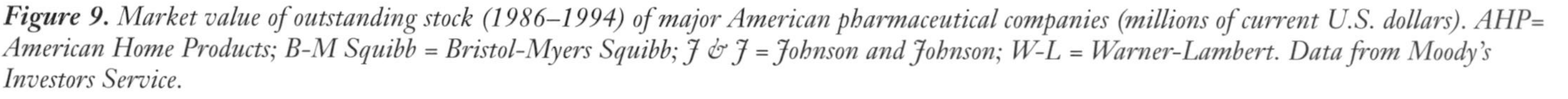

Figure 9. *Market value of outstanding stock (1986–1994) of major American pharmaceutical companies (millions of current U.S. dollars). AHP= American Home Products; B-M Squibb = Bristol-Myers Squibb; J & J = Johnson and Johnson; W-L = Warner-Lambert. Data from Moody's Investors Service.*

overcome not only technological and commercial challenges but also corporate resistance and inertia.

To ease such difficulties, some companies used acquisitions and mergers as shortcuts to technological and market diversification, often taking over narrowly specialized companies unable to grow further on their own. These acquisitions were a foretaste of the mergers of large companies that began in the late 1980s and continued in the 1990s. The resulting conglomerates each covered a wide spectrum of therapeutic technologies and markets.

Large companies also imitated the successful products of their competitors within sectors of their own specialization. Merck and Ciba took advantage of new technologies in antihypertensive drugs—a sector in which they had already established strong CTTs—to develop the new technology for producing ACE inhibitors, introduced by Squibb in 1977. Merck introduced enalapril (Vasotec [1985]) and lisinopril (Prinivil [1987]), and Ciba introduced benazepril (Lotensin [1990]). Lilly followed a similar pattern with Humulin, the bioengineered human insulin, using the new technology to extend its specialization in antidiabetic drugs. For smaller companies, which saw in the new models an opportunity to capture a niche of these lucrative markets, incremental innovation was not hampered by CTTs and was easier to achieve with modest research expenditures. This strategy led to the numerous imitative new drugs of the 1980s and 1990s, including ten new ACE inhibitors, eight new calcium channel blockers, twenty NSAIDs, seven hypolipidemic drugs, and nineteen benzodiazepine anxiolytic agents.

The significant changes in the business environment of the 1980s and the outcome of the business policies that were adopted are reflected both in the money the major American pharmaceutical companies spent on R&D and in the value of those companies' stocks. As Table 18 illustrates, between 1980 and 1989, research expenditures rose significantly in dollar values and as percentage of turnover. Companies became far more research intensive, in part because they shed many of their less research-intensive consumer product businesses (see below).

For most companies total sales increased only moderately during the period (Table 19), primarily because the increased value and volume of pharmaceutical sales were counterbalanced by the divestitures of the nonpharmaceutical businesses. Absolute and relative pharmaceutical sales increased compared with total sales. The value of pharmaceutical stocks reached unprecedented heights as a result of the general business environment, the increased profitability of the companies, and the introduction by most of them of at least one "blockbuster" drug during the decade (Figure 8).

Policies for Growth

Apart from the emphasis on marketing and sales the pharmaceutical industry experienced some other major structural changes during the 1980s. Reversing

Table 17. Some Blockbuster Drugs of the 1970s and 1980s*

Trade Name	Generic Name	Class	Use	Company	Country	Year of Commercialization
Adalat	Nifedipine	Calcium blocker	Hypertension, angina	Bayer	Germany	1974
Naprosyn	Naproxen	NSAID	Arthritis	Syntex	United States	1976
Minipress	Prazosin	Alpha blocker	Hypertension	Pfizer	United States	1976
Tagamet	Cimetidine	H_2-antihistamine	Gastric ulcers	SKF	United States	1977
Capoten	Captopril	ACE inhibitor	Hypertension, heart failure	Squibb	United States	1977
Lopressor	Metoprolol	Beta blocker	Hypertension	Ciba	Switzerland	1978
Clinoril	Sulindac	NSAID	Arthritis	Merck	United States	1978
Platinol	Cisplatin	Platinum compounds	Cancer	Bristol	United States	1978
Ceclor	Cefaclor	Cephalosporin	Infection	Lilly	United States	1978
Mefoxin	Cefoxitin	Cephalosporin	Infection	Merck	United States	1978
Claforan	Cefotaxime	Cephalosporin	Infection	Hoechst	Germany	1981
Tenormin	Atenolol	Beta blocker	Hypertension	ICI	United Kingdom	1981
Xanax	Alprazolam	Benzodiazepine	Depression	Upjohn	United States	1981
Lopid	Gemfibrozil	Fibric acid derivative	Hyperlipidemia	Warner-Lambert	United States	1981
Zantac	Ranitidine	H_2-antihistamine	Gastric ulcer	Glaxo	United States	1982
Feldene	Piroxicam	NSAID	Arthritis	Pfizer	United States	1982
Cardizem	Diltiazem	Calcium blocker	Hypertension, angina	Tanabe Seiyaku Merrell	Japan/ United States	1982
Halcion	Triazolam	Benzodiazepine	Anxiety	Upjohn	United States	1982
Rocephin	Ceftriaxone	Cephalosporin	Infection	Roche	Switzerland	1984
Seldane	Terfenadine	H_1-antihistamine	Allergy	Merrell	United States	1985
Vasotec	Enalapril	ACE inhibitor	Hypertension, heart failure	Merck	United States	1985
Fortaz	Ceftazidime	Cephalosporin	Infection	Glaxo/Lilly	United States/ United Kingdom	1985
Mevacor	Lovastatin	HMG-CoA reductase inhibitor	Hyperlipidemia	Merck	United States	1987
Retrovir	Zidovudine	Reverse transcriptase inhibitor	AIDS	Burroughs	United Kingdom	1987
Cipro	Ciprofloxacin	Quinolone	Infections	Bayer	Germany	1987
Prozac	Fluoxetine	—	Depression	Lilly	United States	1987
Voltaren	Diclofenac	NSAID	Arthritis	Ciba	Switzerland	1988

* Total revenues of more than $500 million a year.

Table 18. Research and Development Expenditures and Patents by Company in the 1980s (in constant 1995 U.S. dollars)

Company	1980		1989		1980–1989		
	U.S. Dollars (in millions)	Percentage of Sales	U.S. Dollars (in millions)	Percentage of Sales	U.S. Dollars (in millions)	Percentage of Sales	Patents*
Abbott	200	4.8	675	9.2	3,990	7.3	320
American Home Products	205	2.6	400	5.0	3,015	4.0	NA
Bristol-Myers	260	4.0	1,145†	8.5†	4,515	7.1	280
Johnson & Johnson	425	4.3	935	7.4	7,100	6.6	375
Lilly	410	7.8	785	14.4	5,760	10.9	790
Merck	480	8.6	975	11.5	6,930	10.9	1,080
Pfizer	330	5.3	690	9.3	4,705	7.1	566
Schering-Plough	185	5.2	425	10.3	2,920	8.3	167
Searle	145	6.5	175‡	8.5‡	815	8.1	136
Smith Kline and French	330	8.0	700	9.8	5,090	9.5	326
Squibb	160	4.7	395§	11.3	2,275	8.1	450
Sterling	120	3.4	160¶	5.2¶	935¶	4.3¶	284
Upjohn	300	8.3	500	14.1	4,200	12.2	573
Warner-Lambert	210	2.5	400	7.3	3,035	5.5	358

Data from company annual reports and *Chemical Abstracts.*
* Patents up to 1987.
† For Bristol-Myers Squibb.
‡ Takeover by Monsanto.
§ For 1988 before merger with Bristol-Myers.
¶ 1986, takeover by Kodak.

their policies of the 1960s and 1970s, most American pharmaceutical companies divested many of the non–health-related businesses they had acquired because consumer product markets were not growing during much of the 1980s and because the blockbuster drugs were providing the cash flow necessary to sustain corporate growth and to support further research and innovation. These cash flows combined with the promise held by the advances in the life sciences convinced managements that purely pharmaceutical companies could not only generate profits at rates superior to those of most other manufacturing sectors but could also sustain corporate growth. After half a century Merck was finally considered an example to be imitated rather than the exception to the rule.

One outgrowth of that philosophy was a new round of mergers, acquisitions, and joint ventures within the pharmaceutical industry. Many of these involved European companies that wanted to take advantage of the technological and commercial advantages of selling their drugs independently in the United States rather than through licensing arrangements with American companies. Squibb merged with Bristol-Myers and Smith Kline and French with Beecham, while American Home Products acquired A. H. Robins, Rhône Poulenc acquired Rorer, Bayer acquired both Miles and Cutter Labs, and ICI acquired Stuart. (In 1993 ICI split up, forming a new chemical company, ICI, and a pharmaceutical company, Zeneca, whose market value rose sharply.) Such companies as Upjohn and Syntex were so outdistanced in size by the industry leaders that they could no longer stand on their own. Roche acquired Syntex in 1994, and Upjohn merged with the Swedish company Pharmacia in 1995. In the same year Hoechst acquired Dow's Marion Merrell and completed its control over Hoechst-Roussel. Ciba-Geigy and Sandoz combined to form Novartis in 1996. In 1997 American Home Products acquired American Cyanamid (now largely pharmaceutical after spinning off its chemical business). There were other such activities as well, of lesser magnitude.

To take advantage of the potential of biotechnology, most pharmaceutical companies established joint ventures with small biotechnology companies, first for funding R&D projects and then for developing and marketing their products. The pharmaceutical companies eventually began to acquire some of the more successful biotech companies (see Chapter 3).

The American companies also formalized numerous joint ventures and licensing agreements with Japanese companies to develop and market their original products in the United States. Abbott and Takeda entered a joint venture, as did SmithKline Beecham and Fujisawa, and Merck and Banyu. Tanabe Seiyaku licensed its outstanding calcium channel blocker diltiazem (Cardizem) to Marion (now Hoechst Marion Roussel).

The increased profitability of pharmaceuticals attracted chemical companies, which for the most part had stayed away from the industry. Finding their profit margins being squeezed in basic chemicals, fertilizers, plastics, and synthetic fibers, the chemical companies began to divest themselves of those businesses and to concentrate on specialties and high value-added products, including

Table 19. Growth of Total Sales, Ethical Pharmaceutical Sales, and Employment by Company in the 1980s (in millions of constant 1995 U.S. dollars)

Company	1980				1989				1980–1989			
	Number of Employees (in 1,000s)	Total Sales	Ethical Pharm. Sales	Percentage	Number of Employees (in 1,000s)	Total Sales	Ethical Pharm. Sales	Percentage	Total Sales	Ethical Pharm. Sales	Percentage	Profitability*
Abbott	31.0	4,180	1,015	24	41.0	7,290	NA	NA	54,755	NA	NA	13.4
American Home Products	49.8	7,785	3,210	41	45.0	8,000	3,750	47	75,000	35,800	48	15.1
Bristol-Myers†	35.0	6,475	2,250	35	35.2	8,060	3,385	42	63,265	24,610	39	14.8
Johnson & Johnson	74.0	9,915	1,750	18	83.0	12,620	3,450	27	107,310	24,170	23	9.0
Lilly	28.1	5,245	2,620	50	28.2	5,430	3,280	60	53,045	29,665	56	15.6
Merck	31.6	5,600	3,885	69	34.4	8,515	7,500	88	55,205	51,740	94	19.4
Pfizer	41.2	6,210	3,370	54	42.0	7,370	3,555	48	66,345	32,240	49	12.2
Schering-Plough	27.4	3,565	1,855	52	21.3	4,110	3,215	78	35,100	20,700	59	11.6
Searle‡	18.2	2,215	1,210	55	10.2	2,055	965	47	9,990	5,285	53	11.4
Smith Kline and French§	31.6	4,135	2,185	53	34.6	6,275	3,125	50	40,000	20,590	51	17.5
Squibb¶	27.0	3,435	1,785	52	18.0	3,490	2,990	86	27,925	18,735	67	12.2
Sterling∥	26.0	3,485	915	26	28.0	3,085	940	30	21,560	6,115	28	7.5
Upjohn	22.0	3,610	2,335	65	15.0	3,535	3,005	85	34,495	25,810	75	10.3
Warner-Lambert	31.0	7,130	2,450	34	33.0	5,455	1,720	31	55,220	18,540	33	5.1

Data from company annual reports.
* Profitability: After-tax profits as percentage of total sales for the decade.
† Data up to 1988 before merger with Squibb.
‡ Data up to takeover by Monsanto.
§ Data up to 1987 before merger with Beecham.
¶ Data up to 1988 before merger with Bristol-Myers.
∥ Data up to 1986 before takeover by Kodak.

pharmaceuticals. Dow led the way in the 1970s by acquiring Merrell and was followed in the 1980s by Monsanto, which acquired Searle; Kodak, which acquired Sterling; BASF, which acquired Knoll; and Du Pont, which established a joint venture with Merck to develop and market pharmaceuticals after its own efforts at in-house drug innovation failed.

The 1990s: Coping with Slow Growth

The slowdown of the world economy caught up with the pharmaceutical industry in the 1990s. Both public and private efforts to reduce the costs of overall health care resulted in significant reductions in traditional pharmaceutical markets. Cost-cutting steps included stricter controls on the prices of medications, withdrawal of reimbursements from selected products, and increases in the contributions of the insured. In Japan, which had had the highest per capita spending on pharmaceuticals in the world, the average level of reimbursement was cut by 40 percent between 1980 and 1987. As a result sales of drugs rose only 5 percent between 1983 and 1986, compared with a 50 percent increase in the previous three-year period. In Europe, government health insurance agencies placed price controls on drugs and urged the use of generic drugs whenever possible.

American health insurance companies also encouraged the use of generics and cooperated with health maintenance organizations (HMOs) and associations of drugstores and hospitals, which used their buying power to obtain lower prices from the pharmaceutical companies. The annual price inflation of drugs dropped from about 14 percent in the 1980s to about 3 to 4 percent in the early 1990s. Prescriptions for generics rose from about 30 percent of all prescriptions written in 1988 to about 50 percent in 1995. That percentage was expected to increase dramatically by the year 2002, when the patents of another two hundred drugs were scheduled to expire.

These measures slowed the rate of growth of pharmaceutical companies. Between 1992 and 1994 the stock value of the eleven large American companies lost about $40 billion, and the American Pharmaceutical Association reported that its member companies lost a total of $80 billion during the same period. Corporate financial data indicated that the problem, at least up to 1994, was not the erosion of profitability, but the deceleration of growth (Table 20). Profitability in the early 1990s was higher than it had been in any of the previous decades, in part because of the divestitures of the non–health-related businesses. Tables 21 and 22 show fluctuations in net profits and profitability caused by the costs of restructuring and the expiration of patents for key products. In contrast to profitability, however, the rates of increase in annual sales fell substantially behind those of the 1980s.

That decline alarmed companies and stockholders, who worried that a continued decline might eventually damage profitability. So managers took several steps aimed at restoring positive growth patterns. In the short term several companies consolidated their organizations and reduced the size of their work

Table 20. Sales and Annual Growth of Sales by Company from 1988 to 1993 (in billions of constant 1995 U.S. dollars)

Company	1988		1989		1990		1991		1992		1993	
	Total Sales	Percentage of Growth	Total Sales	Percentage of Growth	Total Sales	Percentage of Growth	Total Sales	Percentage of Growth	Total Sales	Percentage of Growth	Total Sales	Percentage of Growth
Abbott	6.6	—	7.3	10.6	7.4	1.4	7.8	5.4	8.6	10.2	8.9	3.5
American Home Products	7.4	—	8.0	8.1	8.0	0	8.0	0	8.6	7.5	8.8	2.3
Bristol-Myers Squibb	8.0	—	8.1	1.2	11.5	4.2*	11.9	3.5	12.1	1.7	12.0	(–0.8)
Johnson & Johnson	11.7	—	12.6	7.7	13.2	4.8	14.0	6.1	15.0	7.1	14.9	(–0.7)
Lilly	5.5	—	5.4	(–1.8)	6.1	13.0	6.5	6.6	6.7	3.0	6.8	1.5
Merck	8.0	—	8.5	6.2	9.1	7.0	9.7	6.6	10.5	8.8	11.1	5.7
Pfizer	7.2	—	7.4	2.8	7.6	2.7	7.9	3.9	7.9	0	7.9	0
Schering-Plough	4.0	—	4.1	2.5	3.9	(–4.9)	4.1	5.1	4.4	7.3	4.6	4.5
SmithKline Beecham	6.7	—	6.3	(–0.6)	10.3	63.5†	9.9	(–3.9)	8.6	7.1	9.6	11.6
Upjohn	3.7	—	3.5	(–5.4)	3.5	0	3.8	8.6	3.9	2.6	3.9	0
Warner-Lambert	5.2	—	5.5	5.7	5.5	0	5.7	3.6	6.1	7.0	6.1	0

Data from company annual reports.
* Merger of Bristol-Myers with Squibb.
† Merger of Smith Kline with Beecham.

force. To reinvigorate growth over the long term, companies adopted several strategies, including merging with, acquiring, or entering into joint ventures with related companies to strengthen their competitive position in their own CTTs and to expand into new ones, to produce generic and over-the-counter versions of major drugs, and to extend the geographic reach of the companies into new markets.

These restructurings were followed by the great bull market in Wall Street, which rose to unprecedented levels by July 1998, with a Dow Jones index of stocks above 9,000. In this boom two classes of securities performed exceptionally well: those related to the information age and pharmaceuticals, which were perceived by investors as having great growth potential. In the case of pharmaceuticals it was the high R&D intensity and the recognition of the increasing demand for improved health care. Thus, by 30 June 1998, some of these market capitalizations were as follows (see Figure 9 for the 1994 values to see the change):

Company	Billions of Dollars
Merck	159,866
Pfizer	149,906
Bristol-Myers Squibb	114,397
Johnson & Johnson	99,905
Lilly	73,367
American Home Products	67,927
Schering-Plough	67,262
Abbott Laboratories	63,286
Warner-Lambert	56,851
Pharmacia & Upjohn	23,431

The failure of one or two drugs can also have a sharp effect on these high-flying capitalizations. Conversely, a successful blockbuster like Viagra (for treatment of sexual dysfunction) can be a big boost for Pfizer. These should be compared with more typical old-line manufacturing companies, which have heavy capital investment and diversified portfolios with much higher sales volumes, like DuPont, with a market capitalization of $84,286, and Dow Chemical, with $21,766. It is probable that this relative market evaluation takes into account the intangible capital that is represented by R&D and patents, assets not usually considered in conventional accounting. General accounting practices, however, do not consider such intangible capital. The market capitalization of a few old-time companies that nevertheless have been very well managed are, for example, General Electric, with $296,073, and Exxon, with $174,140. In the information category some high fliers were Microsoft, with $267,044; Intel, with $125,716; and IBM, with $108,257.

Rationalization and Restructuring

The first step most American companies took was to enhance their competitiveness by trimming their operational costs. Companies consolidated and

Table 21. Profitability of Major American Pharmaceutical Companies by Decade and for the Years 1990–93*

Company	1950s	1960s	1970s	1980s	1990–93
			%		
Abbott	11.3	9.6	8.2	13.4	16.1
American Home Products	8.6	10.2	10.5	15.1	18.4
Bristol-Myers	6.4	8.0	7.8	14.8	17.0
Johnson & Johnson	4.5	7.0	8.5	9.0	10.5
Lilly	13.3	12.6	15.2	15.6	16.0
Merck	10.8	15.5	15.7	19.4	23.4
Parke-Davis†	13.8	11.4	—	—	—
Pfizer	10.7	10.0	8.9	12.2	11.0
Schering-Plough	15.3	11.9	14.4	11.6	17.4
Searle†	21.0	21.5	9.6	11.4	—
Smith Kline and French	15.5	16.1	12.9	17.5	14.5
Sterling†	7.9	9.0	8.0	7.5	—
Upjohn	12.9	13.1	9.0	10.3	13.0
Warner-Lambert	7.4	9.0	7.2	5.1	7.0

Calculated from company annual report data.
* Profitability: After-tax profit as percentage of total sales.
† Incomplete data because companies merged with other companies.

restructured their manufacturing and distribution networks, reduced their work forces, and in some cases closed down manufacturing plants. Companies paid particular attention to their western European operations, which were the hardest hit by government controls on the prices of drugs and other medications. They also sold off most of their remaining businesses that were unrelated to health care.

Companies incurred considerable costs in implementing these policies in the form of one-time charges for plant write-offs, cash termination benefits, and pension enhancements for employees opting for early retirement. Merck, for example, incurred a pretax charge of $775 million in 1993 when it streamlined its manufacturing and distribution operations and rationalized its European operations. Pfizer sold off fourteen businesses between 1988 and 1993, including Minerals Technology Inc. and Coty Cosmetics and Toiletries. Restructuring its administrative and operating units and reducing its workforce by about three thousand cost Pfizer an after-tax charge of $525 million in 1993, which was covered by the proceeds of the divestitures; Pfizer's restructuring was expected to generate annual savings of about $130 million.

Lilly divested its Medical Devices Division, spun off its pesticide business as a joint venture with Dow Chemicals (Dow-Elanco), and restructured its remaining manufacturing and distribution operations, laying off 4,000 employees. The restructuring cost Lilly $564 million in 1992 and $1.2 billion in 1993. Warner-Lambert divested its polymer manufacturing business and some of its confectionery businesses, closed down seven manufacturing plants, and reduced overall employment by 2,800. Upjohn pared its work force by 1,500 and

Table 22. Net Income and Profitability (1988–93)*

Company	1988		1989		1990		1991		1992		1993	
	$M	%	$M	%	$M	%	$M	%	$M	%	$M	%
Abbott	752	15.3	860	15.8	927	14.8	1,088	15.8	1,240	15.8	1,400	16.7
American Home Products	995	15.5	1,102	16.3	1,230	18.1	1,375	19.4	1,460	18.5	1,470	17.7
Bristol-Myers Squibb	1,202	15.0	688	8.0	1,691	17.4	1,991	18.8	1,538	13.8	1,960	17.1
Johnson & Johnson	974	10.8	1,082	11.0	1,143	10.2	1,461	11.7	1,030	7.5	1,787	12.6
Lilly	761	21.1	940	22.5	1,127	21.7	1,314	23.0	709	11.5	480	7.4
Merck	1,207	20.3	1,495	22.8	1,781	23.2	2,121	24.7	2,446	25.3	2,166	20.5
Pfizer	791	14.7	681	12.0	801	12.5	722	10.4	811	11.2	657	8.8
Schering-Plough	390	13.1	471	14.9	565	17.0	646	17.9	720	17.8	731	16.8
SmithKline Beecham	—	—	—	—	774	8.9	931	10.6	605	7.7	942	10.4
Upjohn	353	14.3	176	6.6	456	15.4	537	16.0	324	9.0	392	10.7
Warner-Lambert	340	8.7	413	9.8	485	10.3	35	.7	644	11.5	331	5.7

Data from company annual reports.
* $M = millions of U.S. dollars; % = percentage of turnover.

eliminated or reduced manufacturing capacity at fourteen plants. Bristol-Myers Squibb sold off its household products business, restructured its European operations, and cut employment by about 4,000. Johnson and Johnson merged the operations of its two American pharmaceutical companies, Ortho and McNeil, except for their sales and marketing departments, enabling them to lay off about 3,300 workers.

A Changing Marketplace

The advent of managed care, designed to hold down the costs of health care, was by the 1990s inducing the most dramatic changes in the marketing and sales practices of the American pharmaceutical industry since the 1950s. By the mid-1990s half of all Americans covered by health insurance were enrolled in managed-care plans; that proportion was expected to reach 90 percent by the year 2000. Large customers of managed care, including government agencies and group benefit managers, traditionally left drug pricing and prescribing decisions to the marketplace and individual physicians. In the late 1980s that changed as governments, insurance providers, and employers sought ways to cope with spiraling health care costs, a slower-growing economy, and budget constraints. These groups now negotiate lower drug prices from manufacturers and manage the drug benefits portion of their insurance packages in ways that reduce their overall costs. Physicians still decide what medications to prescribe, but they are under increasing pressure to choose from among a list of drugs and formularies approved by insurance providers for their cost-effectiveness as well as their efficacy.

Under these conditions pharmaceutical companies maintain their traditional relations with physicians and pharmacies, which still represent some 60 percent of their customer base, while also marketing to HMOs, mail-order pharmacies, hospital associations, and group-purchasing organizations. The pharmaceutical companies must take into account physicians' concerns about the efficacy and safety of a particular drug but also insurance providers' concerns about the economic advantages of that drug and the economic and therapeutic value of pharmaceuticals compared with such alternatives as hospitalization and surgery. A similar situation pertained in western European countries, where national health insurance systems accounted for 60 to 90 percent of the health insurance market.

This changing marketplace forced pharmaceutical companies to restructure their marketing and sales organizations to reach this important new group of customers. Schering-Plough was one of the few companies to anticipate this development, forming a business unit in 1987 expressly to target the managed-care industry in the United States. Other companies soon followed suit. American Home Products signed agreements for its products with two of the largest hospital purchasing groups, while Bristol-Myers Squibb signed a five-year agreement for its pharmaceutical, nutritional, and other products with American

Healthcare Systems, an alliance of about one thousand nonprofit hospitals. Bristol-Myers Squibb also formed a corporate hospitals account group in 1992 to develop similar arrangements with other hospitals and entered into a joint venture with the Oncology Therapeutic Network, a leading distributor to the expanding office-based chemotherapy market. SmithKline Beecham joined with the Voluntary Hospitals of America to establish the Advanced Practice Institute at Georgia's Auburn University to study methods and practices for coordinated medical care.

The increasing trend toward managing drug benefits led to the establishment of about forty specialized companies, called pharmacy benefit managers (PBMs). By the late 1980s PBMs served about half the insured population by marketing their services to employers, insurance companies, managed-care groups, and Medicaid. Major pharmaceutical companies recognized the commercial importance of these new companies and in the early 1990s acquired or merged with the top five PBMs, accounting for more than 75 percent of that market. In 1992, for example, Merck paid $6 billion to MEDCO, which managed $4 billion in prescription drug benefits and dispensed 72 million prescriptions annually. The newly formed Merck-MEDCO U.S. Managed Care Division markets Merck medicines, among others, and MEDCO's pharmaceutical care services.

Vertical Integration into Generics and OTC Drugs

In the 1960s, when the American pharmaceutical companies faced a similar slowdown, they adopted horizontal diversification outside the industry in less profitable but fast-growing manufacturing sectors. In the 1990s such a strategy could not be used because the decline, contraction, and restructuring of the manufacturing industry severely limited such opportunities. Growth, however, could be achieved by horizontal diversification within the pharmaceutical sector, either through mergers and acquisitions or through vertical integration in the manufacture and distribution of generic and OTC drugs.

Generic drugs are prescription drugs whose active ingredient patents have expired and may be manufactured by the original innovating company and any other manufacturers; they are traded under the generic name, such as chlorpromazine, penicillin, or propranolol. Over-the-counter drugs do not require a physician's prescription and are sold in drugstores and supermarkets by their generic and trade names (such as ibuprofen or Motrin and diphenhydramine or Benadryl).

Generics. In the past research-intensive companies had deliberately stayed away from manufacturing generics because discovering and developing new drugs and improving existing ones were far more profitable endeavors. Moreover, legislation enacted in many foreign countries in the 1950s prohibited anyone but the drug's originator from producing and marketing generic versions when the patent expired. Therefore innovating companies were able to continue to market their products exclusively. Once demand began to drop in the face of competition from new products, the manufacturer would begin to phase out production, leaving that function to companies that specialized in making and distributing generic drugs and to pharmaceutical companies in some European and developing countries, notably Italy, Hungary, India, Egypt, and Korea. This division of labor made sense: Only new drugs could generate the cash flows necessary to sustain profitability at high enough levels to justify continued research and marketing expenditures, while generic manufacturers could accommodate lower profitability because their R&D and marketing expenditures were lower.

The strong innovative record of the 1950s, 1960s, and 1970s created a large pool of generic drugs; and some pharmaceutical companies, recognizing the growing potential of this market, acquired producers of such drugs. But it was in the 1990s that vertical integration became a necessity for even the most innovative companies: Patents for several effective drugs with very large markets were expiring at the same time that public and private health insurers were advocating the use of cheaper generics. Furthermore, in the late 1970s, the United States, Canada, Germany, the United Kingdom, and other countries adopted new legislation to encourage the development of a secondary market supplying cheaper generic products. Federal legislation in the United States made it easier to win Food and Drug Administration approval of generic drugs by requiring only that generics demonstrate blood levels comparable to those of the original products, under the assumption that similar blood levels would deliver comparable clinical efficacy and safety. By the mid-1980s the FDA had approved 8,000 generic versions of 170 brand-name drugs.

These were strong incentives for the pharmaceutical companies to begin to manufacture generic versions of their innovative drugs, particularly those drugs that enjoyed strong demand, in anticipation that the company name, its manufacturing and marketing experience, and the goodwill of the medical profession would ensure the company a smaller but still substantial segment of a more competitive market. The technological maturity of some sectors made lower profit margins acceptable, and companies began to acquire generic manufacturers that produced and marketed generic versions of their own and their competitors' off-patent drugs.

Merck, for example, established West Point Pharma in 1992 and began to market generic versions of eleven of its off-patent drugs. In 1989 Bristol-Myers Squibb formed a generic drug subsidiary called Apothecon, whose volume of sales in the 1990s would have made it the seventh largest American pharmaceutical company. Upjohn acquired Greenstone in 1992 to manufacture generics and signed a supply and distribution agreement with Ciba's generics manufacturer, Geneva Pharms Inc., to market generic versions of Xanax, Halcion, Cleocin, and Micronase—Upjohn's most important drugs of the 1970s and 1980s. Abbott produced Anaquest's off-patent anesthetics isoflurane and enflurane, taking advantage of its tradition in this market.

OTC Drugs. Another market of high-volume, lower-priced products, but one more closely associated with traditional pharmaceutical companies, is that of OTC proprietary drugs. Bristol-Myers, Warner-Lambert, American Home Products, Johnson and Johnson, Beecham, and Boots all were established to market OTC preparations and developed strong traditions in this area. Other large pharmaceutical companies were also active in the OTC business, selling analgesics, cough syrups, antiemetics, laxatives, antacids, nasal decongestants, topical antifungals, and so on. Companies sought to widen these markets by soliciting regulatory approval of OTC versions of some of their prescription drugs whose patents were about to expire. Companies that market their own OTCs have important advantages in that they can promote their established trade names and capitalize on the marketing and sales investments already made and the goodwill already established for the patented product. OTCs do not require a physician's prescription, so they can help reduce health care costs. To that end regulatory agencies in the United States and elsewhere began to approve lower strength and easily self-administered OTC forms of some major prescription drugs, for example, Zantac and Tagamet.

Among the early shifts from prescription drugs to OTC preparations were antihistamines in numerous cough syrups and NSAIDs, notably ibuprofen (Motrin, Advil, Nuprin, Rufen). In the 1990s several pharmaceutical companies rushed OTC versions of their blockbuster anti–peptic ulcer drugs to market as their patents neared expiration. These included SmithKline Beecham's Tagamet, Merck's Pepsid, Glaxo's Zantac, and American Home Products' Axid.

Several companies sought to strengthen their OTC businesses with joint ventures and acquisitions. Schering, a leader in this field, acquired White Labs in the 1970s and by 1993 had completed several prescription-to-OTC shifts, including the decongestant oxymetazoline (Afrin), the antihistamine chlorpheniramine (Chlor-Trimeton), and the antifungal agent clotrimazole (Lotrimin). Glaxo and Wellcome, who merged in 1995, formed a joint venture with Warner-Lambert to market OTC versions of the antiulcer drug Zantac, the antiviral agent Zovirax, and the cholesterol lipid-lowering drug Lopid. American Home Products built an OTC manufacturing plant in Ireland to serve the European markets, while its Whitehall-Robins Division specialized in developing and marketing OTC versions of generic drugs. Merck and Johnson and Johnson formed a joint venture, the J&J-Merck Consumer Pharmaceuticals, and acquired OTC companies in France, Germany, and Spain to manufacture products for the European market. A Pfizer subsidiary, Roerig, introduced OTC versions of its topical antifungal agent fluconazole (Diflucan) and its antihistamine cetirizine (Zyrtec) long before the expiration of its patents.

Global Companies

To widen their markets and ensure growth, pharmaceutical companies during this period aggressively pursued horizontal and geographic expansion. The knowledge that a considerable segment of their markets would be served in the future by technologically mature products, including OTCs and generics, strengthened their desire to create worldwide oligopolistic markets that could generate the cash flows sufficient to support the large research and marketing expenditures needed to discover and develop new innovative medicines. Mergers, acquisitions, and joint ventures with the appropriate competitors could also help a company overcome constraints on growth imposed by its own CTTs and marketing specialization. Moreover, to attain "blockbuster" status, any new drug had to be produced and launched simultaneously throughout the world.

In the early 1990s Roche acquired Syntex and Genentech, American Home Products acquired American Cyanamid/Lederle, Glaxo acquired Wellcome, Beecham merged with Smith Kline, and Ciba merged with Sandoz to form Novartis. The global companies resulting from these and other important acquisitions put pressure on their competitors to imitate them if they wanted to stay competitive. A further concentration of the pharmaceutical industry thus seemed likely before the end of the century. The merger of the Swedish Astra and Britain's Zeneca (1998) and that of Hoechst and Rhône (1999) are the most recent developments.

Joint ventures were another strategy companies pursued to complement their advantages or eliminate their shortcomings, whether scientific, technical, commercial, or geographic. In addition to its joint venture with Johnson and Johnson to market OTC products worldwide, Merck entered several other joint ventures: The Du Pont/Merck Company, for research in medicinal chemistry, manufacturing processes, and the development of cardiovascular, cancer, and AIDS drugs; the Merck/Pasteur-Merrieux/Connaught Laboratories/Chemo-Sero Therapeutic Institute of Japan joint venture, to create a worldwide oligopoly in the development and marketing of vaccines; and the Merck-Astra joint venture, to market Astra's innovative products in the United States. DuPont bought out the Merck interest in the joint venture in 1998.

Bristol-Myers Squibb cooperated with Sterling-Sanofi (the company created when Sanofi, the pharmaceutical division of Elf-Aquitaine, acquired Sterling's prescription drug business) to develop the anticlotting agent clopidogrel and with the biotechnology firm Ixsys to develop the monoclonal antibodies essential for Bristol-Myers's cancer drugs. American Home Products formed joint ventures with two biotechnology firms: It was working with Cygnus Therapeutics Systems to develop transdermal hormones and with Oncogene Science Inc. to develop drugs for asthma, osteoporosis, diabetes, and immunosuppression. American Home also acquired Genetics Institute and redirected its research effort toward cancer therapies.

Geographic expansion was attractive to many companies because the traditional U.S., European, and Japanese markets appeared to have reached maturity. The slowing rates of economic growth coupled with price-cutting regulatory measures caused pharmaceutical sales to level off in the United Kingdom and actually to decline in Germany and Italy in the early 1990s. The geographical expansion of the American pharmaceutical industry into Asia and eastern Europe strongly resembles its expansion in the 1960s into western Europe. Both expansions were caused by shrinking traditional markets, a slowdown in innovation, normalization of the political climate, and improved rates of economic growth in the countries targeted for expansion.

The maturity of many of these traditional markets and the growing importance of generics and OTC drugs forced companies to reevaluate their policies and to invest in less profitable but potentially large markets. American pharmaceutical companies turned primarily to the expanding markets in developing countries, many of which were recording high rates of economic growth: China, with double-digit growth in its gross national product and a pharmaceutical market valued at about $4.2 billion annually; the countries of the Pacific Rim, some Latin American countries; and Russia and eastern European countries, which, despite the upheaval in their economies, held considerable promise for the future.

Upjohn, for example, invested in a facility in China to manufacture the contraceptive pill version of Depo-Provera and built manufacturing plants in India, Poland, and Singapore. Bristol-Myers Squibb formed the Sino-American Shanghai/Squibb Company and intends to introduce a full line of its products in China. Johnson and Johnson formed Johnson and Johnson–China Ltd., the first consumer products company in China wholly owned by a foreign interest; Johnson and Johnson also formed the Xian-Janssen Company to produce most of its drugs for the China market locally.

Research Spending

The rationalization and restructuring programs affected labor, manufacturing, and administrative expenses but not R&D expenditures. Those expenditures, both in dollar values and as a percentage of turnover, increased steadily between 1988 and 1993 (Table 23). Companies recognized that treatments and cures for many widespread diseases are at a high-cost stage in the cycle of innovation, and their commitment to continued research and innovation in these potentially lucrative areas is shown by their investments in biotechnology and other risky, long-term projects.

Even so, research investment decisions depend on cash flow. If cash flow is further constrained by government regulation, price controls, and the changing marketplace, managers will come under increasing pressure to cut their R&D budgets. This concern has been expressed by all companies, particularly those that do not expect to launch innovative products in the immediate future.

A LOOK INTO THE FUTURE

Apart from its contributions to the fight against disease, each generation of pharmaceuticals has been marked by changes in the markets, structure, and business practices of the pharmaceutical industry. The fifth generation is no exception, as the events of the 1980s and 1990s prove. But this generation has

Table 23. Research and Development Expenditures by Company in the 1990s*

Company	1988		1989		1990		1991		1992		1993	
	$M	%	$M	%	$M	%	$M	%	$M	%	$M	%
Abbott	615	9.2	675	9.2	679	9.2	753	9.6	838	9.7	930	10.4
American Home Products	355	4.8	400	5.0	415	5.2	487	6.1	600	7.0	700	8.0
Bristol-Myers Squibb	510†	6.3†	1,145	8.5	1,030	9.0	1,111	9.3	1,176	9.7	1,191	9.9
Johnson & Johnson	875	7.5	935	7.4	1,102	8.3	1,107	7.9	1,224	8.2	1,248	8.3
Lilly	675	12.3	785	14.4	830	13.6	867	13.3	1,004	15.0	1,008	14.8
Merck	824	11.2	975	11.5	1,008	11.1	1,126	11.5	1,233	11.5	1,277	11.2
Pfizer	640	8.8	690	9.3	755	9.9	854	10.8	937	11.9	1,029	13.0
Schering-Plough	400	10.0	425	10.3	448	11.5	481	11.7	567	12.9	610	13.3
SmithKline Beecham	700	10.4	700	9.8	NA	NA	NA	NA	NA	NA	910	9.4
Upjohn	515	14.1	500	14.1	538	15.4	591	15.5	632	16.2	678	17.4
Warner-Lambert	350	6.6	400	7.3	447	8.1	478	8.4	514	8.4	491	8.0

Data from company annual reports.
* $M = millions of constant 1995 U.S. dollars; % = percentage of total sales; NA = not available.
† Only Bristol-Myers.

not yet run its full course, and so the duration of its scientific, technological, therapeutic, and business consequences cannot be predicted with certainty. Technological forecasting can be a frustrating exercise for those who venture to look into a crystal ball. A breakthrough in biotechnology, for example, could accelerate dramatically the pace of innovation and reverse current technological and commercial trends. Radical changes have happened in the past and can be expected in the future. In the absence of such change, however, the trends of the last fifteen years in the context of the industry's technological and commercial history provide some clues about what might be expected in the foreseeable future.

The revolutionary potential of biotechnology and the protracted stages of youth and growth that sector is experiencing suggest that the time horizon of the fifth generation will be longer than the previous postwar generations. Because biotechnology and the life sciences appear to have the potential for developing drugs to treat or cure a host of previously untreatable illnesses, the pharmaceutical industry will remain research-intensive despite occasional or even protracted adversities in the marketplace.

At least for the near future a growing segment of the pharmaceutical market is likely to be served by drugs, including generics and OTC medications, that are already on the market. Pharmaceutical companies will continue to seek competitive advantages in this less profitable marketplace through vertical and horizontal diversification and continued geographic expansion. The ongoing trend of mergers of large, research-intensive companies and their acquisition of the most successful biotechnology firms are likely to reduce the number of companies in the industry.

The intellectual challenges posed by the life sciences and their potential, particularly in fighting disease and improving agriculture, will continue to attract scientists. Financial support will be forthcoming from government and private sources. The smaller number of research-intensive competitors and reduced cash flows may invite much stronger contributions on the part of academic and government research institutions and a further strengthening of their traditional relations with industrial laboratories.

Governments worldwide will play a more important role in the future than they have in the past, not only in funding medical research but also as regulators and customers. Economic upheaval in the countries of the former Soviet Union as well as poverty in both developed and developing countries have left many people unable to afford health care. Widespread poverty and malnutrition have also contributed to the reappearance of such contagious diseases as tuberculosis, diphtheria, cholera, and plague that had been all but eradicated for half a century. Governments will have to take steps to redress these problems, first by fostering economic growth and employment. Dealing with issues of access and public health care will also be important. The pharmaceutical industry is bound to respond by applying new scientific and technological knowledge to developing effective and affordable medicines. The societal need to treat and eliminate disease will continue to be, as it has been, an important driving force for technological innovation in the pharmaceutical industry.

Sources

B. Achilladelis. "The Dynamics of Technological Innovation: The Sector of Antibacterial Medicines." *Research Policy* 22:4 (1993), 279–308.

B. Achilladelis; A. Schwarzkopf; M. Cines. "The Dynamics of Technological Innovation: The Case of the Chemical Industry." *Research Policy* 19:1 (1990), 1–34.

E. D. Adrian. "Sir Henry Dale's Contribution to Physiology." *British Medical Journal* 1 (1955), 1355–1356.

R. P. Ahlquist. "A Study of Adrenotropic Receptors." *American Journal of Physiology* 153 (1948), 586.

American Home Products. Annual Reports, 1940–90. New York, N.Y.

American Medical Association, Department of Drugs. *AMA Drug Evaluations.* Chicago: American Medical Association, 1980.

Anonymous. *World Pharmaceuticals Directory.* Chatham, N.J.: Pharmaco-Medical Documentation, 1991.

Z. M. Bacq. "Chemical Transmissions of Nerve Impulses." In *Discoveries in Pharmacology*, vol. 1. Edited by M. J. Parnham and J. Bruinvels. Amsterdam: Elsevier Science Publishers, 1984, pp. 49–103.

W. A. Bain. "Xylocholine." In *Ciba Foundation Symposium on Adrenergic Mechanisms.* Edited by G. E. W. Wolstenholme and M. O'Connor. London: J. and A. Churchill, 1960.

E. Baumler. *A Century of Chemistry.* Düsseldorf: Econ Verlag, 1968.

———. *In Search of the Magic Bullet.* London: Thames and Hudson, 1965.

T. Beardsley. "Big Time Biology." *Scientific American* (Nov. 1994), 90–97.

Beckman Center for the History of Chemistry. "An Interview with Max Tishler" (mimeo). Philadelphia: Chemical Heritage Foundation, 1984.

Beckman Center for the History of Chemistry. "An Interview with Tadeus Reichstein" (mimeo). Philadelphia: Chemical Heritage Foundation, 1985.

J. J. Beer. *The Emergence of the German Dyestuffs Industry.* Urbana: University of Illinois Press, 1959.

R. Behnisch. "From Dyes to Drugs." In *Discoveries in Pharmacology*, vol. 3. Edited by M. J. Parnham and J. Bruinvels. Amsterdam: Elsevier Science Publishers, 1986, pp. 226–281.

J. D. Bernal. *Science in History.* New York: Hawthorne Books, 1954.

———. *The Social Function of Science.* London: Routledge and Kegan Paul, 1939.

Bernstein Research. *The End of an Economic Cycle in Pharmaceuticals; The Beginning of a Technological One.* New York: Sanford C. Bernstein & Co., 1992.

Biotechnology Medicines in Development. Washington, D.C.: Pharmaceutical Manufacturers Association, 1993.

J. W. Black; A. F. Crowther; R. G. Shanks; L. H. Smith; A. C. Dornhorst. "A New Adrenergic Beta Receptor Antagonist." *Lancet* 1 (1964), 1080–1081.

A. L. A. Boura; A. F. Green. "Peripherally Acting Hypertensives." In *Discoveries in Pharmacology*, vol. 2. Edited by M. J. Parnham and J. Bruinvels. Amsterdam: Elsevier Science Publishers, 1984, pp. 73–105.

Bristol-Myers. Annual Reports, 1940–90. New York, N.Y.

J. F. J. Cade. "Lithium." In *Discoveries in Biological Psychiatry.* Edited by F. J. Ayd and B. Blackwell. Philadelphia: Lippincott, 1970, pp. 218–229.

A. Carlsson. "Antipsychotic Agents: Elucidation of Their Mode of Action." In *Discoveries in Pharmacology*, vol. 1. Edited by M. J. Parnham and J. Bruinvels. Amsterdam: Elsevier Science Publishers, 1984, pp. 197–207.

D. Chapman-Huston; E. C. Cuipps. *Through a City Archway—The Story of Allen and Hanburys, 1715–1954.* London: John Murray, 1954.

Ciba Ltd. *The Story of Chemical Industry in Basel.* Olten/Lausanne, Switzerland: Urs Graf Publishers, 1959.

W. S. Commanor. "The Political Economy of the Pharmaceutical Industry." *Journal of Economic Literature* 24 (1986), 1187–1217.

A Corporation and a Molecule: The Story of Research at Syntex. Palo Alto, Calif: Syntex Corporation, 1966.

G. Cotzias; P. Papavassiliou. *Progress in Neurogenetics.* Edited by A. Barbeau and J. Brunette. Amsterdam: Excerpta Medica, 1969, esp. p. 357.

H. H. Dale. *Adventures in Physiology with Excursions into Auto-Pharmacology.* London: Pergamon Press, 1953.

R. P. T. Davenport-Hines; J. Slinn. *Glaxo: A History to 1961.* Cambridge: Cambridge University Press, 1992.

P. Deniker. "Qui a Inventé les Neuroleptiques?" In *Confrontations Psychiatriques: Les Neuroleptiques, 20 Ans Aprés,* vol. 13. Paris: Specia, 1975, pp. 7–17.

———. "Discovery of the Chemical Use of Neuroleptics." In *Discoveries in Pharmacology*, vol. 1. Edited by M. J. Parnham and J. Bruinvels. Amsterdam: Elsevier Science Publishers, 1984, pp. 164–182.

C. Djerassi. "A Steroid Autobiography." *Steroids* 43 (1984), 351–361.

G. Dosi. "Technological Paradigms and Technological Trajectories." *Research Policy* 11 (3 June 1982), 147.

Drug Amendments of 1962, 21 U.S.C./ /321(p)-(1), 351 (a) (2) (B), 360 et seq., PL 87-781, 76 Stat. 780, 10 October.

D. Dunlop. "Medicines, Governments, Doctors, and Pharmacists." *Chemistry and Industry* (Feb. 1973), 127–131.

P. Ehrlich. "Über den jetzigen Stand der Chemotherapie." *Berichte der Deutschen Chemischen Gesellschaft* 42 (1909), 17–47.

L. Engel. *Medicine Makers of Kalamazoo.* New York: McGraw-Hill, 1961.

M. Ennis; W. Lorenz. "Histamine Receptor Antagonists." In *Discoveries in Pharmacology*, vol. 2. Edited by M. J. Parnham and J. Bruinvels. Amsterdam: Elsevier Science Publishers, 1983, pp. 623–642.

P. Erni. *The Basel Marriage: History of the Ciba-Geigy Merger.* Zurich: Neue Zuricher Zeitung, 1979.

Federal Food, Drug and Cosmetic Act, 21 U.S.C./3016392, 52 Stat. 1040, 25 June 1938.

C. A. Fleeger, editor. *USAN and the USP Dictionary of Drug Names.* Rockville, Md.: United States Pharmacopoeial Convention Inc., 1992.

A. Fleming. *Penicillin: Its Practical Applications.* Philadelphia: Blakiston, 1946.

H. W. Florey et al. *Antibiotics*, vol. 1. Oxford: Oxford University Press, 1949.

Food and Drug Administration's process for approving new drugs. A report prepared by the Subcommittee on Science, Research and Technology of the Committee on Science and Technology, U. S. House of Representatives, 96th Congress, November 1980.

C. Freeman. *The Economics of Hope.* London: Pinter Publishers, 1992.

———. *The Economics of Industrial Innovation.* Harmondsworth, U.K.: Penguin Books, 1974.

C. Freeman; J. Clark; L. Soete. *Unemployment and Technical Innovation.* Westport, Conn.: Greenwood Press, 1982.

J. Fruton. *Molecules and Life.* New York: Wiley-Interscience, 1972.

H. S. Gasser. "Sir Henry Dale: His Influence on Science." *British Medical Journal* 1 (1955), 1359–1361.

Geigy (J. R.) A.G. *200 Years of Geigy.* Bern, Switzerland: Berl and Cie, 1958.

H. G. Grabowski. *Drug Regulation and Innovation: Empirical Evidence and Policy Options.* Washington, D.C.: American Enterprise Institute for Policy Research, 1976.

H. Guerlac. *Antoine-Laurent Lavoisier, Chemist and Revolutionary.* New York: Charles Scribner's Sons, 1975.

C. D. Haagensen; C. Wyndham; E. B. Lloyd. *A Hundred Years of Medicine.* New York: Sheridan House, 1943.

L. F. Haber. *The Chemical Industry during the 19th Century.* Oxford: Oxford University Press, 1958.

———. *The Chemical Industry, 1900–1930.* Oxford: Oxford University Press, 1971.

W. Haefely. "Alleviation of Anxiety: The Benzodiazepine Saga." In *Discoveries in Pharmacology,* vol. 1. Edited by M. J. Parnham and M. Bruinvels. Amsterdam: Elsevier Science Publishers, 1984, pp. 270–304.

R. Hare. *The Birth of Penicillin and the Disarming of Microbes.* London: Allen and Unwin, 1970.

M. Harrington, editor. *Hypotensive Drugs.* New York: Pergamon Press, 1956.

K. Hashimoto; E. Kimura; T. Kobayshi, editors. *Proceedings of the First International Nifedipine "Adalat" Symposium.* Tokyo: University of Tokyo Press, 1975.

G. L. Hobby. *Penicillin: Meeting the Challenge.* New Haven, Conn.: Yale University Press, 1985.

R. Holmstedt; G. Liljestrand, editors. *Readings in Pharmacology.* New York: Pergamon Press, 1963.

B. James. *The Global Pharmaceutical Industry in the 1990s: The Challenge of Change.* London: The Economist Intelligence Unit, Special Report 2071, 1990.

Johnson and Johnson. Annual Reports, 1940–90. New Brunswick, N.J.

E. J. Kahn Jr. *All in a Century: The First 100 Years of Eli Lilly.* Indianapolis: Eli Lilly, 1975.

S. J. Klein; N. Rosenberg. "An Overview of Innovation." In *The Positive Sum Strategy.* Edited by Ralph Landau and Nathan Rosenberg. Washington, D.C.: National Academy Press, 1986, pp. 275–305.

T. Koeppel. "An Interview with Vladimir Prelog" (mimeo). Philadelphia: Beckman Center for the History of Chemistry, 1984.

———. "An Interview with Leo H. Sternbach" (mimeo). Philadelphia: Beckman Center for the History of Chemistry, 1986.

H. Kogan. *The Long White Line: The Story of Abbott Laboratories.* New York: Random House, 1963.

N. D. Kondratiev. "The Long Waves in Economic Life." *Review of Economics and Statistics* 17 (6 Nov. 1925), pp. 105–115.

R. Kuhn. "The Imipramine Story." In *Discoveries in Biological Psychiatry.* Edited by F. J. Ayd and B. Blackwell. Philadelphia: Lippincott, 1970.

T. Kuhn. *The Structure of Scientific Revolutions.* Chicago: Chicago University Press, 1962.

R. Landau; N. Rosenberg, editors. *The Positive Sum Strategy.* Washington, D.C.: National Academy Press, 1986.

D. S. Landes. *The Unbound Prometheus.* Cambridge. Cambridge University Press, 1970.

J. E. Lesch. "The Discovery of M & B 693 (sulfapyridine)." In *The Inside Story of Medicines.* Edited by G. J. Higby and E. C. Stroud. Madison, Wis.: American Institute of the History of Pharmacy, 1997, pp. 101–117.

J. A. Levy; K. B. Menander. *Research in the Pharmaceutical Industry: Major Contributions to Biomedical Science and Implications for Public Policy.* Reston, Va.: National Pharmaceutical Council, 1991.

J. Liebenau. *Medical Science and Medical Industry: The Formation of the American Pharmaceutical Industry.* London: Macmillan Press, 1987.

R. D. Mann. *Modern Drug Use (An Inquiry on Historical Principles).* Lancaster, U.K.: MTP Press, 1984.

J. F. Marion. *This Fine Old House: Smith, Kline's First 150 Years.* Philadelphia: SmithKline, 1980.

L. G. Mattheus. *History of Pharmacy in Britain.* Edinburgh: E. & S. Livingstone Ltd., 1962.

M. S. May; W. M. Wardell; L. Lasagna. "New Drug Development during and after a Period of Regulatory Change: Clinical Research Activity of Major U.S. Pharmaceutical Firms, 1958 to 1979." *Clinical Pharmacology and Therapeutics* 33:6 (1983), pp. 691–700.

A. S. McNalty. "Fifty Years in Public Health Legislation." In *British Medical Association, Fifty Years of Medicine. A Symposium from the* British Medical Journal. Margate, U.K.: Thanet Press, 1950, pp. 226–244.

Merck and Co. "Cardiovascular Research: Improving Prospects for Longer Life through Scientific Innovation." *Merck Review* 45 (Summer 1989), p. 1.

S. Mines. *Pfizer and Co.: An Informal History.* New York: Pfizer, 1978.
M. Moas. *The Scientific Renaissance 1450–1630.* New York: Harper and Row, 1962.
J. Mokyr. *The Lever of Riches.* Oxford: Oxford University Press, 1990.
W. B. Murphy. *Science and Serendipity: A Half Century of Innovation at SYNTEX.* White Plains, N.Y.: Benjamin Company, 1994.
R. R. Nelson; S. G. Winter. "In Search of a Useful Theory of Innovation." *Research Policy* 6:1 (Jan. 1977), pp. 36–76.
M. A. Ondetti; D. W. Cushman; B. Rubin. "Captopril." In *Chronicles of Drug Discovery,* vol. 2. Edited by J. S. Bindra and D. Lednicer. New York: John Wiley and Sons, 1983, pp. 1–31.
Opportunities and Challenges for Pharmaceutical Innovation: Industry Profile. Washington, D.C.: Pharmaceutical Manufacturers Association, 1996.
K. Pavitt. "Sectoral Patterns of Technical Change: Towards a Taxonomy and a Theory." *Research Policy* 13:6 (1984), pp. 343–373.
S. Peltzman. *Regulation of Pharmaceutical Innovation: The 1962 Amendments.* Washington, D.C.: American Enterprise Institute for Policy Research, 1974.
PMA Statistical Fact Book. Washington, D.C.: Pharmaceutical Manufacturers Association, 1989.
PMA Communications Division. *Biotechnology Medicines in Development.* Washington, D.C.: Pharmaceutical Manufacturers Association, 1993.
M. E. Porter. *The Competitive Advantage of Nations.* New York: Free Press, 1990.
J. E. F. Reynolds, editor. *Martindale: The Extra Pharmacopoeia.* Third edition. London: Pharmaceutical Press, 1989.
N. Rosenberg. *Perspectives on Technology.* Cambridge: Cambridge University Press, 1976.
St. M. Scala; R. C. Hodgson; I. C. Sanderson. *Industry Strategies: Pharmaceutical Industry Portfolio Perspectives.* Boston: Gowen & Co., 1994.
J. A. Schumpeter. *Capitalism, Socialism, and Democracy.* Second edition. New York: Harper and Row, 1947.
———. *The Theory of Economic Development.* Oxford: Oxford University Press, 1961. (First edition, 1934.)
D. Schwartzman. *Innovation in the Pharmaceutical Industry.* Baltimore: Johns Hopkins University Press, 1976.
J. C. Sheehan. *The Enchanted Ring: The Untold Story of Penicillin.* Cambridge, Mass.: MIT Press, 1982.
T. Y. Shen. "The Proliferation of Non-Steroidal Anti-Inflammatory Drugs (NSAIDs)." In *Discoveries in Pharmacology,* vol. 2. Edited by M. J. Parnham and J. Bruinvels. Amsterdam: Elsevier Science Publishers, 1984, pp. 524–553.
J. S. Smith. *Patenting the Sun: Polio and the Salk Vaccine.* New York: Anchor Books, 1990.
W. Sneader. *Drug Discovery and Evolution of Modern Medicines.* Chichester, U.K.: J. Wiley, 1985.
B. Sokoloff. *The Story of Penicillin.* Chicago: Ziff-Davis, 1945.
J. L. Sturchio; A. Thackray. "An Interview with Carl Djerassi" (mimeo). Philadelphia: Beckman Center for the History of Chemistry, 1985.
J. L. Sturchio, editor. *Values and Visions: A Merck Century.* Rahway, N.J.: Merck & Co., 1992.
J. P. Swann. "Universities, Industry, and the Rise of Biomedical Collaboration in America." In *Pill Peddlers: Essays on the History of the Pharmaceutical Industry.* Edited by J. Liebenau, G. H. Highby, E. C. Stroud. Madison, Wis.: American Institute for the History of Pharmacy, 1990, pp. 73–87.
M. Tauch. "Endocrine Hormones." In *Discoveries in Pharmacology,* vol. 2. Edited by M. J. Parnham and J. Bruinvels. Amsterdam: Elsevier Science Publishers, 1983.
J. Thuillier. *Les dix ans qui ont change la folie,* vol. 1. Paris: R. Laffont, 1980.
J. Tréfouël; F. Nitti; D. Bovet. Activité du p-aminophenylsulfamide sur les infections streptococciques expérimentales de la souris et du capin. *Comptes Rendus Hebd. Seances Soc. Biol.* 120 (1935), pp. 756–758.
U.S. Department of Health and Human Services, Public Health Service, Food and Drug Administration, Offices of Drug Evaluation. *Statistical Report 1987–1988.* Publication no. PB89-233530. Springfield, Va.: National Technical Information Service, August 1989.

R. Vallery-Radot, editor. *Correspondence de Pasteur*, vol. 3. Paris: Flammarion, 1951.
J. R. Vane. "Progress towards New Drugs." In *The Chemical Industry*. Edited by D. Sharp and T. F. West. Chichester, U.K.: Ellis Horwood Publishers, 1982.
S. W. Waksman. *The Conquest of Tuberculosis*. Berkeley: University of California Press, 1964.
Warner-Lambert. Annual Reports, 1940–90. Morris Plains, N.J.
P. G. Wingate. *The Colorful DuPont Company*. Wilmington, Del.: Serendipity Press, 1982.
H. B. Woodruf; R. W. Burg. "The Antibiotic Explosion." In *Discoveries in Pharmacology*, vol. 3. Edited by M. J. Parnham and J. Bruivels. Amsterdam: Elsevier Science Publishers, 1986, pp. 304–351.

Chapter Two

Discovery and Development of Major Drugs Currently in Use

ALEXANDER SCRIABINE

This chapter deals with the discovery and development of drugs and vaccines currently available for the prevention or treatment of human diseases. With these weapons scientists have conquered, and in some cases even eliminated, many infections and learned to control a multitude of cardiovascular, metabolic, and central nervous system diseases. The drugs were chosen for discussion because of their commercial success, chemical novelty, or effectiveness in treating diseases resistant to previously available drugs.

The drugs are subdivided into different classes in accordance with their therapeutic use or the organ systems that they affect. First discussed are the vaccines, which have saved so many lives through immunization against virulent diseases, followed by sections on antibacterial agents, antiviral and antifungal agents, cardiovascular drugs, drugs affecting the central nervous system, drugs affecting the autonomic nervous system, nonsteroidal anti-inflammatory drugs (NSAIDs) and immunosuppressants, drugs used to treat metabolic diseases, and cancer chemotherapeutics.[1] Trade names (American, if the drugs are marketed in the United States) are listed in parentheses. Definitions of the scientific terms can be found in the glossary at the end of the book.

Vaccines

The best and cheapest way to prevent a disease is to increase the endangered population's resistance to it through immunization. A dozen or so major vaccines are widely used today to inoculate populations against what were once

[1] The classification system is similar to that used in the ninth edition of Goodman and Gilman's *Pharmacological Basis of Therapeutics* (McGraw-Hill, 1996).

devastating diseases. Most of these vaccines consist of killed or live attenuated microorganisms (bacteria or viruses) whose capacity to produce disease has been either eliminated or greatly diminished. The vaccines contain proteins, called antigens, that induce the formation of antibodies (the body's own proteins), which are capable of neutralizing not only antigens in the vaccine but also disease-producing microorganisms.

Vaccination is termed *active* immunization because vaccines contain killed or attenuated microorganisms and the recipients produce their own antibodies. Immunization is termed *passive* when antibodies (usually produced in animals) are administered instead of microorganisms.

The principle of disease prevention by vaccination is an ancient discovery. Smallpox inoculation was allegedly practiced by the Chinese in 1000 A.D., and the first written reference attributes inoculation technique to a Buddhist nun from the eleventh century. In the early eighteenth century, in England, Lady Mary Wortley Montagu introduced the variolation technique after having learned it in Constantinople; the technique involved treatment of patients with dried pus from smallpox pustules. The terms *vaccine* and *vaccination* were coined by English physician Edward Jenner (1749–1823), who derived the terms from the Latin word *vacca* (cow). Jenner first observed that milkmaids did not get smallpox and then inoculated a man with a cowpox virus, thus protecting him from smallpox.

The next important discovery in the field of vaccines was not made until the 1870s, when Louis Pasteur discovered that weakened chicken cholera vibrio can protect humans from virulent cholera.

ANTHRAX AND RABIES VACCINES

In 1881 Pasteur developed the anthrax vaccine and performed his famous controlled experiment. He vaccinated twenty-four sheep, one goat, and six cows with attenuated anthrax bacilli and followed the first vaccination with a second twelve days later. Another group of animals was not vaccinated and served as controls. Fourteen days after the first group received its second vaccination, both groups were inoculated with virulent anthrax bacilli. All of the unvaccinated animals died within three days, while all of the vaccinated animals remained healthy. This first controlled experiment established the value of vaccination, at least for animals. The introduction of anthrax vaccine became an important milestone in the history of vaccination.

Four years later, in 1885, Pasteur took another historic step when he vaccinated two humans with a rabies vaccine he had developed. Pasteur's colleagues sharply criticized him for introducing an actual virus into human beings: By today's standards his viruses would be considered only "partially inactivated." These first two patients survived, and vaccination was considered successful. Pasteur subsequently saved many more patients from rabies, but some died, possibly from vaccination.

DIPHTHERIA IMMUNIZATION

The recipient of the first Nobel Prize in physiology or medicine, in 1901, was Emil Adolf von Behring, a German physician and bacteriologist. He received the prize for the discovery and development of the diphtheria antitoxin, an antibody to the diphtheria toxin, used for passive immunization. The antitoxin is still used not only to prevent but also to treat the disease. It is curative, although it does not always completely eradicate the microorganisms. Diphtheria can be cured by adding penicillin or erythromycin to the antitoxin therapy.

Currently an active immunization against diphtheria is popular; it involves stimulating the production of endogenous antitoxin by administering a toxoid, a chemically modified toxin that has lost its toxicity but still retains the ability to stimulate production of antitoxin. In 1923 A. T. Glenny and Barbara E. Hopkins, from the Wellcome Research Laboratories in the United Kingdom, prepared the first diphtheria toxoid for immunization by treating diphtheria toxin with formalin. The currently used diphtheria toxoid was introduced by a French bacteriologist, Gaston Ramon, the same year. In the 1940s the diphtheria toxoid was combined with tetanus toxoid and pertussis vaccine to form a DPT vaccine, which is widely used today. Because of the high rates of immunization (96 percent) of American children entering school, the DPT vaccine has practically eradicated diphtheria in the United States. In 1921 two hundred thousand cases of diphtheria were reported, while only fifteen cases of respiratory diphtheria were reported from 1980 to 1983.

TETANUS VACCINE

The second component of the triple vaccine DTP is the tetanus toxoid. It is a chemically inactivated toxin of *Clostridium tetani*, a spore-forming, anaerobic bacillus that causes tetanus in animals and humans.

Tetanus is a noncommunicable disease; the bacillus enters the body usually through contaminated puncture wounds. The disease is characterized by a spasm of skeletal muscles caused by action on the central nervous system of one of the toxins produced by *C. tetani.* This toxin, a large protein called tetanospasmin, is one of the most toxic substances known; it can kill a mouse at a dose as low as one nanogram per kilogram of body weight. The toxin was identified and purified in 1890 and was shown to produce in surviving animals antibodies capable of neutralizing the toxin. The toxoid, a related but nontoxic protein capable of inducing similar antibodies, was first prepared and shown to prevent tetanus in 1924.

Passive immunization with antibodies derived from horse serum was initially used to prevent tetanus as well as to treat it. Because of incomplete efficacy and frequent side effects, equine antitoxin was replaced by toxoid. The efficacy of tetanus toxoid was confirmed during World War II. The incidence of tetanus in U.S. soldiers immunized with tetanus toxoid was 0.44 per 100,000, whereas during World War I, when no toxoid existed, the incidence of tetanus was 13.4 per 100,000—thirty times higher.

The American Academy of Pediatrics recommended the routine use of toxoid for active immunization of children in 1944 and of the triple vaccine in 1951. On the U.S. market tetanus toxoid is available either alone (two generic formulations), in combination with diphtheria toxoid (two generic formulations), or as a triple vaccine with diphtheria toxoid and pertussis vaccine (Tri-Immunol from Lederle).

PERTUSSIS VACCINE

Pertussis, or whooping cough, is caused by a gram-negative bacillus, *Bordetella pertussis*. Epidemics of this disease were described in France and England as early as the sixteenth century. It is a severe and long-lasting disease with a high mortality rate that affects mostly infants and children.

Development of the pertussis vaccine was initiated early in the twentieth century. First a whole-cell vaccine was developed, but it suffered from the lack of a useful methodology for assessing its efficacy. Investigators were unable to produce a consistent pertussis infection in animals, so the vaccine could be tested only in clinical trials. Not until the 1950s was the so-called "mouse protection test" developed; it involved the intracerebral inoculation of mice with *Bordetella pertussis*. The test greatly facilitated the development of a more effective vaccine. In 1962 Eli Lilly and Company introduced a cell-free pertussis vaccine that was just as effective as the whole-cell vaccine but produced fewer local reactions and less fever.

POLIO VACCINES

Poliomyelitis is an acute viral disease that occurs either sporadically or as an epidemic. Three types of polio are known. The first type is abortive polio, a mild febrile disease lasting two to three days with no involvement of the central nervous system. The second type is characterized by additional aseptic meningitis, but recovery is still rapid and complete. The third type is a paralytic polio that sometimes follows the initial mild illness. The paralysis, at first flaccid and asymmetric, is more severe in adults than in children. The death rate from paralytic polio is 2 to 5 percent in children and 15 to 30 percent in adults. There is no specific therapy for polio.

The vaccines for preventing polio were introduced in 1955. Two kinds of polio vaccines are currently available: vaccine made of noninfectious, inactivated virus particles, developed by Jonas E. Salk; and vaccine made from infectious virus particles, attenuated in their virulence and developed by Albert B. Sabin of the University of Cincinnati.

The earliest reports on the development of vaccines for polio preceded World War II by a few years. Maurice Brodie and William H. Park, both from New York University, as well as John A. Kolmer, from Temple University, attempted to immunize humans in 1936 with a suspension of monkey brain tissue containing chemically treated viruses. Unfortunately, these trials led to paralytic polio in some patients. In 1948 Isabel Morgan, from Johns Hopkins, demonstrated that suspensions of monkey brain tissue treated with formaldehyde and containing chemically treated viruses induced formation of antibodies to polio in monkeys. Howard E. Howe, also at Johns Hopkins, demonstrated that tissue suspensions free from infectious activity can induce antibody formation in primates as well as humans; this finding was of considerable importance for the development of a noninfectious poliovirus vaccine. It was subsequently discovered that virus propagated in monkey kidney tissue cultures could induce antibody formation. The inactivated poliovirus vaccine was developed by American virologist Jonas E. Salk. The field trial of this vaccine involved a population of 1,829,916 children, and its favorable results made the licensing of the vaccine possible in 1955.

The first successful immunization of humans with a live poliovirus vaccine was reported by Hilary Koprowski and associates at Lederle Laboratories and the Wistar Institute in Philadelphia. Sabin was also intensively pursuing the development of an oral poliovirus vaccine. After clinical testing it appeared to be superior to Koprowski's vaccine in both efficacy and safety and was therefore chosen for licensing in the United States. Sabin's vaccine consists of three types of polioviruses grown in cell cultures and is still widely used today.

The currently available inactivated poliovirus vaccine is manufactured by Pasteur Merieux in France and is distributed in the United States by Connaught Laboratories as an injectable formulation under the trade name of Ipol. The live oral trivalent poliovirus vaccine is available in the United States from Lederle under the trade name of Orimune. The vaccine is used in infants, children, and adolescents up to age eighteen for prevention of poliomyelitis caused by the three types of polioviruses. Routine use of the vaccine in adults is not recommended. It should be used, however, in adults who are at greater risk of exposure to polio than the general population. The vaccine is contraindicated for all patients with immunodeficiency diseases. The ingestion of the vaccine has led on rare occasions to polio infection and even paralytic polio (one case per 1.2 million doses administered). In extremely rare instances a person in contact with a recently vaccinated individual can be infected since poliovirus is excreted in feces of the vaccinated individuals.

The availability of both vaccines has practically eradicated poliomyelitis in the United States. Before 1955 there were twenty-five new polio cases every year per 100,000 people; in 1955 there were only three in 100,000 and in 1968 fewer than one in 100,000.

MEASLES VACCINE

Measles is a viral disease characterized by fever and eruption. It is usually benign and self-limited, but it may lead to several complications involving the heart (myocarditis), abdomen (peritoneal inflammation), kidney (glomerulonephritis), and even the brain (encephalitis). The last complication is rare, but it occurs in 0.1 percent of patients with measles. The death rate of patients with measles encephalomyelitis is as high as 10 percent. Measles can also lead to bleeding from the mouth, intestines, or genitourinary tract. There is no specific therapy for measles; treatment is symptomatic or directed to prevention of secondary bacterial infections.

Measles is primarily a childhood disease, but infants under six months are seldom affected, probably because of the persistence of maternal antibodies. The disease can also occur in unvaccinated adults who have not previously been exposed to measles. Measles is highly contagious; it is spread by droplets from the nose or mouth of infected individuals two to four days before the rash and during the early rash stage of the disease.

Large fatal epidemics occurred in the nineteenth century. The epidemic on the Fiji Islands in 1875 was typical: A few months after a British ship brought the disease to the islands, 40,000 of 150,000 islanders were dead. An effective vaccine was not developed for another eight decades. The inactivated measles virus vaccine initially developed was considered unsatisfactory because resistance to infection declined within a year of vaccination. During the 1950s Samuel L. Katz and John F. Enders from Harvard University developed the Edmonston strain of live measles virus vaccine, which was further attenuated by Maurice R. Hilleman and his associates at Merck.

Today most measles vaccination in the United States is accomplished with the use of a triple vaccine that also inoculates against mumps and rubella viruses (MMR II from Merck Sharp & Dohme). The same company also makes a vaccine containing measles virus only (Attenuvax) or measles and rubella viruses (MRVax II). Vaccination with any of the live virus measles vaccines leads to lifelong protection in more than 95 percent of the recipients. The vaccines are generally safe, but occasional anaphylaxis or anaphylactoid reactions have been reported. Live measles vaccine may cause a very serious complication that affects the brain (subacute sclerosing panencephalitis). Fortunately, it is extremely rare; its frequency was estimated as one case per million of vaccine doses distributed, far lower than the incidence of measles infection itself.

MUMPS VACCINE

Mumps (epidemic parotitis) is an acute viral disease that causes enlargement of the salivary glands, particularly the parotid gland. The virus is spread by saliva and is contagious. Children of five to fifteen years of age are most commonly affected, but the disease has also been described in elderly individuals. During World War I, 170,000 cases of mumps were recorded among American soldiers. Prognosis in younger children is usually excellent. One relatively common complication of mumps in postpuberal males is orchitis, an inflammation of the gonads. It is estimated to occur in 20 percent of postpuberal male patients with mumps and may lead to the loss of fertility. The virus can also enter the brain and cause meningoencephalitis, which can occur without involvement of the salivary glands, making it more difficult to diagnose. Another serious complication of mumps is pancreatitis, which can in very rare cases cause diabetes mellitus.

Before the development and introduction of mumps live virus vaccine, gamma globulin from the plasma of individuals with a high level of mumps virus antibodies was used to prevent the disease in patients exposed to mumps. It had to be given intravenously within a week of exposure. Initial studies with inactivated virus vaccine or with live virus attenuated by passage in the chick embryo were not promising. Not until 1967, with the help of tissue culture techniques, was a successful live mumps virus vaccine developed and marketed. Today three vaccines containing live mumps virus (Jeryl Lynn strain) are available in the United States from Merck Sharp and Dohme. They are Mumpsvax, a lyophilized preparation of mumps virus only; Biavax II, rubella and mumps virus vaccine; and MMR II.

RUBELLA VACCINE

Rubella, or German measles, is characterized by a red eruption on the chest and face, which eventually spreads over the whole body, and by swelling of the lymph nodes at the back of the neck. The rubella virus was isolated in 1962 from the nasal secretion of affected individuals.

Before a vaccine was introduced in 1969, rubella occurred most frequently in five- to nine-year-old children. Now it is more common in young, unimmunized adults. Rubella causes fewer complications than measles, but recurring joint pain for as long as a year after infection and a very rare brain involvement (encephalomyelitis) have been reported. A greater problem is congenital rubella caused by the transmission of the virus through the placenta. The virus can be present in the fetus during intrauterine life and is detectable for six to thirty-one months after birth. Congenital rubella, or so-called expanded rubella syndrome, involves cardiovascular and central nervous system malformations, eye and even bone lesions, as well as growth retardation. Pneumonia, heart disease (myocarditis), enlargement of the liver and spleen, and deafness are other manifestations of congenital rubella. These complications of rubella were recognized during a rubella epidemic in the United States in 1964 and led to the development and introduction of the rubella vaccine. The live virus vaccine produced mild symptoms of the disease, particularly in women. With the advent of cell culture techniques in the production of the vaccines, the risk of developing mild disease symptoms after vaccination was reduced from 25 percent to 2 percent. Because the risk of developing a mild form of rubella after vaccination still exists, pregnant women should not be vaccinated, and vaccinated women should avoid getting pregnant for three months after vaccination.

Live rubella virus vaccine is available in the United States from Merck, Sharp and Dohme under the trade name of Meruvax. It leads to the development of antibodies in 97 percent of vaccinated individuals, and the protection persists for at least ten years. Another Merck Sharp and Dohme vaccine, Biavax II, contains the same rubella virus and the Jeryl Lynn strain of mumps virus. Neither of the two vaccines should be given to pregnant women because the rubella virus is secreted in human milk; the virus therefore can be passed on to nursing infants and can cause a mild rubella infection. Immunization of lactating mothers should also be avoided.

VARICELLA VACCINE

Chickenpox, or varicella, is an acute viral disease characterized by an eruption that appears in patches. Following the initial rash, clear fluid-containing vesicles typically appear. Childhood chickenpox is usually benign, but it is much more severe in children with leukemia or in those receiving corticosteroid therapy. The most common complications in childhood are bacterial superinfection of the skin and cerebellar ataxia. The disease is more severe in adults than in children; fever is higher, the rash is more profuse, and complications are more frequent. Primary varicella pneumonia was reported to occur in 14 percent of adult patients with chickenpox. The varicella-zoster virus causes chickenpox as well as herpes zoster, or shingles. Although varicella is a childhood disease, herpes zoster usually does not develop until a more advanced age. After varicella infection the virus may remain latent, until reactivation, when it appears as herpes zoster, which is characterized by a vesicular eruption in the cutaneous areas supplied by the peripheral sensory nerves.

Both diseases, varicella and herpes zoster, are much more severe in immunodeficient than in normal individuals. Immune prophylaxis with zoster immune globulin, varicella-zoster immune globulin, or zoster immune plasma can be used. To be effective, these preparations have to be given within seventy-two hours of exposure. The live varicella vaccine initially used the Oka strain of varicella, which was isolated in primary human embryonic cell cultures.

Researchers at Pasteur Merieux developed a live attenuated varicella vaccine for the prevention of chickenpox in immunocompromised children. It was launched in France and is available in the United Kingdom on a restricted basis. Merck Sharp and Dohme has a varicella vaccine, Varivax, that was only recently approved for use in the United States.

HAEMOPHILUS INFLUENZAE VACCINE

Infections caused by a gram-negative bacillus, *Haemophilus influenzae*, are among the most serious diseases of childhood. A German bacteriologist, Friedrich Johannes Pfeiffer, isolated *H. influenzae* from the sputum of patients in 1892 during an influenza epidemic. Pfeiffer assumed that this bacillus was responsible for the influenza and therefore named it *H. influenzae*. There are sixteen different species of *Haemophilus* and six types of *H. influenzae*, named "a" through "f," based on the composition of their capsules (the envelope surrounding the bacillus). Nearly 95 percent of systemic *H. influenzae* diseases are caused by *H. influenzae* type b.

The *H. influenzae* infection can start in the upper respiratory tract without any specific symptoms. It tends to spread, however, and can affect sinuses, middle ear, bronchi, lungs, heart, joints, and brain. Young children, two weeks to six years old, are particularly susceptible. The risk of developing an infection is higher among the poor or patients suffering from other diseases, such as sickle-cell anemia or Hodgkin's disease. For unknown reasons the *H. influenzae* infection rate is particularly high among Navajo children. It is the most common cause of bacterial meningitis, which, despite antibiotics, still has a 5 percent mortality rate. In addition, 35 percent of surviving children develop neurological deficits, including seizures, deafness, and mental retardation. *H. influenzae* was originally susceptible to many antibiotics, including ampicillin and tetracyclines. Today, however, many strains are resistant to these drugs, and chloramphenicol or quinolones have to be used.

Because of the severity of the infection and resistance of some strains of *H. influenzae* to antibiotics, attempts were made in the early 1970s to develop a vaccine. Initially favorable results were obtained with a vaccine containing polyribose ribitol phosphate, a major component of the *H. influenzae* type b capsule. The vaccine protected children older than eighteen months for four years or longer, but it failed to protect younger infants. To improve immunogenicity—the ability of the vaccine to stimulate antibody production—polyribose ribitol phosphate was linked to a protein carrier (e.g., diphtheria toxin, tetanus toxoid, or a nontoxic protein from another microorganism). The *Neisseria meningitidis* protein carrier used today in PedvaxHIB is a conjugate vaccine available from Merck Sharp and Dohme. This vaccine is highly effective in all groups studied, including Navajo infants. Another conjugate vaccine containing a capsular antigen of *H. influenzae* type b is HibTITER, available from Lederle. This vaccine is conjugated with the nontoxic variant of diphtheria toxin. It is safe and effective in infants of one to six months of age, and high antibody levels have been maintained for at least two years.

HEPATITIS A VACCINE

Hepatitis A is an acute disease caused by infection with hepatitis A virus. Outbreaks, as well as sporadic cases of hepatitis A, were traced to contaminated food, water, milk, or shellfish. Its incubation time is fifteen to forty-five days, and it affects primarily children and young adults. The fatality rate is approximately 0.1 percent, but probably higher in patients of advanced age or those with various medical disorders. The diagnosis of hepatitis A is made during acute illness by demonstrating high titers of anti–hepatitis A virus antibodies.

Recovery is usually complete. The hepatitis A vaccine (VAQTA) was developed by Merck & Company and approved by the Food and Drug Administration in 1996. Chiron has another vaccine for the prevention of hepatitis A in the advanced stages of development.

HEPATITIS B VACCINE

Hepatitis B is a serious world health problem. It has been estimated that there are 280 million carriers of hepatitis B virus throughout the world and that 20 million new infections occur annually. Hepatitis B was originally called serum hepatitis, based on the assumption that it was transmitted by blood or serum transfusion. It is now known that hepatitis B accounts for less than 10 percent of post-transfusion hepatitis and that it is transmitted primarily through sexual contact and from mother to child, either during the third trimester of pregnancy or during the early postpartum period. Blood, serum, semen, and saliva are infectious, but feces are not. Health workers who handle body fluids, such as surgeons, pathologists, or laboratory technicians, can be infected through contact with any of the above-named fluids.

The complete recovery rate from acute hepatitis B is approximately 90 percent. Patients of advanced age or with other serious illnesses are more likely to experience a severe form of hepatitis B and may have to be hospitalized. The mortality rate for all hepatitis B cases is only 0.1 percent, but for hospitalized patients with the more severe form of the disease, it is 1 percent. The therapy is largely symptomatic. Large doses of corticosteroids are sometimes administered to patients with severe forms of hepatitis B, but their efficacy has not been proved. A rare but dangerous complication of hepatitis B is a fulminant disease leading to massive hepatic necrosis and an 80 percent mortality rate. Another dangerous complication is chronic hepatitis—persistence of the disease for more than six months—which can take one of three forms. So-called chronic active hepatitis leads to liver cirrhosis and has the poorest prognosis.

Because there is no specific therapy for hepatitis B, the development of vaccines has been given high priority. The first vaccines were prepared from the plasma of infected individuals. These vaccines were effective in preventing the hepatitis B infection, but their source was not uniform and extensive purification procedures were required. Two recombinant hepatitis B vaccines are currently available in the United States: Recombivax HB from Merck Sharp and Dohme and Engerix-B from SmithKline Beecham. Both vaccines were developed through genetic engineering techniques that incorporate the hepatitis B virus gene or its fraction into yeast cells and then use cultures of this recombinant yeast strain to produce the vaccine. Neither of the two recombinant vaccines has any association with human blood or blood products. The vaccines are noninfectious and highly effective, and they represent the best example of the application of new biotechnology for the prevention of human diseases.

HEPATITIS C VACCINE

Hepatitis C was originally called non-A, non-B hepatitis and was first identified among recipients of transfused blood. Its incubation period and modes of transmission were consistent with an infectious disease; there was, however, no serologic evidence of either hepatitis A or B infection. Its pathology, similar to that of hepatitis A or B, affects primarily the liver, causing infiltration of liver tissue with mononuclear cells, hepatic cell necrosis, and a variable degree of cholestasis (suppression of bile flow). It is responsible for most cases of blood transfusion–induced hepatitis. Soon after discovery of the identity of the virus, attenuated live vaccines were developed by Pasteur Merieux and by Chiron. The Chiron vaccine is recombinant and in 1997 was still in the preclinical stage of development.

PNEUMOCOCCUS VACCINES

The pneumococcus (also known as *Streptococcus pneumoniae*) is a gram-positive coccus that grows in pairs or short chains. Eighty-one types of this organism have been identified, all of which can cause diseases in humans. Seven types of pneumococci are frequently encountered and are known to cause pneumonia. The infection can also extend to other organs, causing sinus, ear, or eye infections. Most severe is type 3 pneumococcal pneumonia, which occurs in aged and debilitated patients. Despite the effectiveness of penicillin and other antibiotics in the treatment of pneumococcal diseases, the mortality rate in treated adults with pneumococcal pneumonia caused by type 3 organisms exceeds 18 percent.

The initial vaccine against fourteen serotypes of pneumococci was developed in the late 1970s. It was followed by a vaccine against twenty-three serotypes of pneumococci. Two such vaccines are currently available in the United States: Pneumovax 23 from Merck Sharp and Dohme and Pnu-Immune 23 from Lederle. They are recommended for individuals fifty-five years of age or older or for patients with debilitating diseases. Either vaccine consists of highly purified polysaccharides from the twenty-three most prevalent types of pneumococci. The vaccines are effective in the elderly, but children under the age of two have a poor response. Neither of the two vaccines is indicated for younger children or for individuals with impaired immune responses.

PERSPECTIVES FOR NEW VACCINES

The clinical success of currently available vaccines may create the false impression that no further research in this field is required and that available vaccines can protect people from all or most infectious diseases. Unfortunately this is not the case. Scientists are currently at the threshold of a new era in vaccine development. Many new vaccines are needed to combat AIDS, malaria, new strains of influenza viruses, some forms of cancer, and other diseases.

Biotechnology opens new approaches to vaccine development and manufacturing. The use of a disease agent in the vaccine is, at least theoretically, no longer needed because protection can be obtained with individual proteins from viral or bacterial particles. These proteins can now be produced by noninfective cells or organisms; they can even be synthesized. Therefore product liability issues should present less of a problem for manufacturers. New delivery systems, safer components, use of antigen genes instead of the antigens themselves, and other innovative technological developments will further contribute to the safety of the vaccines.

Genetic engineering made it possible to use transgenic plants as the source of vaccines. Bananas that express hepatitis B antigen and potatoes that express components of rotavirus (cause of intestinal disease in infants) have already been developed so that "edible" vaccines may eventually reach the market. New adjuvant mixtures were shown to provide an improved rate of protection compared with the aluminum salts previously used as adjuvants. New DNA vaccines were found capable of encoding a number of different antigens so that protection against many diseases with one vaccine is possible. It has been recently discovered that some noninfectious diseases can also be prevented by vaccines. In the future vaccines containing T-cells might be capable of preventing rheumatoid arthritis and other autoimmune disorders.

These and other scientific discoveries led to an explosion of research and development in the field of vaccines. The June 1997 issue of *R&D Directions* lists eighty-eight different potential vaccines in various stages of development, including twelve in large-scale (phase 3) clinical trials. Among the eighty-eight vaccines are ten for influenza, four for meningitis, and four for gastroenteritis. Six potential vaccines are designed to prevent three or more diseases. At least a dozen new vaccines will reach the market before the end of this century.

Antibacterial Drugs

Many classifications of antibacterial, or antimicrobial, drugs have been proposed, but none of them is entirely satisfactory. Textbooks usually differentiate between antibacterial drugs, which are synthetic chemicals more toxic to bacteria than to mammals, and antibiotics, substances that are produced by living cells. This classification is used here as well, although many derivatives of antibiotics are either partially or completely synthesized today and therefore should not be labeled as such.

SULFONAMIDES

The first antibacterial drug was ethylhydrocupreine (Optoquine, Numoquin). Its activity was discovered at the Charité Hospital in Berlin in 1911 by Julius Morgenroth, who was studying the effects of quinine in experimental infections of mice with *Trypanosoma brucei*. Morgenroth's assistant accidentally found that bile acids dissolve pneumococci and trypanosomes in a similar manner. This observation prompted Morgenroth to try quinine and some of its derivatives in mice inoculated with pneumococci. Quinine had little if any effect, but its derivative, ethylhydrocupreine, prevented the infection not only in mice but also in guinea pigs and rabbits. It was subsequently marketed for the treatment of pneumonia by Chinin Fabrik Zimmer in Frankfurt and patented in the United States in 1913.

Further development of synthetic antibacterials was facilitated by John Churchman of Johns Hopkins University and other scientists who observed that some dyes can selectively destroy bacteria. In 1912 Churchman published a report in the *Journal of Experimental Medicine* on the selective bactericidal activity of a dye, gentian violet. This and other similar observations prompted Carl Browning at the University of Glasgow to test many different dyes for antibacterial activity. His work led to the discovery of the antibacterial properties of acriflavine in 1913. Unlike antiseptics, acriflavine (trypaflavine, Panflavin) was not deactivated by body fluids. Further search for synthetic antibacterials at Hoechst in Germany led to the synthesis of ethacridine and its marketing as Rivanol. Unfortunately, neither acriflavine nor ethacridine was powerful enough to cure systemic bacterial infections in patients, even though both drugs were effective in some experimental infections of mice.

The systematic search for synthetic antibacterials was initiated in 1927, when Gerhard Domagk joined IG Farben in Wuppertal, Germany, and started using, as an experimental model, mice infected with highly virulent bacteria—hemolytic streptococci. Initially Domagk found that gold and other heavy metal compounds had antibacterial activity, but renal toxicity prevented their clinical use. His persistence paid off in 1932, when he discovered the antibacterial activity of sulfamidochrysoidine (Streptozon, Prontosil) and developed it for the treatment of streptococcal infections. By 1933 this drug was reported to have saved the life of a ten-year-old boy and in 1935 the life of Domagk's own daughter. Domagk's discovery was not immediately accepted by his skeptical colleagues in other countries. His work was repeated in France, England, Switzerland, and the United States before all agreed that his findings were correct. Domagk was awarded the 1939 Nobel Prize in medicine. (The Nazi government in Germany persuaded Domagk to reject the prize, so it was only after World War II that he went to Stockholm to accept it.)

Domagk's discovery stimulated the search for other antibacterials. Of considerable importance for the further development of sulfonamides was the finding from Ernest Fourneau's laboratory in France that sulfamidochrysoidine is metabolized in the body to another antibacterial sulfonamide, sulfanilamide. Because sulfanilamide had already been synthesized in 1908, it was not patentable and therefore became widely available in various formulations. In the United States it is still being used in cream or suppositories for the treatment of vaginal infections.

Many sulfonamides were synthesized and tested during World War II and in the early 1950s. Because of the advance of antibiotics only a few of them

Gerhard Johannes Paul Domagk (1895–1964), German physician, biochemist, and director of research at Bayer AG, was recipient of the 1939 Nobel Prize in medicine for the discovery of the antibacterial activity of sulfonamides and specifically for the discovery of sulfamidochrysoidine (Prontosil), the first drug to cure streptococcal infections. His discovery stimulated the search for other antibacterial drugs. Courtesy Bayer AG.

survived. Among them is sulfasalazine (Azulfidine), which is poorly absorbed from the gastrointestinal tract and was therefore used in intestinal infections. It was found to be useful in ulcerative colitis, possibly because of its anti-inflammatory rather than its antibacterial activity. Sulfasalazine is metabolized in the intestines to sulfapyridine and 5-aminosalicylate, which has anti-inflammatory properties.

Another antibacterial sulfonamide still widely used today is sulfisoxazole (Gantrisin), discovered in 1947 at Hoffmann–La Roche. The main advantage of this drug is its very high solubility; it is not deposited in the kidneys as other sulfonamides are and consequently does not cause renal toxicity. Sulfisoxazole is used in urinary tract infections and meningococcal meningitis and as an ophthalmic solution for superficial ocular infections. It is marketed in various formulations as well as in combination with phenazopyridine, an antiseptic dye, for the treatment of urinary tract infections (Azo Gantrisin) and with erythromycin for the treatment of middle ear infection (Pediazole).

Of considerable theoretical and practical importance is the current use of another sulfonamide, sulfamethoxazole, in combination with trimethoprim (Bactrim, Septra). This is a rational combination based on two observations: In 1940 Donald Woods of Oxford University found that sulfonamides interfere with bacterial metabolism by inhibiting the formation of an important nutritional factor—folic acid; and in 1948 George Hitchings of Wellcome

Laboratories in the United States discovered that trimethoprim is an extremely potent and selective inhibitor of folic acid formation in bacteria (but not in humans or other mammals) and that its mechanism of inhibition is different from that of the sulfonamides. Hitchings proposed, therefore, a combination of a sulfonamide and trimethoprim as a potentially powerful antibacterial formulation. Sulfamethoxazole was selected because its duration of action is similar to that of trimethoprim. The management of Burroughs Wellcome Laboratories was initially unconvinced by Hitchings's proposal, but when Hoffmann–La Roche expressed an interest in licensing trimethoprim from Burroughs Wellcome, it decided to co-market the proposed combination with Hoffmann–La Roche. This simple idea paid off handsomely. By 1992 the combined sales figures for Bactrim and Septra had reached $5 billion.

The introduction of sulfonamides as antibacterials led to the discovery of other potential uses for this class of drugs. Sulfonamyl diuretics were developed to treat heart failure, hypertension, kidney, and liver diseases, while sulfonylureas, oral hypoglycemic agents, were developed for diabetes mellitus.

Antibiotics

Antibiotics are chemical substances produced by living cells and antagonistic to other forms of life. The concept of antibiosis can be traced to Pasteur, who observed that the presence of other bacteria can slow the growth of anthrax bacilli. At the end of the 1800s another French scientist, Paul Vuillemin, used the term *antibiosis* to describe the inability of one organism to thrive in the presence of another.

PENICILLIN AND SEMISYNTHETIC PENICILLINS

The reports that *Penicillium* molds had antibiotic activity preceded the discovery of penicillin by more than thirty years. The first scientific report on the ability of molds to destroy bacteria came from John Burdon at St. Mary's Hospital in London in 1870. In 1895 Vincenzo Tiberio of Naples published his extensive work on the antibacterial properties of molds.

Despite these reports, the discovery of penicillin by Alexander Fleming in 1928 was accidental. As Walter Sneader from Glasgow's University of Strathclyde tells in his well-known book, *Drug Discovery: The Evolution of Modern Medicines*, Fleming noticed that a mold, which had contaminated his bacterial culture plate, caused bacteria in its immediate vicinity to undergo lysis (gradual decomposition). Fleming had left the bacterial culture on his workbench instead of placing it as usual in an incubator to enhance bacterial growth. The mold most likely came through the air from the laboratory of his colleague, who used molds in his work on vaccine development. Nevertheless, the fact that Fleming noted the lysis and properly interpreted the significance of his finding makes him the discoverer of penicillin and points out the importance of careful observations in the discovery process. Fleming derived the name *penicillin* from the name of the mold in his bacterial culture, *Penicillium notatum*.

Sir Alexander Fleming (1881–1955), a Scottish bacteriologist, shared the 1945 Nobel Prize in physiology or medicine for the discovery of penicillin with Ernst Chain and Howard Florey. His observation that a mold in his cultures was capable of killing bacteria led to the development of penicillin and subsequently to other antibiotics. The discovery revolutionized the treatment of bacterial infections. Courtesy Nobel Foundation.

Immediately after discovering the lysis of bacteria in the filtrate of broth in which the molds had been grown, Fleming and his colleagues Frederick Ridley and Stuart Craddock prepared active extracts and evaluated their activity against a variety of bacteria. They found that penicillin was active against staphylococci, streptococci, meningococci, gonococci, and diphtheria bacilli. Early attempts to purify penicillin failed, primarily because researchers lacked resources and because scientific administrators and many scientists failed to recognize the potential value of penicillin. Even the Royal Society did not find the discovery of penicillin important enough to admit Fleming as a member.

Fortunately, some very capable scientists were impressed with the potential of penicillin. Among them was Ernst Chain, who in 1938 began to study penicillin under the directorship of Howard Florey at Oxford. With Florey's help Chain obtained a small grant from the Medical Research Council (£300 a year for three years) and then a much larger grant from the Rockefeller Foundation (£5,000 a year for five years). These grants made it possible for Chain to isolate and purify penicillin. In May 1940 Chain had one hundred milligrams of penicillin powder, enough for Florey to start biological studies. He injected eight mice with a lethal amount of streptococci; one hour later two of the mice each received ten milligrams of penicillin, and two other mice five milligrams each. These last two mice received four similar injections over the next ten hours. All four mice not treated with penicillin died as did one of the two mice that had only a single injection of penicillin. The others survived. This experiment encouraged the scientists to increase their efforts, and in 1941 they had enough penicillin for clinical investigations. The first attempt to treat a patient with severe infection was made in February 1941. The patient, a policeman with severe staphylococcal and streptococcal infection, improved initially but still died one month later. The next two patients who received penicillin completely recovered from their infections.

The next major problem was manufacturing sufficient quantities of penicillin to meet demand. Neither Imperial Chemical Industries (ICI) nor Boots Company in England was willing to manufacture penicillin by fermentation, the only process available at that time. The manufacturers assumed that penicillin would soon be synthesized and were hesitant to invest in its production by fermentation. Florey went to the United States and convinced Merck, Pfizer, and Lederle to start penicillin production. Soon thereafter a consortium of midwestern pharmaceutical companies—Abbott, Lilly, Parke-Davis, and Upjohn—began to manufacture penicillin as well.

The efforts to synthesize penicillin continued in the United Kingdom as well as in the United States. Oscar Wintersteiner and his group at Squibb Laboratories in New Jersey and Sir Robert Robinson's group in Oxford crystallized penicillin and discovered to their surprise that their materials were not identical. The American penicillin became known as penicillin G, or benzylpenicillin, while British penicillin was 2-pentenylpenicillin, or penicillin F. Soon thereafter five more penicillins were discovered. But the precise structure of penicillin remained controversial until 1945, when Dorothy Hodgkin at Oxford firmly identified it using X-ray crystallography. Penicillin was not synthesized until 1957, when John Sheehan, a chemistry professor at the Massachusetts Institute of Technology and former Merck chemist, synthesized phenoxymethylpenicillin, or penicillin V, which is more stable in the acidic medium than penicillin G and therefore better absorbed from the gastrointestinal tract.

The recognition of penicillin's importance in the treatment of bacterial infections, as well as of some of its drawbacks, led to the development of many semisynthetic derivatives of penicillin. Their advantages over penicillin G included better oral absorption, slower excretion, and resistance to penicillinase, an enzyme that destroys penicillin; there were also minor differences in the spectrum of antimicrobial activity.

An improvement in oral absorption was achieved with the development of ampicillin (Polycillin, Amcill, Omnipen). The first patent for ampicillin was assigned to Beecham in 1961. The drug is effective against some bacteria resistant to penicillin G, but many bacterial strains, such as *Escherichia coli*, *Salmonella*, and *Shigella*, which were originally sensitive to ampicillin, have developed resistance. Like penicillin G, ampicillin is sensitive to penicillinase.

In 1964 Beecham Laboratories was granted a patent for another semisynthetic penicillin, amoxicillin (Amoxil, Larotid, Polymox). Its advantages over ampicillin are twofold: even more complete oral absorption and a reduced tendency to produce diarrhea. The spectrum of its antimicrobial activity is, with a few exceptions, similar to that of ampicillin.

To synthesize many new and stable penicillins, Beecham chemists used two important discoveries in process chemistry. The first was made by MIT's Sheehan. In addition to synthesizing penicillin V, he developed a new method for converting 6-aminopenicillanic acid to therapeutically useful penicillins, while Beecham chemists discovered the presence of the acid in mold juice and devised methods of producing it in large quantities by fermentation. Sheehan's research was supported by Bristol Laboratories. The initial cooperation between Beecham and Bristol did not last long, and legal disputes between the two companies continued until 1979, when the U.S. Board of Patent Interferences ruled in favor of Sheehan (and Bristol). The development of new semisynthetic penicillins continued, however, at both companies.

In 1959 Beecham introduced methicillin (Staphcillin), which is resistant to penicillinase but poorly absorbed orally and readily destroyed by the acid in the stomach. Its development was made possible when Beecham scientists discovered that bulky substituents on the penicillin side chain produced so-called steric hindrance and protected penicillins from destruction by penicillinase. In 1961 Beecham introduced the orally active derivatives oxacillin (Prostaphlin, Bactocill), which was initially marketed in the United States by Bristol Laboratories, and cloxacillin (Cloxapen, Tegopen), marketed by Beecham. These two semisynthetic penicillins do not differ substantially in their efficacy or oral absorption, and they remain drugs of choice for the treatment of staphylococcal infections. Among many other semisynthetic penicillins, Beecham's nafcillin (Nafcil, Unipen) is the most active penicillinase-resistant penicillin. It enters the central nervous system in sufficient quantities to be effective in the treatment of bacterial meningitis, but unfortunately its oral absorption is variable and unpredictable.

Attempts to expand the antimicrobial spectrum of semisynthetic penicillins led to the discovery and development at Bayer AG in Germany of mezlocillin (Baypen, Mezlin). It is effective against many gram-negative microorganisms, including *Pseudomonas* and *Klebsiella*. Mezlocillin is poorly absorbed orally and is given either intramuscularly or intravenously.

Another more recent development was based on the observation of H. C. Neu and K. P. Fu in 1978 that a substance called clavulanic acid, which is produced by an organism called *Streptomyces clavuligerus* and has little or no antibacterial activity on its own, can inhibit the enzyme β-lactamase, which can destroy penicillins and other antibiotics containing a β-lactam nucleus. Clavulanic acid was combined successfully with amoxicillin and marketed as Augmentin. This combination product was found to be effective in vivo against β-lactamase–producing strains of staphylococci, gonococci, *H. influenzae*, and *E. coli*.

STREPTOMYCIN AND OTHER AMINOGLYCOSIDES

Most antibiotics discovered between 1940 and 1960 were natural products isolated from soil samples. Many of them were found through large-scale screenings of soil samples for antibacterial activity, which were conducted by various industrial laboratories. The idea to test soil samples came primarily from Selman Waksman, a microbiologist at Rutgers University, who became convinced in the early 1940s that his extensive knowledge of soil microbes could be useful in the search for new antibiotics. The first antibiotic, which he obtained from *Actinomyces antibioticus*, was actinomycin A, which was too toxic for systemic use. Waksman was determined to continue his search, although his grant request was rejected by the newly established Committee on Medical Research in Washington, D.C. With private funds (a grant from the Commonwealth Fund) Waksman isolated three more antibiotics: clavacin, fumigacin, and streptothricin. Unfortunately they all had to be abandoned because of toxicity. But in 1943 his persistence paid off, when, in searching for a new antitubercular drug, he isolated from *Streptomyces griseus* an antibiotic named streptomycin. This substance was used for many years, particularly in the therapy of tuberculosis, but also in rather unusual infections, such as tularemia, brucellosis, and plague. Because of its side effects, such as ototoxicity, which can lead to deafness, and renal toxicity, as well as the availability of safer antibiotics, the use of streptomycin has been largely abandoned, except in a combination therapy for tuberculosis.

Streptomycin was one of the first antibiotics belonging to the family of aminoglycosides. These antibiotics consist of two or more amino sugars joined with another carbohydrate molecule, either hexose or aminocyclitol. Another aminoglycoside, gentamicin (Garamycin, Cidomycin, Garasol), was first described in 1963. It consists of at least four components, which differ from each other in the composition of one of the sugar molecules. Gentamicin is a broad-spectrum antibiotic produced by an organism called *Micronospora* and is useful in treating severe infections caused by gram-negative bacteria. It is, however, being replaced by quinolones for this indication. The patents for gentamicin are assigned to Schering Corporation.

Another aminoglycoside, tobramycin (Nebcin, Tobrex), is produced by another microorganism, *Streptomyces tenebrarius*. It was first described in 1967 by C. E. Higgins and R. E. Kastner. The spectrum of its antimicrobial activity is similar to that of gentamicin. All known aminoglycosides suffer from the same side-effects profile. Like streptomycin they are capable of producing ototoxicity and nephrotoxicity, although in experimental animals tobramycin appears to be less toxic than streptomycin.

TETRACYCLINES

The use of soil samples in the search for antibiotics led to other important discoveries. Among them are chlortetracycline (Aureomycin), oxytetracycline

(Terramycin), and tetracycline (Tetracyn, Achromycin). These three first-generation tetracyclines have good oral activity and a broad spectrum of antimicrobial efficacy. According to Sneader, chlortetracycline was discovered by Benjamin Duggar at Lederle Laboratories. Duggar had retired as professor of botany at the University of Wisconsin and joined Lederle in 1943 at the distinguished age of seventy-one. Lederle asked Duggar to establish a screening operation to find an antibiotic for treating tuberculosis that would be safer than streptomycin; Duggar in turn asked his academic friends to send him soil samples. From a sample he received in 1945 from William Albrecht at the University of Missouri, Duggar isolated a new organism, *Streptomyces aureofaciens*. Chlortetracycline, the antibiotic produced by this organism, was useless against tuberculosis but had an otherwise broad spectrum of antibacterial activity.

The discovery of chlortetracycline, which was marketed in 1948, was rapidly followed by the development of oxytetracycline at Pfizer by a team of sixty scientists directed by Alexander C. Finlay. Pfizer scientists were collecting soil samples from all over the world but found antibacterial activity in a sample obtained on the company's own property in Terre Haute, Indiana. The organism that produced the antibiotic was named *Streptomyces rimosus*, and the drug was called oxytetracycline. The drug was aggressively marketed and by 1951 had captured one-fourth of the U.S. market for antibiotics.

The search for new antibiotics continued. Lloyd Conover, another scientist at Pfizer, discovered that the removal of a chlorine atom from chlortetracycline led to another broad-spectrum antibiotic, tetracycline. Scientists at Lederle and Bristol Laboratories found that tetracycline was produced by other *Streptomyces* species as well and obtained patents for their production methods. A legal dispute followed but was eventually settled by cross-licensing agreements among the companies involved.

The three tetracyclines represented a major advance in the development of antibiotics, primarily because of their wide range of antibacterial activity against both gram-positive and gram-negative organisms. They are also effective in rickettsial infections, which include epidemic typhus and Rocky Mountain spotted fever. *Rickettsiae* are intracellular parasites, in size similar to bacteria, but they are transmitted by insect bites. Rickettsial infections are resistant to most other antibacterial drugs, except chloramphenicol.

Tetracyclines are still drugs of choice in the treatment of *Chlamydia* infections, which cause eye diseases, including trachoma, or conjunctivitis, and a venereal disease, lymphogranuloma venereum. *Chlamydia* are intracellular parasites originally thought to be large viruses but now classified as bacteria. The wide spectrum of tetracycline activities includes other venereal diseases, such as gonorrhea and syphilis; mycoplasma or bacillary infections, such as brucellosis; tularemia; and cholera. Tetracyclines were widely used to promote growth in cattle, exposing meat or milk consumers to trace amounts of tetracyclines.

Widespread use led to the development of bacterial resistance to the drugs, which, like resistance to penicillin, is considered to be an inducible trait. In other words the bacteria can become resistant after exposure to low concentrations of an antibiotic. Some bacterial strains, however, are intrinsically resistant to certain antibiotics because of the structural differences in the bacterial enzymes that are targets of the antibiotic therapy.

The three original tetracyclines (chlortetracycline, oxytetracycline, and tetracycline) are only partially absorbed after oral administration. Their presence in the intestines leads to substantial changes in the intestinal flora and often to diarrhea. In patients with renal infections the first-generation tetracyclines also tend to accumulate in plasma because of the kidneys' inability to excrete them properly.

The efforts to develop a tetracycline without these drawbacks led to the discovery of doxycycline (Vibramycin, Doryx) by Manfred Schach von Wittenau and his colleagues at Pfizer. Doxycycline is eliminated by pathways different from those of other tetracyclines and is therefore less likely to produce diarrhea or to accumulate in the plasma of patients with renal disease. The U.S. patent for doxycycline was issued in 1965 and assigned to Pfizer.

Among other tetracyclines, minocycline (Minocin) and demeclocycline (Declomycin) should be mentioned. Both compounds were developed and patented by scientists at Lederle Laboratories—minocycline in 1964 and demeclocycline in 1969. Both are effective antibiotics with indications and side effects similar to those of other tetracyclines.

CEPHALOSPORINS

The discovery of cephalosporins is attributed to Giuseppe Brotzu in Cagliari, Sardinia, who, inspired by the discovery of penicillin, was looking for antibiotic-producing organisms. He was intrigued by the seawater near the local sewer that appeared to have the capacity to purify itself, and he subsequently succeeded in isolating a mold from it—*Cephalosporium acremonium*. After discovering that the mold produced an antibiotic with a broader spectrum than penicillin, Brotzu initiated primitive clinical studies by applying the mold juice to abscesses and by injecting it in patients with typhoid fever.

Nobody in Italy appeared to be interested in Brotzu's work, so in 1948 he sent one of his cultures to Sir Howard Florey at Oxford, where Norman Heatley extracted an antibiotic from Brotzu's extract and named it cephalosporin P. It turned out to be a steroid and, unlike Brotzu's antibiotic, had only a narrow spectrum of antibacterial activity. Continuing their isolation attempts, Oxford's scientists found cephalosporin C, which was chemically related to penicillin but was resistant to penicillinase and less toxic than benzylpenicillin. Large-scale production methods were developed in Oxford and licensed to Glaxo Laboratories. Cephalosporin C was not used as a drug but became the main starting material for synthesizing the semisynthetic cephalosporins, which turned out to be much more potent than cephalosporin C.

The first semisynthetic, clinically useful cephalosporin was cephalotin (Keflin, Seffin), discovered in 1962 by Robert Morin and his associates at Lilly Laboratories. It had a broader spectrum of antibacterial activity than penicillin G but was poorly absorbed orally and caused pain by intramuscular injection; therefore it had to be given intravenously. Soon after introduction of cephalotin Glaxo introduced cephaloridine (Ceporin), which was chemically more stable than cephalotin and had a slightly different antibacterial spectrum. Among many first-generation cephalosporins that followed, cephalexin (Keflex, Ceporex) and cefazolin (Kefzol, Ancef) should be mentioned. Cephalexin was introduced by both Lilly and Glaxo in 1967; it is orally active and has a broad spectrum of antibacterial activity but is somewhat less effective than other cephalosporins against penicillinase-producing staphylococci. Cefazolin is not active orally but has a longer half-life than other first-generation cephalosporins. It was developed at Fujisawa Laboratories in Japan and patented in 1967.

The so-called second-generation cephalosporins, developed in the 1970s, are characterized by a broader spectrum of antibacterial activity, which includes many gram-negative organisms, but they are less potent against gram-positive organisms than are the first-generation cephalosporins. Among them is cefoxitin (Mefoxin), a semisynthetic derivative of cephamycin C. Cefoxitin was synthesized by Burton Christensen and his associates at Merck and patented in 1971. It is not active orally but is highly resistant to β-lactamase, which is produced by gram-negative bacteria, and is particularly useful in treating such anaerobic infections as pelvic inflammatory disease and lung abscesses. Of interest is another second-generation cephalosporin, cefaclor (Ceclor), developed and marketed by Lilly Laboratories. It is active orally and is particularly effective in pediatric respiratory infections caused by *H. influenzae*.

The first cephalosporin of the third generation to become available in the United States was cefotaxime (Claforan). Developed and marketed by Hoechst-Roussel Pharmaceuticals in the early 1980s, it is highly resistant to β-lactamases and has a broad spectrum of antibacterial activity, including many gram-negative organisms. It is effective in treating meningitis caused by gram-negative organisms. One of its metabolites, a desacetyl derivative, also has antibacterial activity, so its therapeutic effect lasts longer than its own half-life. Another metabolite, cefprozil (Cefzil), is commonly used to treat streptococcal throat or ear infections. Marketed by Bristol Laboratories, this cephalosporin is active in vitro against a wide spectrum of microorganisms (gram-positive and gram-negative) and is well absorbed after oral administration (up to 95 percent). Its half-life is rather short, on average 1.3 hours. Another cephalosporin of the same generation, developed in the early 1980s at Hoffmann–La Roche is ceftriaxone (Rocephin). It is not active orally, but its half-life is eight hours; and it is capable of curing gonorrhea, even that caused by penicillinase-producing gonococci, in a single dose.

MACROLIDES

Macrolides are antibiotics that contain one or more deoxysugars attached to a large lactone ring. The most important among them is erythromycin (E.E.S., E-Mycin, Erythrocin, Ilosone). Erythromycin was isolated in 1952 by Robert Bunch and James McGuire at Lilly Laboratories from a culture of *Streptomyces erythreus* found in a soil sample from the Philippines. The spectrum of its antibacterial activity is similar to that of penicillin, but it is the preferred drug in certain infections, such as Legionnaires' disease, which is caused by *Mycoplasma pneumoniae*, and urogenital chlamydial infections. It is also the drug of choice for patients allergic to penicillin.

Azalides are derived from erythromycin and are considered a subclass of macrolide antibiotics. The best known among them is azithromycin (Zithromax), marketed in the United States by Pfizer. It differs chemically from erythromycin in that methyl-substituted nitrogen is incorporated into the lactone ring. Azithromycin, which has a wide spectrum of antimicrobial activity, is well absorbed orally and is rapidly distributed into the tissues. Of clinical significance is the fact that its levels in tissue are usually higher than its concentrations in plasma.

Other Antibacterials

Despite successful development of antibiotics, attempts to synthesize new chemicals, structurally unrelated to known antibiotics, continued. These efforts led to the synthesis of new antitubercular drugs, nitroimidazoles, and more recently quinolones.

ANTITUBERCULAR DRUGS

The development of sulfonamides as antibacterials stimulated the search for other synthetic antibacterials. In 1946 at the Squibb Institute in New Jersey a screening program for antitubercular drugs was initiated under the direction of Frederick Wiseloge. In this program about five thousand compounds were synthesized and tested in mice infected with tuberculosis. One of the chemists in this program, Harry Yale, synthesized isonicitinaldehyde thiosemicarbozone from an intermediate, isonicotinic acid hydrazide. To his surprise the intermediate was found to be much more effective than the desired product. It was given the generic name isoniazid, tested in New York hospitals, and found highly effective, even in patients with severe tuberculosis. It was introduced worldwide and is listed in the Merck Index under seventy-five different trade names. It is currently available in the United States as INH from Ciba or as a generic from Duramed.

The discovery of isoniazid was complicated by the fact that its effectiveness as an antitubercular drug was discovered independently and nearly simultaneously by two other groups. Two weeks before the publication of the Squibb report, Robert Schnitzer and his associates at Hoffmann–La Roche announced the discovery of the antitubercular activity of isoniazid. Similar work was also pursued in Domagk's laboratory at Bayer in Germany, and Bayer's announcement appeared shortly after those of the two American groups.

Isoniazid's mechanism of action is still unknown, although several hypotheses have been offered. One hypothesis is based on the selectivity of isoniazid action for the mycobacteria that cause tuberculosis. Unlike other bacteria, mycobacteria contain mycolic acids in their cell walls, and isoniazid may inhibit the synthesis of those acids. Not all mycobacteria are isoniazid sensitive; some resistant strains have emerged. The sensitivity of the microorganisms to isoniazid should therefore be tested before the drug therapy is initiated.

The most common adverse effect of isoniazid is peripheral neuritis, inflammation of peripheral nerves, which can be prevented by pyridoxine. The most dangerous adverse effect is hepatitis; its incidence is higher with advanced age and increased alcohol consumption. A U.S. Public Service Surveillance study reported 174 cases of hepatitis among 13,838 patients receiving isoniazid.

The discovery of isoniazid led to the development of many related antitubercular drugs. Soon after its introduction, two companies, Lederle and Merck, announced simultaneously the discovery of pyrazinamide (Zinamide, Aldinamide). Still on the market, pyrazinamide is used primarily as a component of a short-term multidrug therapy. Its major adverse effect is also hepatitis, which is at least as common as with isoniazid. Another isoniazid derivative, ethionamide (Nisotin, Trecator-SC) was introduced in 1956 by Theraplix in France and is distributed in the United States by Wyeth-Ayerst. It can produce hepatitis, severe postural hypotension, depression, and anorexia and is used only when other drugs have failed or are contraindicated.

Attempts to develop an antitubercular agent with fewer and less severe side effects led Lederle Laboratories in 1961 to the discovery of a chemically unrelated drug, ethambutol (Myambutol). The incidence of adverse effects with ethambutol is lower than with the previously available antitubercular drugs, and hepatitis does not occur. Its side effects include inflammation of the optic nerve, which can lead to visual impairment and an inability to discriminate between red and green. Ethambutol also blocks the excretion of uric acid, therefore elevating its levels in blood. Currently, ethambutol is used only in combination with other antitubercular drugs.

NITROIMIDAZOLES

In 1956 H. Horie in Japan discovered that 2-nitroimidazole is lethal to trichomonas, parasitic protozoa that can cause diarrhea or vaginal infections in humans. This discovery led to the synthesis of many related compounds in various laboratories. In 1959 Charles Cosar and his associates at Rhône Poulenc in

France reported the synthesis and superior activity of a related nitroimidazole, metronidazole (Flagyl). It is still widely used today to treat trichomonal infections as well as amebic dysentery. Metronidazole was used as an antiparasitic for many years before its antibacterial activity was discovered. In 1975 Anthony Chow and his colleagues at the University of California, Los Angeles, reviewed the accumulated evidence for the effectiveness of metronidazole in the treatment of serious infections caused by anaerobic bacteria (bacteria capable of living without air), which were notoriously difficult to treat. In 1980 the FDA approved use of metronidazole in severe infections caused by anaerobic bacteria; its use for this purpose has saved many lives.

The mechanism of selective toxicity of metronidazole for anaerobic bacteria is not yet completely understood. The nitro group in metronidazole is thought to interact with bacterial proteins involved in the iron transport, and a reduced form of metronidazole has been shown to disrupt the helical structure of DNA. Still another use of metronidazole has recently been found: It enhances sensitivity of hypoxic tumor cells to radiation. Metronidazole is not without side effects. The most serious are convulsive seizures and numbness of the extremities.

CHLORAMPHENICOL

Some of the antibacterial drugs that are being completely synthesized today were found originally in soil samples. Among them is chloramphenicol (Chlormycetin), the first broad-spectrum antibiotic. In 1943 Paul Burckholder at Yale University found the substance in a soil sample from Venezuela, and then sent the sample to Parke-Davis, where John Ehrlich and Quentin Bartz isolated chloramphenicol. The same substance was isolated three months later by David Gottlieb and his associates at the University of Illinois. Chloramphenicol was first tested clinically in typhus patients during a 1947 epidemic in Bolivia. Twenty-two patients received chloramphenicol and were cured. Subsequent clinical trials confirmed the broad-spectrum activity of chloramphenicol, and the drug was synthesized by John Controulis and his colleagues at Parke-Davis. By 1949 chloramphenicol had become one of the best-selling drugs. After about eight million patients had been treated with the drug, reports started to appear that chloramphenicol might be responsible for aplastic anemia and the death of some patients. The incidence of chloramphenicol-induced aplastic anemia was estimated to lie between 0.001 and 0.005 percent. Because of the drug's value in treating infections not controllable by other drugs, the FDA left chloramphenicol on the market, but with a warning that it should not be used to treat infections for which other drugs are effective.

QUINOLONES

Another group of antibacterial agents—the quinolones—were also highly successful. The first quinolone discovered was nalidixic acid (NegGram). In 1946

Alexander Surrey and H. F. Hammer of Sterling-Winthrop Research Institute were developing a new method for synthesizing chloroquine, when they accidentally produced a by-product related to nalidixic acid. The compound was included in a general screening and was found to have antibacterial activity. Subsequently, a number of related compounds were tested for antibacterial activity, and nalidixic acid was chosen for development. It was rapidly excreted unchanged in the urine and therefore evaluated in urinary tract infections. Nalidixic acid is effective against most gram-negative organisms, but large amounts—four grams a day—must be administered. Resistance to the drug developed quickly in as many as 25 percent of patients.

In searching for a superior derivative of nalidixic acid, a related quinolone, oxolinic acid (Ossian, Pietil), was developed. It was about four times more potent than nalidixic acid, but its side effects were more frequent. The search continued unsuccessfully until 1981, when ciprofloxacin (Cipro) was discovered at Bayer's laboratories in Germany. This antibacterial agent was found to be one thousand times more potent than nalidixic acid. It was tested against twenty thousand different strains of bacteria, and 98.3 percent of them were sensitive to the drug. Ciprofloxacin was effective against bacteria resistant to all other antibacterial agents available at that time. Clinical studies confirmed the unique effectiveness of ciprofloxacin. Studies of the mechanism of action revealed that ciprofloxacin inhibits a bacterial enzyme, called gyrase, which facilitates DNA folding inside the bacterial cell. Its inhibition leads to the death of bacteria.

Within the next few years many other potent quinolones with antibacterial activity were synthesized and marketed, including norfloxacin (Noroxin), ofloxacin (Floxin), temafloxacin (Omniflox), lomefloxacin (Maxaquin), enoxacin (Penetrex), and cinoxacin (Cinobac). Despite the introduction of these other drugs, ciprofloxacin retained the largest portion of the quinolone market. The confidence of the clinical investigators and the decision by Bayer to undertake studies for many different indications simultaneously ensured its success in the United States. The FDA quickly approved ciprofloxacin, not only for urinary tract infections, but also for gastrointestinal, pulmonary, and skin infections.

MONOBACTAMS

Development of bacterial resistance to many commonly used antibacterial drugs is a continuous problem. Bacteria adapt by producing enzymes that destroy the drugs. Penicillinase and cephalosporinase are enzymes produced by bacteria that can hydrolyze penicillin or cephalosporin, respectively. Fortunately scientists keep finding new drugs that cannot be destroyed by the same enzymes and to which resistance cannot easily develop. One such synthetic antibacterial agent is aztreonam (Azactam), the first drug discovered in the monobactam family. By virtue of its molecular structure aztreonam is highly resistant to bacterial enzymes and is therefore highly effective in the therapy of infections caused primarily by gram-negative bacteria that have developed resistance to other antibacterial agents. Aztreonam was synthesized at Squibb Institute for Medical Research in New Jersey and marketed in injectable form for intravenous or intramuscular administration. It is indicated for treatment of urinary tract, respiratory, abdominal, and gynecological infections. The incidence of side effects is relatively low, but abdominal cramps, jaundice, and seizures were described in less than 1.5 percent of the patients.

AUGMENTIN

Bacterial resistance to penicillins and cephalosporins usually occurs when microorganisms produce a variety of enzymes called β-lactamases, which can inactivate the antibiotics. In 1975 researchers at Beecham Laboratories reported that clavulanic acid inhibits the β-lactamases. This observation was central to the creation of a combination product, amoxicillin and clavulanic acid, which is marketed in the United States by SmithKline Beecham as Augmentin. This combination drug is particularly useful in the therapy of infections caused by β-lactamase–producing strains of *Staphylococcus aureus* and *Haemophilus influenzae*.

PRIMAXIN

Another recently developed, injectable antibacterial product is Primaxin, a fixed combination of cilastatin and imipenem. Cilastatin, which has no antibacterial activity of its own, enhances and prolongs the activity of imipenem by preventing its rapid destruction in the kidney by the enzyme dehydropeptidase I. Imipenem is highly effective against many gram-negative organisms that are resistant to other antibacterial agents. Primaxin, developed at Merck in New Jersey, is indicated in the treatment of low respiratory tract, abdominal, skin, and gynecological infections. Because imipenem belongs to the class of β-lactam antibiotics, its side effects are similar to those produced by other antibiotics. They include a serious anaphylactic reaction, which is likely to require emergency treatment with steroids, oxygen, and epinephrine, as well as severe diarrhea.

NEW ANTIBACTERIALS

Since many microorganisms are capable of developing resistance to the available antibacterial drugs, the medical need for new antibacterials continues. During 1996 the FDA approved five new antibacterial drugs. Among them is levoflaxacin (Levaquin) from Ortho-McNeil, a broad-spectrum quinolone with a 99-percent absorption rate after oral administration. Its indications include sinusitis, bronchitis, pneumonia, and urinary tract infections. Another quinolone, sparfloxacin (Zagam), from Rhône-Poulenc Rorer was approved in the same year, with the indications limited to bronchitis and pneumonia. Still another chemically related antibacterial, grepafloxacin (Raxar) was discovered at Otsuka Pharmaceuticals in Japan and licensed to Glaxo Wellcome. It was approved by the FDA in 1997 with indications for gonorrhea and nongonococcal urethritis in addition to bronchitis and pneumonia. In the June 1997 issue of *R&D Directions*, forty-four antibacterials are listed in the pharmaceutical industry's pipeline for 1996. At the end of 1996 seventeen of them were either awaiting approval or were in phase 3 clinical trials. Among them are balofloxacin (Baloxin) from Chugai Pharmaceutical in Japan, and its U.S. licensee Novartis, and a cephalosporin, cefoxopran (Firstcin), from Takeda Chemical Company. In the 1990s Japanese pharmaceutical companies greatly intensified their research and development efforts not only in the area of antibacterials but also in many other areas of drug development, and as a result introduced drugs with chemical and biological properties similar to or even superior to those discovered in the United States or western Europe. The properties of these drugs were attractive to the multinational companies, so they licensed them for marketing outside Japan. Some of these drugs brought in annual revenues of $250 million or more.

Antiviral Agents

Few drugs are available for the treatment of viral infections, and they have a narrow spectrum of activity. Only a few types of viruses are sensitive to the available therapeutics. One of the first synthetic drugs found to be effective against skin infections caused by herpesvirus is idoxuridine (Herplex). The effectiveness of topical idoxuridine in this indication was discovered in 1959 by William Prusoff of Yale University. Unfortunately idoxuridine is too toxic for systemic use and can only be used locally for skin infections.

In 1975 George Hitchings and Gertrude Elion reported antiviral activity in a series of antileukemic drugs. In an attempt to separate antileukemic and antiviral effects and optimize antiviral activity, Howard Schaeffer of Burroughs Wellcome synthesized acyclovir (Zovirax), which was highly effective in treating herpes infections, such as cold sores and genital herpes. Acyclovir blocks viral DNA synthesis after it is incorporated into DNA.

The major problem in antiviral therapy today is treatment of the acquired immunodeficiency syndrome (AIDS). This disease is caused by the human immunodeficiency virus (HIV) and is manifested by profound depression of cell-mediated immunity. Many substances have been found to inhibit the important viral enzymes reversed transcriptase and DNA polymerase, but many of them failed in the clinic or produced only transient remissions. The first drug approved by the FDA for the treatment of AIDS was zidovudine (Retrovir, also known as AZT), which is manufactured and distributed by Glaxo Wellcome. Zidovudine in vitro blocks up to 90 percent of detectable HIV replication if added a short time after laboratory infection of cells. Clinical studies with zidovudine demonstrated prolongation of life but no cure of AIDS. Viral infections associated with AIDS (cytomegalovirus and herpes simplex) are being successfully treated with foscarnet (Foscavir).

After the discovery and development of zidovudine a series of related inhibitors of the same enzyme, viral reversed transcriptase, were developed and marketed, including didanosine (DDI, Videx), zalcitabine (DDC, Hivid), stavudine (d4T, Zerit), lamivudine (3TC, Epivir), and delavirdine (Rescriptor). After entering the target cells, these drugs are converted into active triphosphorylated forms by cellular enzymes.

A chemically different inhibitor of viral reversed transcriptase is nevirapine (Viramune), which inactivates the enzyme by a different mechanism than previously marketed inhibitors. In combination with zidovudine it reduced viral load in patients with HIV infection more effectively than zidovudine alone. Nevirapine was discovered and developed by Boehringer Ingelheim Pharmaceuticals, Inc., and marketed by Roxane Laboratories. None of the reversed transcriptase inhibitors cure AIDS, but in combination with a new class of drugs, protease inhibitors, these drugs are capable of reducing, often to nondetectable levels, the virus levels in the blood. Protease inhibitors act at a different site in the HIV multiplication chain, destroying the enzyme used by the virus to reproduce itself. The following protease inhibitors are now available: indinavir (Crixivan), saquinavir (Invirase), ritonavir (Norvir), and nelfinavir (Viracept). Despite a dramatic decline of virus levels in the blood, patients are not yet considered cured of AIDS, because the virus can remain dormant in the lymph nodes for long periods. It has been recently estimated that two to three years of viral suppression are needed to rid the body of HIV.

The treatment of influenza (types A and B) with antiviral drugs became a reality in 1997. The results of a well-controlled clinical trial of a Glaxo Wellcome product, zanamivir (Relenza), were published in the *New England Journal of Medicine*. When the therapy was started within thirty hours of onset, the symptomatic period decreased from seven to four days. Zanamivir inhibits an enzyme, neuramidase, that is present on the surface of the flu virus and helps the virus invade host cells. The approval of zanamivir in the United States is imminent; its annual sales are expected to reach $1 billion.

Antifungal Agents

The ability of fungi to cause diseases was recognized in the early nineteenth century. The concept of parasitic infections was promoted by Augustino Bassi de Lodi, who determined in 1835 that silkworm disease is caused by a fungus, *Beauvaria bassiana*. Fungal infections in humans were considered rare and less important than bacterial infections. But the incidence of fungal infections has increased substantially during the last twenty years, and fungal diseases are now recognized as common complications of cancer chemotherapy and AIDS.

The first clinically useful antifungal agent was nystatin (Mycostatin), the activity of which was first reported in 1950. The drug is produced by *Streptomyces noursei* and other streptomycetes. Although it has no antibacterial or antiviral properties, it is active against such fungi as *Candida*, *Cryptococcus*, *Histoplasma*, and *Blastomyces*. Nystatin is too toxic for parenteral use and is used primarily in the treatment of topical *Candida* infections. In 1959 another antifungal, griseofulvin (Fulvicin, Grisaxin), was marketed. It was originally isolated from *Penicillium griseofulvum* in 1939 by Harold Raistrick and his associates at the London School for Hygiene and Tropical Medicine, but its antifungal activity was not recognized until 1957, when J. C. Gentles, a mycologist in Glasgow, demonstrated its activity in guinea pigs infected with *Microsporum canis*. Griseofulvin is still used, primarily to treat ringworm infections of the skin.

One of the most important antifungal agents is amphotericin B (Fungizone). Isolated in 1953 from *Streptomyces nodosus* at E. R. Squibb and Sons in New Jersey, it was marketed in 1958. It is used to treat severe fungal infections, which were invariably fatal before its introduction. Amphotericin B remains the most effective drug in the therapy of fungal and yeast infections, although it has many adverse effects and has to be given intravenously.

Currently the most widely used antifungals belong to the chemical class of azoles—either imidazoles or triazoles. The mechanism of antifungal action of azoles appears to involve inhibition of the synthesis of ergosterol, a steroid, in the fungal cell membrane. The first azole used as an antifungal was chlorimidazole, which was initially introduced as a 5 percent topical cream; it is no longer used. Merck's anthelminthic, thiabendazole (Mentazole), was shown to have antifungal activity in 1961 but was not developed for antifungal indications. Similarly Janssen Pharmaceutica never developed its anthelminthic, mebendazole (Vermox), as an antifungal. The demonstration of the antifungal activity of these two anthelminthic agents, however, led to the development of more specific antifungal imidazoles by other companies. In 1969 Karl-Heinz Büchel and his associates at Bayer synthesized clotrimazole (Lotrimin, Mycelex), which has a broad spectrum of antifungal activity and is active orally. Its indications include athlete's foot, jock itch, and ringworm infections. It was also shown to be safe and effective in the topical treatment of other cutaneous and vaginal fungal infections. In the United States clotrimazole is currently sold by Schering-Plough and Bayer. Another antifungal imidazole, miconazole (Monistat), was synthesized at Janssen Pharmaceutica, also in 1969. It is used in the United States primarily to treat vaginal *Candida* infections and is available as capsules, vaginal cream, or a solution for intravenous infusion from either the Ortho or Janssen division of Johnson and Johnson.

Even more successful commercially than miconazole is another imidazole, ketoconazole (Nizoral), introduced in 1977 by Janssen. Nizoral is active orally and is effective in treating a wide variety of fungal infections. Its side effects include hepatotoxicity and inhibition of synthesis of adrenocortical steroids.

Many more antifungal imidazoles, triazoles, or related agents were introduced during the 1980s and 1990s. Among them are three significant innovations. Pfizer introduced fluconazole (Diflucan), a highly effective treatment for *Cryptococcus neoformans* infections, which are deadly to AIDS patients; meanwhile Janssen introduced itraconazole (Sporanox), which is indicated in the treatment of pulmonary as well as toenail and fingernail fungal infections. For the treatment of fungal nail infections Novartis introduced terbinafin (Lamisil), which is chemically different from itraconazole and is probably less likely to affect the blood levels of other drugs (e.g., cimetidine) that are metabolized by the same liver enzyme. Rare cases of hepatitis were, however, reported in patients receiving either terbinafin or itraconazole.

Fungi tend to develop resistance to the available antifungal drugs, so new drugs are always needed. The mechanism of resistance development has not yet been established.

Cardiovascular Drugs

Cardiovascular drugs are used to treat diseases of the heart and blood vessels, including cardiac weakness (heart failure), irregularities of heart rhythm (arrhythmias), elevated blood pressure (hypertension), and occlusion of blood vessels by cholesterol deposits (atherosclerosis) or blood clots (embolism and thrombosis). Cardiovascular disease is the main cause of morbidity and premature mortality in the Western world.

Like most currently used drugs, cardiovascular drugs act at their receptors (proteins on the cell surface or sometimes in the cell). The drugs can either mimic or block the effects of the body's own chemicals (transmitters). Drugs that mimic the effects of transmitters are called agonists; those that block their effects are called antagonists. Most transmitters act on more than one receptor. The receptor families and their subclasses are identified by Greek letters and subscript numerals (for example, α_1- or α_2-adrenoceptors). These concepts are discussed in greater detail in the section on the autonomic nervous system.

ANTIHYPERTENSIVE DRUGS

Hypertension, or abnormally high blood pressure, is a genetic (inherited) disease that can be aggravated by diet and such environmental factors as stress. If not treated, hypertension can lead to stroke, heart failure, or both. Since the early 1950s it has been possible to control hypertension with the class of drugs called antihypertensives.

Centrally Acting Antihypertensives

Among the drugs of plant origin still used in modern medicine is reserpine (Serpasil), an alkaloid derived from *Rauwolfia serpentina*, an Asian snakeroot. The plant was used in ancient Indian medicine primarily as a sedative. Reserpine was isolated in 1954 by L. Dorfman and his associates at Ciba Research

Laboratories in Basel, Switzerland, and was found to have sedative and hypotensive properties. Its mechanism of hypotensive action was discovered many years after its introduction and is associated with the depletion of norepinephrine and serotonin from their storage sites in the central and peripheral nervous system. This depletion reduces the ability of the sympathetic nervous system to maintain peripheral vascular tone and heart rate. As a result blood pressure falls, and the heart rate is lowered. Reserpine is seldom used in the United States today, primarily because of its sedative activity, which is severe and may even lead to depression. But it is still used in many other countries, mostly in combinations with diuretics, hydralazine, or other antihypertensive drugs.

Another centrally acting antihypertensive drug, which was used for many years without a clear understanding of its mechanism of action, is methyldopa (α-methyldopa, Aldomet). It was synthesized in 1952 by Karl Pfister and his associates at Merck in Rahway, New Jersey, and patented in 1959 as an inhibitor of an enzyme, dopa decarboxylase, which controls the formation of norepinephrine from its precursor dihydroxyphenylalanine (dopa). Consequently, methyldopa was thought to interfere with the synthesis of norepinephrine. The drug was marketed and successfully promoted as an antihypertensive by virtue of its dopa decarboxylase inhibitory activity. Surprisingly, no other inhibitor of dopa decarboxylase was found to have antihypertensive activity. It was subsequently discovered that methyldopa has only a transient inhibitory effect on dopa decarboxylase, and its antihypertensive effect is not caused by the inhibition of this enzyme.

The next proposed mechanism of action for methyldopa was the "false transmitter" hypothesis. A metabolite of methyldopa, α-methylnorepinephrine, was found to accumulate in the tissue, replacing norepinephrine in its storage sites. Stimulation of the sympathetic nervous system by methyldopa was thought to release α-methylnorepinephrine instead of norepinephrine. Because α-methylnorepinephrine is a less potent vasoconstrictor, it was argued that the vascular tone and therefore blood pressure should be reduced. The release of α-methylnorepinephrine from the sympathetic nerve endings after treatment with methyldopa was demonstrated in 1963 by Erich Muscholl and his associates at the University of Mainz in Germany.

This hypothesis did not survive for long either. It was found that in rats or dogs α-methylnorepinephrine is as potent as norepinephrine, although methyldopa has antihypertensive activity in these species as well. This and other findings led to the conclusion that another mechanism must be responsible for the antihypertensive action of methyldopa. In 1968 Pieter A. van Zwieten and his associates at the University of Amsterdam demonstrated that methyldopa, administered directly to the brain, lowers the blood pressure of anesthetized cats at much lower doses than if it is administered intravenously. This finding led to the current concept, that methyldopa lowers arterial pressure by a central mechanism involving stimulation of the α_2-adrenoceptors by its metabolite α-methylnorepinephrine. The agonists at these receptors are thought to reduce the release of neurotransmitters, including norepinephrine and acetylcholine.

Even while scientists argued about the specific mechanisms responsible for the antihypertensive activity of methyldopa, the drug became widely accepted because of its effectiveness and less severe side effects in comparison with previously available antihypertensive drugs. Methyldopa was used either alone or in combination with the diuretic hydrochlorothiazide (Aldoril). In the 1960s and 1970s methyldopa was one of Merck's major products, playing a significant role in the establishment of the company as the leading American pharmaceutical manufacturer.

Among other centrally acting antihypertensive drugs of particular interest is clonidine (Catapres). In the early 1960s Wolfgang Hoefke at the laboratories of Boehringer Ingelheim in Germany studied various imidazolines with vasoconstrictor activity for potential use as nose drops. He noticed that after an initial pressor effect (elevation of blood pressure) these substances lowered arterial pressure and heart rate. One of these imidazolines, clonidine, was selected for further evaluation as a potential antihypertensive agent. Like methyldopa, clonidine lowered arterial pressure at much lower doses if given directly to the brain rather than intravenously. Its effect was attributed to the stimulation of central α_2-adrenoceptors and the subsequent reduction of vascular tone. More recently more specific receptors for clonidine—imidazoline receptors—were found in the human brain; their activation also reduces nerve impulses in the peripheral sympathetic nerve fibers. Other therapeutic uses were also found for clonidine: It appears to be effective in managing withdrawal reactions from opiates and alcohol.

α-Adrenoceptor Antagonists

The functions of most of our internal organs are not under our voluntary control; these organs—including the heart, intestines, and blood vessels—are controlled by the autonomic nervous system, which consists of sympathetic and parasympathetic nerves. Chemicals (transmitters) released at the endings of sympathetic nerves are sympathomimetic amines: norepinephrine (noradrenaline) and epinephrine (adrenaline). At the parasympathetic nerve endings the transmitter is acetylcholine. The receptors that are excited by sympathomimetic amines are called adrenergic, or adrenoceptors, and drugs that block them are adrenoceptor antagonists. The adrenoceptors are subdivided into α- and β-adrenoceptor families. The excitation of α-adrenoceptors leads to vasoconstriction and an increase in blood pressure; that of β-adrenoceptors leads to an increase in heart rate. Blood pressure is lowered by α-adrenoceptor antagonists, which prevent norepinephrine and epinephrine from interacting with α-adrenoceptors.

The first adrenoceptor antagonists discovered were ergot alkaloids, isolated from the fungus ergot, or *Claviceps purpurea*, which contaminates rye. Its ingestion caused many deaths in the Middle Ages. Ergot intoxication was known at that time as "St. Anthony's Fire." Vascular spasms caused by ergot alkaloids led to gangrene of the extremities and eventual death. The association of gangrene with the ingestion of the fungus was not recognized until the seventeenth century. Small amounts of ergot were used by midwives to speed delivery. The first ergot alkaloid, ergotinine, was isolated in 1875 by a French pharmacist, Charles Tanret. The antagonism of epinephrine by ergot extracts was first demonstrated by Sir Henry Dale in 1905. It could not be used therapeutically because of the many different pharmacological effects produced by ergot alkaloids, particularly their ability to constrict vascular smooth muscle.

One of the first synthetic compounds shown to antagonize the effects of epinephrine was piperoxan, discovered in 1933 by Ernest Fourneau and Daniele Bovet at the Pasteur Institute in Paris. It was marketed in Europe, but its duration of action was too short for any meaningful therapeutic use. In 1939 Ciba marketed tolazoline (Priscol), a vasodilator and adrenoceptor antagonist. It had only transient effects on arterial pressure, and its peripheral vasodilator effects were also unimpressive. In 1950 Ciba introduced another adrenoceptor antagonist, phentolamine (Regitine). Because of the short duration of its action and pronounced cardiac and gastrointestinal stimulant effects, phentolamine is not used to treat hypertension but is used as a diagnostic tool in patients with hypertension. It is highly effective in reducing arterial pressure in patients with pheochromocytoma, an epinephrine-secreting tumor of the adrenal gland, but not in patients with other types of hypertension.

Based on the Ciba experience with tolazoline and phentolamine, it was assumed that α-adrenoceptor antagonists would not be effective in treating hypertension. This assumption proved incorrect with the introduction of prazosin (Minipress) in 1968, which was synthesized by Hans-Jürgen Hess, a medicinal chemist at Pfizer. The goal of his program was to develop an antihypertensive with "peripheral vasodilator" properties but little or no cardiac acceleration. The antihypertensive and α-adrenoceptor–blocking effects of prazosin were discovered by Alexander Scriabine and Jay Constantine. Initially the α-adrenoceptor–blocking activity of prazosin was not considered its major mechanism of action. The decision to develop the drug was based on the assumption that it dilated blood vessels by a novel mechanism, at a site peripheral to the adrenergic receptors. Without this erroneous assumption the drug would not have been tested clinically because at the time the available α-adrenoceptor antagonists were not useful in treating hypertension. In many subsequent studies prazosin was clearly shown to interact directly with α-adrenoceptors. The drug became the standard ligand for these receptors and the first α-adrenoceptor antagonist routinely used in the treatment of hypertension.

Sir Henry Hallett Dale (1875–1968) shared the 1936 Nobel Prize in physiology and medicine with Otto Loewi. They were honored for the discovery of chemical transmission of nerve impulses. During his scientific career, which included ten years at Wellcome Laboratories (1904–14) and twenty-eight years with the National Institute for Medical Research, Dale established a scientific basis for the development of drugs affecting the autonomic nervous system. Many well-known pharmacologists of the twentieth century were either educated or otherwise influenced by Dale. Courtesy Nobel Foundation.

The development of prazosin was followed in 1977 by the discovery of terazosin (Hytrin) by Jaroslav Kyncl and his associates at Abbott Laboratories in Chicago. Terazosin has longer duration of action than prazosin; its half-life is approximately twelve hours. In addition claims were made that terazosin was better and more predictably absorbed than prazosin following oral administration. The mechanism of action and other pharmacological properties of the two drugs are otherwise very similar. More recently Pfizer introduced doxazosin (Cardura), another chemically related α-adrenoceptor antagonist with a long duration of action, which was discovered at Pfizer Laboratories in the United Kingdom.

All three α-adrenoceptor antagonists—prazosin, terazosin, and doxazosin—have favorable effects on blood lipids. Their major drawback is occasional fainting after the first dose of the drug. To avoid this effect, the therapy is initiated at the lowest possible doses, which can then be gradually increased. Terazosin is now being promoted for a new indication, to control urine flow in elderly men with enlargement of the prostate gland. Prazosin and doxazosin appear to have similar effects.

β-*Adrenoceptor Antagonists*

The separation of adrenoceptors into α and β families was made by Raymond P. Ahlquist at Emory University in Atlanta, Georgia; in 1948 he observed that the available adrenoceptor-blocking drugs did not block the increase in heart rate caused by sympathomimetic amines. Ahlquist thus suggested that there were two types of adrenoceptors, which he called α- and β-adrenoceptors. Other investigators confirmed his findings and discovered many subgroups of adrenoceptors.

Beta-adrenoceptor antagonists (or β-blockers) are used in the treatment of hypertension and coronary heart disease. The first β-adrenoceptor antagonist was discovered in 1957 by Irwin Slater at Lilly Laboratories in Indianapolis. He tested various analogs of isoproterenol for bronchodilator activity and found accidentally that dichloroisoproterenol (DCI) antagonized the cardiac stimulant effects of isoproterenol. Unfortunately DCI had many other pharmacological effects, including some agonist effects at β-adrenoceptors, so it was not considered useful therapeutically. Slater's findings were confirmed by Neil Moran at Emory University in Atlanta. James Black, at that time with Imperial Chemical Industries (ICI), learned about Slater's and Moran's findings and asked John Stevenson at ICI to synthesize new DCI derivatives in an attempt to develop useful therapeutic agents.

The tremendous therapeutic potential for β-adrenoceptor antagonists was not immediately recognized. In 1962, when Black presented the initial pharmacological findings with his first β-adrenoceptor antagonist, pronethalol (nethalide, Alderlin), at the International Congress of Pharmacology in Holland, he was asked what therapeutic indications he considered likely for such a drug. He responded, with some hesitation, that the drug might conceivably be useful in treating abnormal heart rhythm. Antony C. Dornhorst and B. F. Robinson in London had some success when they tested pronethalol clinically in the therapy of angina pectoris (chest pain caused by coronary artery spasm or occlusion). Its hypotensive action in patients with angina pectoris was discovered in 1964 by Brian N. C. Prichard, also in London. Further development of pronethalol was precluded by chronic toxicological studies in mice, which revealed its carcinogenicity (it was found to produce cancer of the thymus gland).

Imperial Chemical Industries followed the development of pronethalol with that of propranolol (Inderal), which was synthesized by Leslie Smith and demonstrated to have an antihypertensive effect in the clinic by Prichard and Peter M. S. Gillam in 1964. Propranolol was the first β-adrenoceptor antagonist marketed successfully. Its initial indication was angina pectoris, but hypertension rapidly became its major market. Many more indications were discovered, and propranolol is currently approved in the United States for use in hypertension, angina pectoris, arrhythmias, myocardial infarction (heart attack), glaucoma, and migraine. The β-adrenoceptor antagonists—drugs "in search of a disease"—found many diseases to treat.

The success of propranolol led almost every large pharmaceutical company to mount an intensive search for other β-adrenoceptor antagonists. Merck

Sir James Whyte Black (1924–), a British pharmacologist, is one of the recipients of the 1988 Nobel Prize in physiology or medicine. His pioneering work in analytical pharmacology led to the discovery of β-adrenoceptor and histamine type 2 antagonists. He discovered both propranolol (Inderal) and cimetidine (Tagamet). Both drugs had a major effect on the practice of medicine and are highly successful commercial products. Courtesy Nobel Foundation.

scientists developed timolol (Blockadren, Timoptic); it is more potent than propranolol and has no local anesthetic activity, whereas propranolol is comparable to lidocaine as a local anesthetic. This property is of particular importance for treating glaucoma (elevation of intraocular pressure), because local anesthesia is not desirable for locally applied antiglaucoma drugs. Merck pioneered the use of β-adrenoceptor antagonists in glaucoma and established timolol as the therapy of choice for this condition.

Other companies exploited the drawbacks of propranolol. First was the fact that propranolol is not selective for β_1-adrenoceptors. By inhibiting β_2-adrenoceptors (in addition to β_1), propranolol and similar unspecific antagonists are likely to cause bronchospasms, an effect particularly dangerous for patients suffering from bronchial asthma. In 1970 ICI patented atenolol (Tenormin), and in 1971 Hässle in Sweden developed metoprolol (Lopressor). These two β-adrenoceptor antagonists are selective for β_1-adrenoceptors and are therefore less likely to cause bronchospasms. They are as effective as propranolol for treating hypertension and angina pectoris. A second disadvantage of propranolol is its ability to enter the brain and to produce vivid dreams in some patients. Bristol Laboratories marketed nadolol (Corgard), a nonselective β-adrenoceptor antagonist that does not enter the brain and does not interfere with normal sleep, as an alternative to propranolol. Propranolol can also produce in some patients, by virtue of its β-adrenoceptor–blocking action, a pronounced lowering of heart rate, which may be undesirable, particularly in patients with heart failure. Pindolol (Visken), a drug with some agonist (in addition to antagonist) activity at the cardiac receptors, is less likely to lower heart rate to the same extent as propranolol. It was patented in 1969 by Sandoz and introduced in the United States in 1977.

The development of new β-adrenoceptor antagonists continued during the 1980s and 1990s, but at a somewhat slower pace. The patents on most of the older drugs expired, allowing companies that manufacture generic drugs to market them under generic or new trade names. The identification of specific medical needs led to the discovery and marketing of new antagonists of β-adrenoceptors. Among them was labetalol (Trandate, Normodyne), which combines α_1- and β-adrenoceptor blocking effects in one molecule. It has vasodilator properties and little if any effect on the heart rate. These properties are desirable in the treatment of hypertensive patients with a weak heart. The discovery of the antiarrhythmic activity of another β-adrenoceptor antagonist, sotalol (Betapace), led to its marketing for the treatment of life-threatening ventricular arrhythmias refractory to other antiarrhythmic drugs. Cardiac surgeons desired a short acting β-adrenoceptor–blocking drug with a rapid onset of action that can be given during surgery by infusion to control the sudden occurrence of ventricular arrhythmia and hypertension. This need was satisfied by esmolol (Breviblock). Robert R. Ruffolo and his associates at SmithKline Beechman in King of Prussia, Pennsylvania, developed another β-adrenoceptor antagonist with multiple actions, carvedilol (Coreg). This drug combines, like labetalol, α_1- and β-adrenoceptor–blocking effects, but in addition it has Ca^{2+}-channel blocking and tissue protective properties. It was found to be clinically useful in the treatment of heart failure and was approved by the FDA for both this indication and hypertension.

Many more β-adrenoceptor antagonists were developed and marketed throughout the world. In 1998 twelve different β-adrenoceptor antagonists were on the U.S. market; at least twice as many were available in most European countries. The impact of β-adrenoceptor antagonists on the treatment of hypertension and coronary heart disease was impressive. These drugs increased the life expectancy of many cardiac or hypertensive patients and substantially changed the practice of medicine. In 1988 James Black received the Nobel Prize in physiology or medicine for his part in the discovery of propranolol and cimetidine.

New Adrenoceptor Subtypes

In the early 1970s scientists began to realize that many more subtypes of adrenoceptors existed than originally anticipated. In 1974 Klaus Starke at the University of Freiburg in Germany and Solomon Langer at Synthelabo in Paris proposed the existence of two subtypes of α-adrenoceptors: α_1 and α_2. This classification was based on the discovery of adrenoceptors at the nerve endings (α_2). These receptors were found to control the release of norepinephrine and had different characteristics than previously recognized adrenoceptors. Norepinephrine was an agonist at either subtype of receptor, but the affinity (selective attraction) of other agonists and of antagonists for the two subtypes of α-adrenoceptors differed substantially.

Researchers currently distinguish at least three subtypes of α_1-adrenoceptors: α_{1A}, α_{1B}, and α_{1C}; the existence of an α_{1D} subtype is questionable. Four subtypes of α_2-adrenoceptors have been identified: α_{2A}, α_{2B}, α_{2C}, and α_{2D}; the existence of further subtypes is considered possible. The search for β-adrenoceptor subtypes in addition to β_1- and β_2-adrenoceptors led to only one more subtype, β_3. Many of the adrenoceptor subtypes have been isolated, structurally identified, and even cloned. Their identification and characterization, as well as that of any other receptors, is of more than academic interest. Some highly selective agonists or antagonists for the receptor subtypes that have already been discovered are likely to replace many current drugs. These new drugs will be much more selective in modulating body functions and will have fewer side effects.

Angiotensin-Converting Enzyme Inhibitors

The development of angiotensin-converting enzyme (ACE) inhibitors has had a major impact on cardiovascular medicine. In the 1970s ACE inhibitors became the drugs of choice in the treatment of two major cardiovascular diseases: hypertension and heart failure. The development of the first clinically useful ACE inhibitors is often cited as an example of the type of rational drug design that is likely to become routine in the next century. It is also an example of the importance of basic research for drug development.

The discovery of ACE inhibitors was made possible by a hundred years of basic research in the control of arterial blood pressure. Robert Tigerstedt in Germany made the first relevant observation in 1898, when he reported that kidney extracts contain a vasoconstrictor substance, which he named renin. His discovery generated little interest until 1934, when Harry Goldblatt at Western Reserve University in Ohio produced sustained elevation of blood pressure in a dog by constricting the renal artery. In 1940 Eduard Braun-Menendez in Argentina discovered that renin catalyzes the formation of a potent vasoconstrictor substance, which he called hypertensin. At the same time Irvine Page and his associates at the Cleveland Clinic independently discovered a vasoconstrictor substance they called angiotonin, which turned out to be identical to hypertensin. In 1958 both investigators agreed to call it angiotensin.

In the mid-1950s Leonard T. Skeggs and colleagues recognized the existence of more than one angiotensin. An enzyme, renin, was found to activate angiotensinogen, an angiotensin precursor, to form angiotensin I; another enzyme was found to convert angiotensin I to angiotensin II. Angiotensin II was synthesized in 1957 by F. Merlin Bumpus and his associates at the Cleveland Clinic. In 1958 Franz Gross at Ciba in Switzerland recognized the physiological role of the renin-angiotensin system in the maintenance of water and salt balance. But the importance of this system in maintaining elevated arterial pressure in all forms of hypertension was not suggested until the early 1970s, by John Laragh and his group in New York.

The development of ACE inhibitors would not have been possible without the seemingly unrelated observation of Sergio T. Ferreira and his colleagues in São Paulo, Brazil, that the venom of a Brazilian pit viper contains factors that enhance the vasodilator effect of another endogenous peptide, bradykinin. Ferreira called these factors bradykinin-potentiating factors, or BPFs. These factors (which were also peptides) were shown to inhibit an enzyme that degrades bradykinin. In 1968 Erwin G. Erdös and his associates in Chicago demonstrated that the enzyme responsible for degrading bradykinin is also responsible for forming angiotensin II. Ferreira's BPFs were, therefore, the first identified ACE inhibitors.

One of Ferreira's peptides, BPF_{9a}, or teprotide, was tested in animals and eventually in humans and was found to lower arterial pressure. David W. Cushman and his associates at Squibb, in Princeton, analyzed the structural requirements for the ACE inhibitory activity of teprotide and identified the active site of the enzyme. On the basis of this analysis they synthesized a series of orally active nonpeptides with ACE inhibitory activity. Captopril (Capoten) was selected for development as one of the most potent compounds from this series. The advantages of captopril and other ACE inhibitors in treating hypertension include their effectiveness in all types of hypertension and freedom from the side effects that attended earlier antihypertensive agents—for example, increased heart rate, headache, edema, and salt retention. The only occasionally disturbing side effect of ACE inhibitors is a cough caused by the bronchoconstrictor effects of bradykinin.

Angiotensin-converting enzyme inhibitors were also shown to have beneficial effects in heart failure. Because of their vasodilating action, they improve cardiac performance, probably by reducing heart work. Another more recently discovered advantage is organ protection: ACE inhibitors can protect the heart, kidneys, and possibly the brain from experimentally induced hypoxic tissue damage. The mechanism of this protective action is not clearly understood but is probably determined by inhibition of angiotensin II formation, the major pharmacological action of ACE inhibitors.

Since the development of captopril many other ACE inhibitors have been created and introduced on the market. Arthur Patchett and his associates at Merck synthesized enalapril (Vasotec), whose pharmacological properties were evaluated by Charles S. Sweet and his associates. Enalapril has a longer duration of action than captopril and lacks a so-called sulfahydro group, which is thought to be responsible for some of the side effects of captopril. Enalapril was followed by lisinopril (Prinivil, Zostril), marketed by Merck as well as Stuart Pharmaceuticals. Lisinopril does not offer any significant advantages over enalapril, except once-a-day dosing. With its two ACE inhibitors Merck dominated the field, even though other pharmaceutical companies introduced their own versions. Parke-Davis introduced quinapril (Accupril), while Hoechst developed ramipril (Altace). Ciba-Geigy developed benazepril (Lotensin), and Rhône-Poulenc Rorer introduced perindopril (Aceon), licensed from Servier Laboratories in France. According to the 1998 *Physicians' Desk Reference* (*PDR*), ten ACE inhibitors are marketed in the United States. They have a similar mechanism of action and side-effects profile but differ in such areas as their relative potency, pharmacokinetics, duration of action, and tissue distribution. New ACE inhibitors, currently under development, have more than one mechanism of antihypertensive action. A new ACE inhibitor from Schering-Plough, SCH 54470, inhibits two other enzymes—endothelin-converting enzyme and neutral endopeptidase—and is likely to be important in cardiovascular control as well.

Angiotensin II Antagonists

In 1982 Y. Furukawa and his associates at Takeda Chemical Industries in Osaka reported that their hypotensive derivatives of imidazole-5-acetic acid inhibit the pressor effect of angiotensin II. On the basis of this report the DuPont Merck Company in Delaware initiated a research program that succeeded in optimizing this effect with the company's own imidazole derivative, DUP 753, or losartan. In 1991 David J. Carini and associates reported the synthesis of losartan, and Pancras C. Wong reported the pharmacology. Losartan was found to be an effective antihypertensive agent in various animal models of hypertension. Its activity was apparently mediated by its metabolite (EXP 3174), which is a specific inhibitor of a subtype of angiotensin receptors (AT_1). (For sites of action of ACE inhibitors and losartan, see Figure 1.)

Losartan was subjected to extensive clinical studies and was shown to be as effective an antihypertensive as the ACE inhibitors, but less likely to produce cough. The FDA approved losartan in 1995, and it was marketed by Merck as Cozaar. Many other angiotensin II antagonists followed in development.

Scientists at Ciba-Geigy in Switzerland developed valsartan. Unlike losartan, valsartan is not a "pro-drug": It does not have to be metabolized in the liver to an active principle and is therefore less likely to interact with other drugs metabolized by the same enzyme. In addition it generally is not contraindicated in patients with liver disease. In the United States valsartan was licensed to Novartis and marketed under the trade name Diovan.

In 1997 two more angiotensin II antagonists were approved in the United States: irbesartan (Avapro), codeveloped by Sanofi-Winthrop and Bristol-Myers Squibb, and eprosartan (Tevetan), developed by SmithKline Beecham. Hyper-

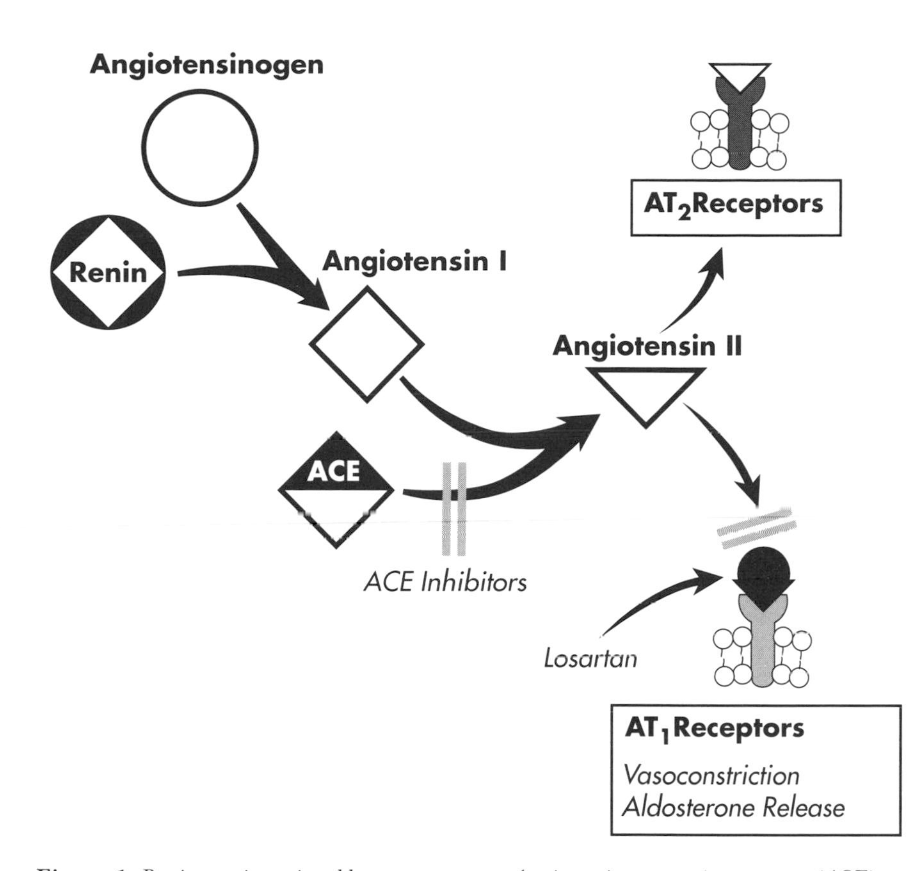

Figure 1. *Renin-angiotensin-aldosterone system. Angiotensin-converting enzyme (ACE) inhibitors block (||) the formation of angiotensin II from angiotensin I, while losartan blocks angiotensin (AT1) receptors and, therefore, prevents angiotensin II–induced vasoconstriction and aldosterone release.*

tension remained the major indication for angiotensin II antagonists, but ongoing clinical trials suggest that their indications may be expanded to heart failure and even to vascular complications of diabetes. The market for angiotensin II antagonists became so lucrative that every major pharmaceutical company attempts to develop or to license a product in this field.

At least two more "sartans"—tasosartan (American Home Products) and telmisartan (Boehringer Ingelheim)—are in advanced clinical trials. The investigators are trying desperately to find clinically significant differences among these compounds. It was reported at the 1998 annual meeting of the American Hypertension Society that patients on tasosartan (but not on other angiotensin II antagonists) can even skip a dose or two and still maintain control over their blood pressure.

Ca^{2+} Channel Antagonists

The importance of calcium for muscular contractions was first recognized by British physiologist Sydney Ringer in 1882. He was perfusing isolated frog hearts and recording their contractile activity. One day his assistant accidentally omitted calcium from the perfusion medium, and the hearts would not contract. After discovering that calcium was missing, Ringer concluded that it was essential for cardiac contractility.

Since this original observation many studies on the role of calcium in muscle contraction have been conducted. It is now established that constant blood calcium levels are required to maintain not only a normal heart beat, but also normal functioning of nerves and muscles, blood coagulation, hormone secretion, and many other processes in the human body. Excessively high calcium levels in blood and other extracellular fluids, however, can be detrimental for normal cellular functions. If calcium is allowed to enter vascular smooth muscle cells in excessive amounts, it can lead to vasospasms and hypertension; excessive amounts of calcium can also lead to cell death. Preventing excessive amounts of calcium from entering the cells represents, therefore, an attractive goal for drug therapy.

Calcium enters the cells through specific proteins in the cell membrane, which are called channels. In the 1980s calcium channels were studied extensively by Richard Tsien and his associates, initially at Yale and later at Stanford University; by Rodolfo Llinas and his group at New York University; and by many other investigators. Calcium channels are classified according to their physiological properties, as voltage- or receptor-operated channels. Voltage-operated channels open in response to depolarization of the cell membrane by potassium or other means, while receptor-operated channels open in response to various chemical transmitters.

There are at least six different types of voltage-operated calcium channels, three of them named according to the duration of their open state: L (for long-lasting), T (for transient), and N (for neither). When N channels were discovered to occur primarily in neurones, N came to indicate "neuronal." P channels were named for Purkinje cells in the brain, where they were first discovered. Q and R channels followed in alphabetical order. P and Q channels are now considered by some investigators to be identical and are often referred to as P-Q channels. The binding of drugs at certain sites of the channels can either block or enhance calcium entry into the cells. Most of the calcium-blocking drugs that are used therapeutically today inhibit L-type calcium channels.

The history of the discovery of calcium-blocking drugs was recorded by Albrecht Fleckenstein, a professor of physiology at the University of Freiburg, in his book *Calcium Antagonism in Heart and Smooth Muscle*, published in 1983. According to Fleckenstein, he was the first to report, in 1964, evidence that drugs could mimic the effects of calcium withdrawal from the medium surrounding the cells. The first drugs shown to have such activity were verapamil

Albrecht Fleckenstein (1917–1992), German physician and physiologist, made major contributions to the discovery and development of calcium channel antagonists. His discovery that drugs can reduce the entry of calcium ions into the cells and, by doing so, cause coronary and systemic vasodilation, established the basis for the use of these drugs in the therapy of angina pectoris and hypertension. His findings also suggested the cytoprotective and anti-atherosclerotic effects of these drugs. Courtesy G. Fleckenstein-Grün.

(Isoptin, Calan, Verelan) and prenylamine (Segontin). Some of the other investigators were not willing to give all the credit to Fleckenstein. Theophile Godfraind of the University of Louvain, Belgium, claimed that the concept of calcium antagonism, as a mode of action of drugs, was derived from his studies with lidoflazine and cinnarizine, published earlier.

Verapamil was originally developed at Knoll Pharmaceutical Company (part of the BASF group) in Germany. The drug was developed as a coronary vasodilator without any theory on its mechanism of action. One of the Knoll executives was Fleckenstein's brother, who provided him with a sample of verapamil and asked him to study its effects. Fleckenstein not only described the mechanism of verapamil's action but also established the concept that drugs can dilate blood vessels and protect vital organs by reducing the entry of calcium into the cells.

Fleckenstein's findings attracted the interest of other pharmaceutical companies. Hans-Günther Kroneberg, at that time head of the Pharmacological Institute of Bayer, took a sample of Bayer's coronary vasodilator, nifedipine, to Fleckenstein. The company was not anxious to market this drug, but a team of two stubborn scientists—Friedrich Bossert, a chemist, and Wulf Vater, a pharmacologist—were so intrigued by its properties that they continued to work with nifedipine and related substances, disregarding management's suggestion to stop. Fleckenstein saved the drug as well as Bossert and Vater. Soon after receiving the sample, he called Kroneberg and announced that nifedipine was the most potent and most specific inhibitor of calcium entry into the cells he had seen. By "specific" he meant that the other pharmacological effects of the drug were negligible. At the same time this work was being done at Bayer, Bernard Loev and Stewart Ehrreich were synthesizing and evaluating a series of closely related compounds at Smith Kline and French Laboratories in Philadelphia. They even synthesized nifedipine without patenting it. The first compound evaluated pharmacologically was not active orally, and they had no hypothesis for its mechanism of action. Their project was discontinued.

Clinical trials with nifedipine in coronary artery disease were initiated in Europe and Japan in 1970 and were highly successful; nifedipine significantly reduced the incidence of anginal attacks (chest pain) in patients with coronary heart disease, and the effect was attributed to its coronary vasodilator action.

In the United States at that time coronary vasodilation was not considered a promising approach to the therapy of coronary artery disease, and the concept of drug interference with Ca^{2+} entry into the cells was not well accepted. But American cardiologists and pharmaceutical executives began to change their opinions after an Italian cardiologist, Atilio Maseri, published a series of papers in leading American medical journals on coronary arterial spasm as a probable cause of not only variant angina but also myocardial infarction. Gerald Laubach, then the president of Pfizer, flew to Germany to negotiate with Bayer to license nifedipine (Adalat) to Pfizer for sale in the United States. Marketed by Pfizer as Procardia, nifedipine was exceptionally successful.

The second major indication for calcium antagonists was hypertension. Initially, the lowering of arterial pressure was considered an undesirable side effect of calcium antagonists, although verapamil, by intravenous administration, had been shown in 1970 to lower arterial pressure in emergencies with hypertensive patients. In 1978 O. Lederballe Pedersen and E. Mikkelsen in Denmark and Kyuzo Aoki and his associates in Japan published the first clinical data on the usefulness of nifedipine in the acute and chronic treatment of hypertension. Many other reports in subsequent years confirmed their findings. Nifedipine was found to be dramatically effective even in cases of severe hypertension that could not be managed by other drugs used at that time, such as diuretics, methyldopa, or β-adrenoceptor antagonists. Verapamil found additional applications in the treatment of disturbances of cardiac rhythm and certain other heart diseases.

During the late 1960s Masanori Sato and his associates at Tanabe Seiyaku Company in Japan developed diltiazem (Cardizem). Its chemical structure differed substantially from that of either nifedipine or verapamil, but the mechanism of action was similar. Diltiazem interfered with the entry of calcium into vascular smooth muscle or myocardial cells through the same ion channel (L-type calcium channel) as nifedipine or verapamil but at a different site, and it had a more favorable side-effect profile than the other two drugs. Diltiazem has been very successful in the United States, where it is distributed by Marion Merrell Dow (now part of Hoechst).

The commercial success of the first three calcium channel antagonists—verapamil, nifedipine, and diltiazem—led almost every major pharmaceutical company to initiate research projects in this area. Most second-generation calcium channel antagonists belong to the same chemical class as nifedipine (dihydropyridines). Sandoz (now Novartis) developed and marketed isradipine (Dynacirc), and Merck-Astra created felodipine (Plendil). Nicardipine was licensed from Yamanouchi, initially by Syntex, and eventually by American Home Products (Wyeth-Ayerst Division) and marketed under the trade name Cardene. Pfizer developed amlodipine (Norvasc) and Miles (now Bayer) introduced nimodipine (Nimotop). One of Bayer's second generation Ca^{2+} channel antagonists, nisoldipine, was licensed to Zeneca and marketed in the United States as Sular. Many more dihydropyridines were marketed in Europe.

The second-generation calcium channel antagonists offer some advantages over the original three compounds. Amlodipine has a longer duration of action; nicardipine, a water-soluble derivative, is available for intravenous administration; felodipine appears to be more selective than nifedipine for vascular smooth muscle. Nimodipine enters the brain better than nifedipine does and has greater specificity for brain cells. It has, therefore, a different indication: prevention of disability following subarachnoid hemorrhage (a type of stroke). Many more indications for calcium channel antagonists have been discovered. They improve circulation in the extremities and are therefore beneficial in treating peripheral vascular disorders. They are useful in migraine and even in heart failure.

Rapid-release formulations of nifedipine and similar short-acting calcium channel antagonists are now largely replaced by slow-release formulations of the same drugs or by new long-acting derivatives. Rapid lowering of arterial pressure is considered undesirable, particularly in patients with coronary heart disease. Recently published retrospective analyses of some clinical studies indicated that high doses of rapidly acting nifedipine may have even increased the mortality rate among patients with a history of heart disease.

A retrospective analysis of some clinical studies with nifedipine was published in 1996 by Curt Furberg and his associates at the Bowman Gray School of Medicine in Durham, North Carolina. This analysis indicated that high doses of a rapidly acting formulation of nifedipine increased mortality in patients with a history of heart disease. Furberg's conclusion was questioned by many investigators. It was generally agreed, however, that a rapid fall in arterial pressure should be avoided in patients with coronary heart disease. So most physicians switched their patients from rapid-release to extended-release formulations of nifedipine or other Ca^{2+} channel antagonists. These formulations were considered safe since there was no evidence that blockade of L-type calcium channels per se is detrimental to cardiac patients. Extended-release formulations of most Ca^{2+} channel antagonists are now available.

In 1997 eleven Ca^{2+} channel antagonists were marketed in the United States, some of them by two or more companies under different trade names. Among them was a new Ca^{2+} channel antagonist, mibefradil (Posicor). This drug was discovered in the late 1980s at the research laboratories of Hoffmann–La Roche in Basel, Switzerland. Jean-Paul Clozel and his associates were looking for a Ca^{2+} channel antagonist with a high bioavailability and long half-life, and freedom from cardiac depressant or strong vasodilator effects. On the basis of these criteria as well as chemical novelty they selected mibefradil and recommended it for clinical evaluation in the treatment of hypertension and angina. In spite of the favorable initial clinical studies the development of mibefradil proceeded slowly, until American pharmacologist, R. Kent Hermsmeyer, at the Oregon Primate Center, reported a unique effect of the drug on T-type channels in the heart. He found that unlike other available Ca^{2+} channel antagonists mibefradil was inhibiting T-type in preference to L-type calcium channels. This finding was interpreted as an indication of greater potential efficacy and tolerability of mibefradil in comparison to other Ca^{2+} channel antagonists. Hoffmann–La Roche heavily promoted the drug, and many clinical studies appeared to support company claims. In June 1997 mibefradil was approved for the treatment of hypertension in the United States and by the end of that year had been approved in twenty-eight other countries. One year later the FDA requested the drug be withdrawn because of serious, even fatal, interactions with other drugs metabolized in the liver by the same enzymes as mibefradil. At the time of withdrawal three hundred thousand patients in the United States alone were being treated with this drug. The fact that mibefradil inhibits liver enzymes (so-called cytochrome P450 isozymes) was known at the time of approval, but the consequences were apparently not anticipated. Mibefradil's case is a good lesson for both the pharmaceutical industry and the regulatory agencies: They learned that drug interactions should be anticipated and carefully evaluated in the early phases of development. The fate of mibefradil does not affect other Ca^{2+} channel antagonists since mibefradil's toxicity is not related to an effect on Ca^{2+} channels. On the contrary, the search for a specific T-type Ca^{2+} channel antagonist free from an effect on liver enzymes will probably be intensified.

Other Vasodilators

Drugs that lower arterial pressure by dilating peripheral blood vessels are called peripheral vasodilators. Sodium nitroprusside (Nipride) acts by releasing or increasing formation of nitric oxide, a vasodilator gas that forms and acts in the vascular wall and is rapidly destroyed at the site of its action. Sodium nitroprusside is an old compound, known since 1929 to lower arterial pressure. It was not approved until 1974 for use in the United States in treating acute high blood pressure episodes.

Another vasodilator, still being used, is hydralazine (Apresoline), discovered in 1950 by Franz Gross and his coworkers at Ciba during a search for novel antihistaminics. Hydralazine was thought originally to act by blocking α-adrenoceptors in the blood vessels. When that theory was refuted, many other mechanisms were proposed, but none proved correct. Hydralazine is usually not used alone to treat high blood pressure, but is given in combination with diuretics, β-blockers, or both. Its side effects, salt and water retention and an increased heart rate, are antagonized by these drugs.

Minoxidil (Loniten, Rogaine), a long-acting vasodilator and antihypertensive agent, was discovered in 1973 by Donald W. DuCharme and his associates at Upjohn Company. The initial clinical studies demonstrated the ability of minoxidil to control severe hypertension even in patients not responding to other drugs. But when toxicological evaluation of minoxidil was done in dogs, the drug produced lesions in the upper right chamber of the heart. Because of minoxidil's unique effectiveness, the FDA approved the drug but restricted its use to severely ill patients, refractory to other drugs. In subsequent clinical use it was observed that one of minoxidil's side effects is excessive hair growth, which led to its current major indication, treatment of baldness. For this purpose minoxidil is used locally, so that its blood concentrations do not reach levels capable of causing cardiac lesions or even lowering blood pressure. The market for minoxidil as a hair growth stimulant is substantial.

More recently minoxidil was shown to activate ion channels (so-called ATP-dependent potassium channels) in cell membrane. This effect causes the intracellular concentration of calcium ions to decrease and consequently the blood vessels to dilate. Various pharmaceutical companies discovered a large number of new drugs acting by this mechanism, and further new indications, including treatment of urinary incontinence, were proposed. Activators of potassium channels are now under intensive research and development.

Diuretics

Soon after sulfanilamide was introduced as an antibacterial agent, it was noticed that the patients receiving it excreted alkaline urine. In 1940 T. Mann and D. Keilin at the University of Cambridge discovered that sulfanilamide inhibits carbonic anhydrase, an enzyme that catalyzes the decomposition of carbonic acid into carbon dioxide and water. Inhibition of carbonic anhydrase was not a general property of all antibacterial sulfonamides, and structural requirements for the enzyme inhibition were clearly different from those for the antibacterial activity.

The role of carbonic anhydrase in the renal excretion of bicarbonate and sodium was of interest to Robert Pitts at Cornell University, in Ithaca, New York, who used sulfanilamide as a tool to study renal function. He identified the site of carbonic anhydrase inhibitory action of sulfanilamide in the kidney. The inhibition of the enzyme led to an increase in excreted bicarbonate and consequently to alkaline urine. Theoretically this effect should have led to a clinically useful increase in the excretion of sodium and water. In a clinical test sulfanilamide did increase the excretion of sodium and water in a patient with heart failure, but only at a dose that produced unacceptable side effects. Richard Roblin of Lederle Laboratories initiated a program to develop a new synthetic inhibitor of carbonic anhydrase, which led to the discovery by Thomas Maren of acetazolamide (Diamox), a sulfanomyl diuretic and a potent inhibitor of carbonic anhydrase. It is still used today, but seldom as a diuretic. The drug was found useful in the treatment of glaucoma to reduce intraocular pressure, in acute mountain sickness, and in the management of periodic paralysis. Most therapeutic, as well as adverse, effects of acetazolamide result from the inhibition of carbonic anhydrase either in the kidney or in other organs. The inhibition of this enzyme leads to acidosis, accumulation of acid in the body. Excessive acidosis is the major limiting factor in the therapeutic use of acetazolamide and other inhibitors of carbonic anhydrase. It was also the major reason to search for safer diuretics.

Karl Beyer at Sharp and Dohme (later Merck Sharp and Dohme, a division of Merck & Company) noticed that some sulfonamide diuretics increased the excretion not only of bicarbonate but also of chloride anions. This observation led James Sprague and his associates at Sharp and Dohme to synthesize many new sulfonamides for evaluation as diuretics. Among them was chlorothiazide, which was found to excrete more chloride than bicarbonate but was still a weak inhibitor of carbonic anhydrase. An ideal diuretic should excrete ions in the urine in the same proportion as they are present in the body fluids. Chlorothiazide met this requirement better than any other previously known diuretic.

Beyer and the management team at Merck, including Max Tishler, who was head of research, realized the importance of this observation and decided to develop and market chlorothiazide (Diuril) despite the market research data, which did not anticipate the potential importance of diuretics. Market research analysis based its predictions of a very limited market potential on sales figures for mercurial diuretics, relatively toxic compounds with limited indications. Beyer visualized the medical need for safer diuretics and the possible expansion of indications to other diseases. He used to tell his associates that in the search for new drugs it is always good business to satisfy medical needs. If a medical need for the treatment of one disease is satisfied, the indications for a new drug are likely to expand beyond any predictable limits.

Even though Beyer realized the possibility of using diuretics in hypertension, their use was not a recognized medical concept at the time. Within the next five years the antihypertensive activity of diuretics was demonstrated, and the market expanded beyond imagination. Almost every major pharmaceutical company in the United States and Western Europe initiated a program to develop diuretics. Many "holes" were found in Merck's chlorothiazide patent, and more than twenty related sulfonamyl diuretics were marketed. The most

Karl Henry Beyer Jr. (1914–1997) spent thirty years of his professional life in the pharmaceutical industry (Sharp and Dohme, later a division of Merck). Among the many drugs he and his associates developed are diuretics (chlorothiazide, hydrochlorothiazide, ethacrynic acid, and amiloride), the uricosuric agent probenecid, the antihypertensive drug methyldopa, and antiserotonin (cyproheptadine). He described his scientific career in the 1977 issue of the Annual Review of Pharmacology and Toxicology *as well as in the book* Discovery, Development and Delivery of New Drugs, *published in 1978. Courtesy E. S. Vessel.*

successful derivative, which largely replaced chlorothiazide, is hydrochlorothiazide, which both Merck and Ciba claim to have discovered. The settlement between the two companies granted marketing rights to both, so hydrochlorothiazide was marketed by Merck as Hydrodiuril and by Ciba as Esidrix. The advantages of hydrochlorothiazide included higher potency and the virtual absence of carbonic anhydrase inhibitory activity at therapeutic doses. What had been regarded as a desirable mechanism of action became a side effect!

A longer duration of action was an advantage of another sulfonamyl diuretic, polythiazide (Renese). In the United States only a few additional sulfonamyl diuretics, closely related chemically to chlorothiazide (also known as thiazides or benzothiadiazines), survived the intense competition for market share. They are hydroflumethiazide (Saluron, Diucardin), benzthiazide (Exna), bendroflumethiazide (Naturetin), and methyclothiazide (Enduron). Other sulfonamides, unrelated chemically to chlorothiazide, were also found to have diuretic activity. The best known among them are furosemide (Lasix), bumetanide (Bumex), quinethazone (Hydromox), and chlorthalidone (Hygroton).

Of these, the most successful is furosemide; it differs from chlorothiazide-like diuretics primarily in its higher efficacy. Its maximum obtainable salt-excreting ability is much higher than that produced with chlorothiazide, hydrochlorothiazide, or other benzothiadiazine-type diuretics. Only bumetanide has the same high efficacy. Because of this property furosemide and bumetanide are called "high-ceiling" diuretics. Their higher efficacy is explained by a different site of action within the kidney—Henle's loop, where a larger portion of salt and water is reabsorbed. Because of this site of action furosemide and similar diuretics are also called "loop" diuretics. The onset of furosemide's action is more rapid than that of hydrochlorothiazide, and its dose response curve steeper. The maximal effect of furosemide can therefore be reached faster, and a small increment in dosage achieves a greater increment in the effect. Because of its high efficacy furosemide is often effective in patients who do not respond to other diuretics.

Furosemide was discovered at Hoechst in Germany by Roman Muschaweck and his associates and patented in 1962. Its high efficacy surprised its inventors, who were hoping to develop a sulfonamide diuretic with fewer side effects than benzothiadiazines. Its different site of action also surprised many academic investigators, who did not expect a significant therapeutic advance from another sulfonamide diuretic.

The acute toxicity of benzothiadiazines and furosemide is very low, and their therapeutic index (the toxic dose divided by the therapeutic dose) is very high—more than 2,000 in either rats or dogs. When taken over a long period, however, diuretics can produce serious side effects even at therapeutic doses. The earliest recognized side effect was potassium depletion. Benzothiadiazines excrete slightly more potassium than needed to maintain a normal balance of ions in the body fluids. Other side effects are hyperglycemia (high levels of blood sugar) and hyperuricemia (high levels of uric acid in the blood).

Attempts to develop another sulfonamide free from these side effects failed, and investigators concentrated their efforts on other chemical classes. In 1961 Charles M. Kagawa and his associates at Searle discovered spironolactone (Aldactone), an antagonist of aldosterone, a hormone that retains sodium and water and increases the excretion of potassium. As an inhibitor of aldosterone, spironolactone increases the excretion of sodium but tends to retain potassium. Spironolactone was patented in 1961 and has been used since then to treat heart failure, liver cirrhosis, and nephrotic syndrome. Its most common side effect is hyperkalemia (high levels of plasma potassium), an effect opposite to that of benzothiadiazines. The use of spironolactone is limited, however, because of its relatively weak diuretic activity and because of the observation that its long-term use in rats has been shown to increase the incidence of benign tumors.

The search for diuretics that do not deplete potassium led to the development of triamterene (Dyrenium) at Smith Kline and French (now SmithKline Beecham) and to amiloride (Midamor) at Merck Sharp and Dohme. Neither of the two compounds is a sulfonamide and both are potassium retaining. Their site of action in the kidney is different from that of benzothiadiazines, furosemide, or spironolactone. Their major use is in combination with hydrochlorothiazide or other benzothiadiazines. The fixed combination of triamterene with hydrochlorothiazide is marketed as Dyazide and that of amiloride as Moduretic. These combinations have little or no effect on plasma potassium. They are used in patients with hypertension or heart failure, who developed hypokalemia (low levels of blood potassium) while receiving hydrochlorothiazide or another benzothiadiazine.

Among nonsulfonamide diuretics, ethacrynic acid (Edecrin), a high efficacy drug, is worthy of note. It was discovered and developed at Merck in 1962 by Karl Beyer and his associates. Ethacrynic acid is a completely novel chemical structure; no similar compounds were ever used as diuretics. Unlike furosemide, ethacrynic acid was not successful commercially, possibly because of the gastrointestinal side effects (occasional vomiting and diarrhea).

CARDIAC GLYCOSIDES

Cardiac failure is the most debilitating cardiovascular disease. Heart muscle is weakened because of insufficient blood supply or excessive workload (or both), often as the consequence of atherosclerosis, hypertension, or other diseases. It is the most common cause of death in the elderly. The main therapeutic approaches are cardiac stimulants (primarily digitalis glycosides) or drugs capable of reducing cardiac work by peripheral vasodilation or other means.

Although most of the currently used drugs were discovered during the last fifty years, some of the glycosides and alkaloids have been extracted from plants for many centuries. Among them are digitalis glycosides, the discovery of which is usually attributed to William Withering, an English physician who, in 1785, described the use of foxglove for treating dropsy and other diseases. Foxglove, which contains digitalis glycosides, was actually mentioned as early as 1250 in the writings of Welsh physicians. Digitalis glycosides were used in the treatment of heart failure for centuries without a clear understanding of their mechanism of action. Only in 1940 was a definitive proof of their direct action on cardiac muscle provided, when two American pharmacologists, Harry Gold and McKeen Cattell, demonstrated that one of the digitalis glycosides, ouabain, increases the contractile force of isolated cat heart papillary muscles.

At the cellular level, however, the mechanism of the cardiac stimulant action of digitalis glycosides remained speculative. Then Julius Allen and Arnold Schwartz, and subsequently Tai Akera and Theodore Brody, described inhibition of cardiac adenosine triphosphatase, an enzyme controlling sodium and potassium ion transport in the cardiac cell membrane (also called the sodium pump, or Na^+,K^+-ATPase), by ouabain and attributed digitalis-induced increase in the cardiac contractile force to the inhibition of the sodium pump. In the United States digoxin is the most popular cardiac glycoside, in part because of the availability of its reliable formulation, Lanoxin, from Burroughs Wellcome (now Glaxo Wellcome).

ANTIANGINAL DRUGS

Angina pectoris is defined in *Dorland's Medical Dictionary* as a paroxysmal thoracic pain, often radiating to the arms, particularly the left, and sometimes accompanied by a feeling of suffocation and impending death. It is usually caused by coronary artery disease, although the initial description of the clinical manifestations of angina pectoris did not relate them to the coronary artery. Only at the end of the nineteenth century were the consequences of the occlusion of the coronary arteries realized and described. Drugs used in the therapy of angina pectoris are usually referred to as antianginal, which can be subdivided into those used to abort an acute anginal attack and those used continually to prevent or reduce the incidence of anginal attacks. Some nitrites (esters of nitrous acid) and nitrates (esters of nitric acid) are capable of stopping anginal attacks. Amyl nitrite and nitroglycerin are used for this purpose.

Nitrites and Nitrates

Amyl nitrite was synthesized in 1844 at the Sorbonne in France. Its usefulness in treating anginal attacks was not demonstrated until 1867, by Thomas Brunton of Edinburgh, who based his use of amyl nitrite on his experience in treating an anginal attack by venesection. Observing that blood removal reduced the severity of, or even stopped, the anginal attack and lowered arterial pressure, Brunton assumed that the lowering of blood pressure per se would abolish the chest pain. When he learned that amyl nitrite reduced arterial pressure in animals, he expected it to abolish chest pain. As so often happens in drug research, the rationale was wrong, but the drug was effective. Within seconds after inhaling amyl nitrite, the severe chest pain that Brunton's patients were experiencing disappeared. Brunton's experience led to the universal acceptance of amyl nitrite to treat angina pectoris attacks.

Brunton tried other nitrites and nitrates and found that some of them had effects similar to those of amyl nitrite. Glyceryl trinitrate, also called nitroglycerin, was among them. In the initial trial nitroglycerin produced severe headache, so Brunton did not pursue this drug further. In 1877 a British physician from Westminster Hospital, William Murrell, tested nitroglycerin on himself and found that its effect persisted for a longer period than that of amyl nitrite, although the onset of nitroglycerin action was somewhat slower, a few minutes instead of a few seconds. Murrell also found that nitroglycerin produced headache only at doses higher than those required for its therapeutic action. After treating patients with nitroglycerin, Murrell published his findings in *Lancet* in 1879, and nitroglycerin gradually replaced amyl nitrite as the drug of choice for the treatment of anginal attacks.

Today twenty-two different formulations of nitroglycerin are on the U.S. market, including three patented transdermal delivery systems (Deponit, Minitran, Nitro-Dur). The transdermal systems are designed to deliver the drug at a slow and constant rate and are more useful for preventing anginal attacks than for treating them. The development of tolerance to nitroglycerin is the major limiting factor in the use of transdermal delivery systems. To prevent tolerance from developing, the transdermal patches should not be applied for longer than twelve hours each day. When these patented delivery systems were introduced, the market for nitroglycerin expanded significantly.

Because nitroglycerin is capable of dilating coronary arteries, it was thought that arterial coronary vasodilation was the sole mechanism of its antianginal action. But more recently it became clear that nitroglycerin also dilates veins and that venous dilation contributes significantly to the mechanism of its antianginal action. By virtue of dilation of systemic veins the amount of blood returned to the heart is diminished, and cardiac work is reduced. The effects of nitroglycerin are thought to be mediated by the formation of a free radical, nitric oxide, a gas that causes vasodilation by activating an enzyme, guanylate cyclase. This enzyme controls contractile proteins and, therefore, vascular tone in coronary and systemic circulation.

β-*Adrenoceptor Antagonists as Antianginal Drugs*

In addition to transdermal nitroglycerin delivery systems two important groups of drugs are used for the continuous therapy of angina pectoris: β-adrenoceptor antagonists (β-blockers) and calcium channel antagonists (calcium antagonists or calcium blockers). Both groups of drugs are useful in treating hypertension, and some drugs in both groups also have antiarrhythmic properties. Beta-adrenoceptor antagonists are recommended for patients with exertional (exercise-induced) but not vasospastic angina. These antagonists are effective because they reduce myocardial oxygen consumption; they have no coronary vasodilator activity and cannot antagonize vasospasms. The first β-adrenoceptor antagonist used clinically, and still the most popular among them, is propranolol (Inderal). The FDA has also approved nadolol (Corgard) and metoprolol (Lopressor) for use in the chronic therapy of angina pectoris. Nadolol is longer acting than propranolol, while metoprolol is cardioselective (selective for a subgroup of β-adrenoceptors) and is less likely to produce bronchospasms, an occasional side effect of propranolol.

Timolol (Blockadren) was approved for the treatment of myocardial infarction after a Norwegian Multicenter Study demonstrated that the drug reduces mortality and reinfarction in patients who survive the first heart attack. Another clinical study found that timolol reduces the size of myocardial infarction. Similar clinical studies were successfully performed with other β-adrenoceptor antagonists. The fact that the FDA has not approved other drugs for the therapy of angina does not necessarily indicate their ineffectiveness. Either the proper clinical studies have not been performed or the pharmaceutical companies have not applied for approval, most likely because the cost of the required studies and marketing are too high to justify the introduction of still another β-adrenoceptor antagonist without clear-cut clinical advantages.

Other Antianginal Drugs

Among other drugs commonly used for the chronic therapy of angina pectoris are calcium channel antagonists. According to Albrecht Fleckenstein, the first "specific" calcium antagonists were prenylamine (Segontin) and verapamil

(Isoptin, Calan, Verelan). Both drugs were discovered and developed in Germany as coronary vasodilators (at Hoechst and Knoll, respectively) and were recognized by Fleckenstein as being capable of inhibiting calcium entry into the cardiac and vascular smooth muscle cells. Fleckenstein was also the first to recognize the calcium channel antagonist property of nifedipine (Adalat, Procardia), a dihydropyridine-type calcium channel antagonist, first described as a coronary vasodilator. G. Mabuchi and other Japanese cardiologists found nifedipine to be highly effective in preventing attacks of variant angina. Nifedipine was also shown to be effective in treating exertional angina pectoris. For more extensive discussion of calcium channel antagonists and their use in other diseases, see the section on antihypertensive drugs.

ANTICOAGULANTS AND THROMBOLYTICS

Drugs that prevent formation of blood clots are called anticoagulants. Blood clots (also called coagula) are semisolid masses of blood elements that can be formed either in vitro or in vivo. In the body, clots are often formed in veins and can be transported in blood to the heart and lungs. When they occlude smaller lung vessels and interfere with the circulation of blood, they are called pulmonary emboli. Anticoagulants are often used to prevent the formation of pulmonary emboli.

Drugs that prevent the formation of thrombi are called antithrombotics. Thrombi, like clots, consist of blood elements, but these are mostly platelets and fibrin and, unlike clots, adhere to the vessel wall. Thrombi cannot form in vitro because a damaged blood vessel wall is a prerequisite for their formation. Secondary clotting at the site of the thrombus leads to eventual vessel occlusion. Anticoagulants and antithrombotic drugs can prevent it. The drugs that prevent platelet aggregation or their adherence to the vessel wall have antithrombotic properties with or without anticoagulant activity. Some of the more recently developed drugs not only prevent, but also lyse, or dissolve, fibrin and thrombi; they are therefore called fibrinolytic or thrombolytic drugs. Anticoagulants are more likely to cause excessive bleeding than are antithrombotic or thrombolytic drugs, which can prevent interaction of blood elements with the vascular wall without any significant blood loss.

Anticoagulants

For the last thousand years physicians practiced anticoagulant therapy using blood-sucking leeches. Removal of blood was considered therapeutic for a variety of diseases. Not until the first quarter of the nineteenth century was it recognized that leeches, while sucking blood, were releasing an anticoagulant. British physician J. B. Haycraft discovered the anticoagulant, which he named hirudin. In the 1950s it was isolated from the homogenized heads of leeches and characterized by Fritz Markwardt of the University of Erfurt in Germany. It is a small protein (sixty-five amino acids) that inhibits the enzyme thrombin.

Recombinant hirudin can now be produced by various biological systems, and its use is likely to increase.

Currently the most commonly used anticoagulant is heparin. It was discovered and named in 1922 by William Howell of Johns Hopkins University, who isolated the material from the liver. It was only 2 percent pure, and attempts to use it in the clinic failed because of side effects, probably caused by impurities. In 1928 Charles Best at the University of Toronto discovered that an anticoagulant is present in many different tissues and identified bovine lung as its best source. Best also called his material heparin, although it appeared to be different from Howell's. The international standard for heparin was established in 1935, and its clinical evaluation was conducted in Toronto and Stockholm. Heparin was found to consist of polymers of two repeating disaccharide units. It is still produced commercially from bovine lung and porcine intestinal mucosa, although semisynthetic "heparinoids" have been shown to have anticoagulant activity. Heparin acts at multiple sites in the coagulation system; at low concentrations it enhances the activity of an antithrombin factor. In 1937 Best used heparin to prevent thrombosis during blood transfusion; during World War II heparin's use for that purpose saved many lives. Current indications for heparin include treating and preventing thrombosis and preventing clot formation in cardiac and arterial surgery as well as in blood transfusion.

The discovery of anticoagulant activity in a series of chemical substances, known as coumarins, was purely accidental. In the early 1920s a strange disease that caused cattle to bleed to death broke out in North Dakota and Alberta. Alberta veterinarian F. W. Schofield traced the cause of the disease to hay containing spoiled sweet clover. Many attempts to isolate an anticoagulant from sweet clover failed, until 1939 when Harold A. Campbell from the University of Wisconsin isolated crystals of a potent anticoagulant, a derivative of coumarin. This anticoagulant, identified as one of the substances responsible for the hemorrhagic disease of cattle, was named dicumarol (Dicoumarol, Dicumol, Melitoxin). The drug was first marketed in 1942 after successful conclusion of clinical studies at the Mayo Clinic. The mechanism of the anticoagulant activity of dicumarol was subsequently identified as antagonism of vitamin K in its effect on the biosynthesis of clotting factors in the liver. Dicumarol is still used to prevent and treat deep vein and pulmonary thrombosis and after implantation of artificial heart valves. Many related substances were synthesized in the laboratory of Karl Link at the University of Wisconsin. Link named one of the most potent and successful compounds warfarin (for Wisconsin Alumni Research Foundation, which received royalties for the discoveries made in Link's laboratory). Initially used as a rat poison, warfarin was eventually marketed for human use as an anticoagulant by Endo Laboratories under the trade name of Coumadin. Warfarin is indicated for the prophylaxis or treatment of venous thrombosis and pulmonary embolism and as an adjunct in the prophylaxis of systemic embolism after heart attack.

Antithrombotics

In the 1990s low-molecular-weight heparins were introduced as antithrombotics, to prevent deep-vein thrombosis in patients undergoing abdominal or other major surgery. These drugs are heparin fragments or chemically related substances with little or no anticoagulant activity in vitro; they are still capable of preventing thrombosis in vivo, but have less tendency than heparin to cause systemic bleeding. The best known low-molecular-weight heparins are daltepan sodium (Fragmin), enoxoparin sodium (Lovenox), and danaparaoid sodium (Orgaran). In 1997 the United States was the first country to approve the Wyeth-Ayerst product ardeparin sodium (Normiflow) for specific use in the prevention of thrombosis after knee replacement surgery.

Drugs that prevent platelets from aggregating or adhering to vessel walls, called antiplatelet agents, represent still another approach to the prevention of thrombosis. The leading drug in this class is still aspirin. One tablet of aspirin a day can reduce the chances of coronary artery thrombosis. The major problem in the development of new platelet aggregation inhibitors is the need to demonstrate in experimental as well as in clinical studies that the new drug is more effective than aspirin. Two other antiplatelet agents are currently marketed in the United States for specific indications. They are ticlopidine (Ticlid), a product of Roche Pharmaceuticals, indicated for the prevention of thrombotic stroke, and dipyridamole (Persantine), a Boehringer Ingelheim product, indicated as an adjunct to anticoagulants for the prevention of thrombosis after cardiac valve replacement. Another synthetic antithrombotic likely to be marketed in the near future is a Sanofi product, clopidogrel. It is at least as effective as ticlopidine. A new antithrombotic biotechnology product, abciximab (ReoPro), was developed by Centocor and marketed by Eli Lilly. It is an antibody to a specific glycoprotein receptor (known as GPIIb/IIIa) on the platelet surface. It inhibits platelet aggregation and is indicated as an adjunct to aspirin and heparin in the prevention of closure of coronary arteries after angioplasty. DuPont and other companies are currently developing small synthetic inhibitors of this glycoprotein receptor.

Thrombolytics

The currently used thrombolytics or fibrinolytics act by catalyzing the formation of plasmin, an enzyme that degrades fibrin, a protein that forms the essential portion of the blood clot. One of these thrombolytics is urokinase (Abbokinase), a proteolytic enzyme, which was originally isolated from human urine and is normally produced by the kidney. It is primarily used to lyse pulmonary emboli (a mass of clotted blood brought from another vessel), but it is also used in coronary thrombosis to restore patency of coronary vessels and to reduce damage to the heart muscle caused by prolonged vessel occlusion.

Another thrombolytic agent is streptokinase (Streptase, Kabikinase), a bacterial enzyme, which acts on plasmin indirectly by forming a complex with plasminogen, the precursor of plasmin. The complex has proteolytic activity and also catalyzes formation of plasmin. The major indication for streptokinase is coronary artery occlusion. Streptokinase is expected to lyse intracoronary thrombi. Either of the two antithrombotics also lyse fibrin deposits, which occur at the sites of injury or needle puncture. This lysis could lead to excessive bleeding, so fibrinolytics should not be used for ten days after major surgery or in patients with hemorrhagic diseases.

Attempts to develop a more specific thrombolytic agent that would not produce bleeding led to the development of tissue plasminogen activator (tPA), which is released from endothelial cells and preferentially activates plasminogen adsorbed to fibrin clots. The drug thus lyses clots without causing bleeding from minor wounds. The isolation and production of tPA presented many technical problems, which were solved in the early 1980s by modern biotechnology. Dianne Pennica and her associates reported in 1983 in *Nature* the cloning and expression of a human tissue-type plasminogen activator, cDNA, in *Escherichia coli.* A year later Frans Van de Werf and colleagues reported in the *New England Journal of Medicine* the efficacy of recombinant tissue plasminogen activator in lysing coronary thrombi in patients with evolving myocardial infarction.

The recombinant tPA (alteplase, or Activase), marketed by Genentech, is used to treat pulmonary embolism and acute myocardial infarction. The reduction of sudden death in heart attack victims was clearly demonstrated in numerous clinical trials, although some studies failed to demonstrate a significant advantage of alteplase over streptokinase. The development of alteplase is considered to be one of the major accomplishments of the biotechnology industry in the last decade.

In 1989 A. Stern and his associates from Boehringer Mannheim in Germany filed a patent on a new plasminogen activator with a longer half-life than alteplase. It was produced by deletion of certain parts of alteplase molecule (N-terminal domains). In animal studies the new plasminogen activator was more potent and more effective and had more rapid onset of action than alteplase. Three controlled clinical trials confirmed its efficacy and superiority over streptokinase. The plasminogen activator was named retaplase and marketed in the United States under the trade name Retavase. It is indicated in the management of acute myocardial infarction.

HYPOLIPEMIC DRUGS

Atherosclerosis, the leading cause of death in the United States, is a form of arteriosclerosis, a thickening and hardening of arterial wall. Atherosclerosis is a disease of major arteries involving deposition in the arterial wall of plaques containing lipids. The lipids deposited are usually directly proportional to the concentration of lipids—cholesterol and triglycerides—in the plasma. The lipids in the plasma are transported in the form of lipoproteins (complexes of cholesterol, its esters, and triglycerides with proteins). A condition that involves an increased concentration of lipids in the blood is called hyperlipoproteinemia, and drugs that lower plasma lipids are called hypolipemic, hypolipidemic, or hypolipoproteinemic drugs.

Most of the cholesterol in the plasma is transported in the form of a complex with low-density lipoproteins (LDLs). Free cholesterol is transported in the form of a complex with high-density lipoproteins (HDLs), which facilitate the removal of cholesterol from the tissues. Conversely, LDLs control cholesterol's deposition. The elevation of LDL cholesterol leads therefore to atherosclerosis, while the elevation of HDL is viewed as beneficial. The normal levels of plasma lipids are arbitrarily defined; individuals with concentrations of plasma lipids higher than 95 percent of the population are considered hyperlipoproteinemic. The need to lower plasma lipids was a controversial issue until 1984, when the completed Coronary Primary Prevention Trial provided strong evidence that a decrease in blood LDL cholesterol can reduce the risk of coronary heart disease.

Niacin

Drugs can lower plasma cholesterol levels either by decreasing the production of lipoproteins or by facilitating their removal from the plasma. The first drug shown to decrease production of lipoproteins was niacin (nicotinic acid; Niacor, Nicobid, Nicolar). Nicotinic acid was synthesized in 1897 by Alfred Ladenburg of the University of Kiel, Germany. Its physiological role as one of the B vitamins was not recognized until 1937, when Conrad Elvehjem, of the University of Wisconsin, used niacin to cure "black tongue" (vitamin B deficiency) in dogs.

The lipid-lowering property of niacin is not related to its role as a vitamin. This effect was discovered only in 1955 by Rudolf Altschul and his associates. The mechanism of action of niacin as a hypolipemic is not precisely known, but it is likely to be related to its ability to break down lipids in the adipose tissue (lipolytic action) or to interfere with the formation of triglyceride esters in the liver. Niacin's usefulness in treating hyperlipoproteinemia is limited by its side effect: intense flushing and itching of the skin.

Clofibrate

In 1962 J. M. Thorp and W. S. Waring reported in *Nature* that clofibrate (Atromid-S) lowered total lipid and cholesterol levels in the blood of rats. The mechanism of action of clofibrate has not yet been conclusively established. It is thought to inhibit the release of lipoproteins from the liver or to enhance the activity of an enzyme, lipoprotein lipase, that degrades lipids. Clofibrate was found to be tumorigenic (increasing the number of all spontaneous tumors) in rats; it also increases bile stone formation in humans.

In a large clinical study (five thousand patients) sponsored by the World Health Organization, in which the subjects received clofibrate or a placebo for five years, mortality from noncardiovascular diseases was higher in the group treated with clofibrate than in the placebo group. Only in patients with no previous history of heart disease did clofibrate decrease the incidence of subsequent nonfatal heart attack. The FDA thus restricted the use of clofibrate for two indications: one type of hyperlipidemia (type III) that does not respond adequately to diet and to patients with extremely high triglyceride levels (type IV or V hyperlipidemia). These restrictions severely limited the market share of clofibrate, which initially had been highly successful.

Bile Acid–Binding Resins

In 1965 Sami A. Hashim and Theodore B. Van Itallie in New York proposed another approach to lowering blood lipid levels, which involved the use of bile acid–binding resin, or cholestyramine (Questran). Cholestyramine and another resin, colestipol (Colestid), introduced in 1969 by Upjohn, are polymers with high molecular weight. They are not absorbed in the gastrointestinal tract and act by binding bile acids in the intestines. This effect leads to the increased production of bile acids from cholesterol in the liver and consequently to reduction of plasma cholesterol levels.

Bile acids are also required for the intestinal absorption of cholesterol, so the two resins result in larger amounts of cholesterol being excreted in the feces. A more recently proposed mechanism involves increasing the number of hepatic LDL receptors in patients receiving either of the two resins. These receptors tend to retain more LDL cholesterol in the liver and leave less cholesterol for circulation. Because the resins are not absorbed, they represent a safer therapy than clofibrate. They have to be administered, however, at very high doses: twelve to sixteen grams of cholestyramine and fifteen to thirty grams of colestipol a day. Nausea and abdominal discomfort are common side effects. The resins also interfere with the intestinal absorption of many other drugs.

Gemfibrozil

In a search for a safer clofibrate-like compound Parke-Davis screened more than eight thousand related drugs for lipid-lowering activity in animals. One of the compounds found effective in this screening operation was gemfibrozil. It was synthesized in 1968, its evaluation was completed in 1976, and it was finally marketed as Lopid in the United States in 1986 by Parke-Davis. Although its mechanism of action is not completely understood, it seems to differ from that of clofibrate. Unlike clofibrate, gemfibrozil elevates plasma concentrations of HDL cholesterol. It is also much more effective in lowering plasma triglycerides than clofibrate. One theory is that gemfibrozil acts by inhibiting the release of very-low-density lipoproteins from the liver. In a five-year clinical study (the Helsinki Heart Study) gemfibrozil significantly reduced the rate of serious coronary events, but not the total mortality rate from all causes. In a second study the drug was beneficial as a preventive therapy in patients with no coronary disease. Like clofibrate, gemfibrozil may enhance the formation of bile stones.

Inhibitors of Cholesterol Synthesis

During the last thirty years numerous attempts were made to control plasma cholesterol levels by inhibiting cholesterol's biosynthesis. In the early 1960s Merrell and Company developed triparanol (MER-29), which inhibited the synthesis of cholesterol but led to the accumulation of desmosterol (another steroid formed from cholesterol) with many undesirable side effects, including formation of cataracts. Triparanol was withdrawn and is not even mentioned in current textbooks of medicine or pharmacology.

In 1967 Akido Endo and his associates at Sankyo Laboratories in Tokyo discovered a new class of cholesterol biosynthesis inhibitors, which inhibited an enzyme called 3-hydroxy-3-methylglutaryl-coenzyme A reductase, or HMG-CoA reductase. The enzyme controls the rate-limiting step in the biosynthesis of cholesterol. The first reported compound was isolated from cultures of *Penicillium citrinum*. It was shown to be a highly potent inhibitor of the enzyme and to reduce cholesterol biosynthesis in vitro and in vivo. The drug was named metastatin (CS-500, ML-236B, Compactin). Its effectiveness in reducing plasma cholesterol levels in patients with primary hypercholesteremia was reported in 1980 by Akira Yamamoto, Hiroshi Sudo, and Endo. In 1979 Endo had isolated another inhibitor of HMG-CoA reductase from a culture of *Monascus ruber* and named it monacolin K, while Alfred W. Alberts and his associates at Merck had isolated from cultures of *Aspergillus terreus* a compound they called mevinolin (lovastatin, Mevacor), which turned out to be identical to monacolin K. The total synthesis of metastatin and lovastatin has since been accomplished, and many analogs have been synthesized.

Merck gave high priority to the development of lovastatin, which became the first inhibitor of HMG-CoA reductase approved for clinical use. In subsequent studies lovastatin was shown to be a pro-drug that is converted to corresponding β-hydroxy acid form. (A pro-drug is a substance that is converted to an active principle in the body.) The active form of the drug inhibits cholesterol synthesis in the liver; this inhibition triggers a compensatory increase in the synthesis of LDL receptors. The uptake of LDL cholesterol by its hepatic receptors helps reduce plasma cholesterol. The main clinical indication for lovastatin is as an adjunct to the diet of patients with primary hypercholesteremia (types IIa and IIb). At the optimal dose level lovastatin can reduce plasma LDL cholesterol by up to 41 percent.

The scientists at Sankyo Laboratories continued their work with HMG-CoA inhibitors and selected pravastatin (CS-514, Pravachol). Pravastatin is a urinary metabolite of metastatin in dogs and was initially obtained by microbial transformation of metastatin. It has been shown to be tissue selective for liver and intestines, the major sites of cholesterol formation. Its indications are similar to those of lovastatin.

At Merck the development of lovastatin was followed by that of simvastatin (MK-733, synvinolin, Zocor). Simvastatin is twice as potent as lovastatin but very similar in its mechanism of action and clinical efficacy. At least six other inhibitors of HMG-CoA reductase are now in various stages of development by different companies.

In the 1998 *Physicians' Desk Reference* three more "statins" are listed: cerivastatin (Baycor), fluvastatin (Lescol), and atorvastatin (Lipitor). Cerivastatin is the most potent of all available statins. In the American clinical trial of cerivastatin a total dose of 0.3 milligrams per day was shown to be effective in lowering LDL cholesterol, total cholesterol, and triglyceride levels. The next most potent inhibitor of HMG-CoA reductase approved for use in the United States is atorvastatin; its clinical dose is 10 milligrams. The mechanism of action and the clinical indications of all statins are identical, and all marketed compounds of this type have been associated with occasional biochemical abnormalities of liver function tests.

Drugs Affecting the Central Nervous System

The human nervous system is usually subdivided into central and peripheral systems. The central nervous system (CNS) consists of the brain and spinal cord, while nerves in the rest of the body belong to the peripheral system. Sleep, pain, behavior, thought process, motor activity, and many other important human functions are controlled by the central nervous system and can be selectively affected by various drugs. Sleep was first controlled with hypnotics, consciousness with anesthetics, and pain with analgesics. Abnormal thought processes have been controlled more recently with antianxiety, antipsychotic, or antidepressant drugs. Diseases associated with abnormal motor functions, such as Parkinson's disease and epilepsy, can now also be influenced by drugs.

ANESTHETICS

One of the definitions of anesthesia is loss of the ability to feel pain. General anesthesia also involves temporary loss of consciousness or inducement to sleep (hypnosis), while local anesthesia abolishes pain peripherally by acting at the nerve endings without affecting the central nervous system. As discussed below, anesthetics were discovered accidentally. A systematic search for better anesthetics did not begin until the twentieth century, and the scientific understanding of the mechanism of action of anesthetics is still incomplete.

Chloral Hydrate

Chloral hydrate, which was synthesized by Justus von Liebig in 1832 in Germany, was the first clinically useful sleep-inducing (hypnotic) drug. These properties, however, were not recognized until 1861, when the first professor of pharmacology, Rudolf Buchheim, experimented with chloral hydrate in his laboratory in Dorpat (now Tartu, Estonia). He and his colleagues fell asleep promptly after ingesting this substance. Eight years later Oscar Liebreich in Berlin, unaware of Buchheim's experience, noted that chloral hydrate could produce unconsciousness in animals and recommended it for clinical use as a hypnotic, attributing its effect to the formation of chloroform in the blood. This mechanism of action was questioned, and Josef von Mering suggested in 1881 that chloral hydrate may be metabolized in the liver to a hypnotic alcohol, trichloroethanol, rather than chloroform. Not until 1948, however, was this metabolic pathway confirmed experimentally by an American investigator, Thomas Butler. Despite its unpleasant odor and tendency to produce gastric irritation, respiratory depression, and cardiac toxicity at higher doses, chloral hydrate is still used today as a hypnotic for short-term procedures.

The discovery of the hypnotic action of chloral hydrate is of major historical importance because it led to the development of other superior hypnotics. Buchheim's student, Oswald Schmiedeberg, introduced urethane in 1885; A. Heffter discovered chloralose in 1889; and numerous barbiturates were introduced during the first part of the twentieth century.

Barbiturates

The discovery of the hypnotic properties of the substituted derivatives of barbituric acid can be traced to the beginning of the twentieth century, when 5,5-diethylbarbituric acid was shown to have hypnotic activity in animals. It was patented by the famous German chemist Emil Fischer in 1903 and subsequently marketed by Bayer under the trade name of Veronal. During World War I, 5,5-diethylbarbituric acid was given the generic name of barbital in the United States; the same substance was called barbitone in the United Kingdom. In 1911 Fischer synthesized and Bayer marketed another hypnotic derivative of barbituric acid, phenobarbital (phenobarbitone) under the trade name of Luminal. After World War I many new barbiturates were synthesized and marketed, including amylobarbital (Amytal) and secobarbital (Seconal) by Eli Lilly, and sodium pentobarbital (Nembutal) by Abbott Laboratories.

A rapid onset of action is highly desirable for an intravenous anesthetic so that an anesthesiologist can precisely control the dosage of a drug. In response to this need IG Farben introduced hexobarbital under the trade name of Evipan. Further attempts to improve the safety of anesthetics led to the synthesis and introduction by Abbott Laboratories of thiopental sodium (Pentothal), an ultra-short-acting anesthetic. An initial dose lasts only a few minutes, but repeated administration results in tissue saturation and occasional dangerous increases in blood levels. This problem, common in many ultra-short-acting anesthetics, has not yet been overcome.

Other General Anesthetics

Since the 1960s general anesthetics were not of great interest to the pharmaceutical industry, and few compounds were introduced into medical practice. Only two additional anesthetics should be mentioned here. Both of them are approved for use in the United States and are currently marketed. Propofol

(Diprivan), a product of Zeneca, was released for general use in 1989. It is chemically unrelated to any other anesthetic agents. In recent years it has become widely adapted for the induction and maintenance of anesthesia. It does not impair kidney or liver functions but can severely depress respiration and lower arterial pressure. Midazolam (Versed), a product of Roche Laboratories, is a short-acting benzodiazepine, chemically related to diazepam, which is used intravenously for induction of general anesthesia as well as a preanesthetic medication to produce sedation.

Local Anesthetics

The first attempts to develop a local anesthetic for skin surgery were made by Bernard W. Richardson in England during the 1860s. Richardson first experimented with cooling liquids and then with ether ethyl bromide and ethyl chloride sprays. The first local anesthetic was cocaine. In 1884 Carl Koller in Vienna conducted the initial experiments with cocaine for use in ophthalmology. Since then the use of cocaine as a local anesthetic has become widely accepted.

Gradually the toxicity and addictive properties of cocaine were discovered, and extensive structure-activity studies were undertaken in search of new local anesthetics. One of the most active laboratories in this field at the beginning of the twentieth century was that of Albert Einhorn at the University of Munich. His work led to the introduction of several local anesthetics, including procaine, which was patented by Hoechst as Novocain in 1906, although its efficacy and duration of action were not yet optimal for either skin or dental surgery. Another finding of considerable importance was Heinrich Braun's discovery that epinephrine can enhance and prolong the action of local anesthetics.

The continuing search for superior local anesthetics led to the discovery of lidocaine (lignocaine) in 1946 by Nils Löfgren and Bengt Lundqvist at the University of Stockholm. Lidocaine, marketed in 1948 by Astra under the trade name Xylocaine, has a more rapid onset of action and is safer and less irritating than older local anesthetics, including procaine. Because of these advantages lidocaine rapidly dominated the local anesthetic market. Other local anesthetics were introduced later, but they offered only minor advantages over lidocaine. Among them is bupivacaine (Marcaine, Sensorcaine), which has a longer duration of action than lidocaine (five to seven hours) and is occasionally preferred to lidocaine for dental procedures or infiltration anesthesia. Two other long-acting local anesthetics introduced in the 1970s are etidocaine (Duranest) and prilocaine (Citanest). Etidocaine is often preferred for spinal or epidural anesthesia, while prilocaine is used almost exclusively in dental procedures.

Attempts to further improve local anesthetics continued in the 1980s, and Astra introduced two more agents that were eventually approved in the United States and are currently marketed. They are ropivacaine (Naropin) and chlorprocaine (Nesacaine). Ropivacaine was the result of a search for a less cardiotoxic drug than bupivacaine. It is suitable for epidural as well as regional anesthesia and is used in obstetrics for cesarean section and major nerve block. Chlorprocaine is recommended for infiltration and not for epidural anesthesia. Its onset of action is rapid (six minutes) and the duration of action does not exceed sixty minutes. The general tendency in the development of local anesthetics is to design different drugs for each surgical procedure.

Volatile Anesthetics

The first volatile (rapidly evaporating) anesthetics were introduced in the middle of the nineteenth century. The discovery of ether is usually attributed to William G. Morton, a Boston dentist who successfully demonstrated the use of ether at Massachusetts General Hospital on 16 October 1846. The first physician to use ether in surgery was probably Crawford Long from Jefferson, Georgia. He excised a growth from the neck of a patient using ether to prevent pain. Long did not share his experience with the medical community until 1849.

The first use of nitrous oxide in dental surgery is attributed to Horace Wells in Hartford, Connecticut. Wells attended a public lecture, which included a demonstration of nitrous oxide as a laughing gas. A volunteer who had inhaled nitrous oxide as part of the demonstration accidentally injured his leg and felt no pain. Wells invited the lecturer, Gardner Colton, to administer nitrous oxide to him while his molar tooth was extracted by another colleague. This operation was successfully accomplished on 11 December 1844.

The discovery of the anesthetic properties of chloroform is attributed to James Y. Simpson, professor of obstetrics at the University of Edinburgh. He tested chloroform on himself and some of his friends and reported his findings to the Medical-Surgical Society of Edinburgh on 10 November 1847. Simpson found that chloroform had advantages over ether: It had a more rapid onset of action and higher potency, was more pleasant to inhale, and was simpler to administer.

Many other volatile anesthetics were discovered during the next fifty years, including ethylene, propylene, vinyl ether, trichloroethylene, and cyclopropane. None of them was ideal, although all of them were introduced into medical practice.

World War II contributed to the next step in the development of volatile anesthetics. Among the many compounds synthesized during the war for various purposes were stable fluorocarbons, initially made as potential solvents for uranium salts. After the war B. H. Robbins evaluated stable fluorocarbons in the pharmacology laboratory at Vanderbilt University and discovered that some of them had anesthetic activity. The use of cauterization techniques in surgery had increased the danger of explosion in operating rooms, and new nonexplosive anesthetics were urgently needed. Ohio Medical Products synthesized many new fluorocarbons in an attempt to develop a nonflammable volatile anesthetic superior to ether. This work led to the discovery of fluroxene, by John Krantz of the University of Maryland, which was marketed in 1956.

The work on fluorocarbons was also pursued in the United Kingdom by

Charles Suckling and James Raventos at the ICI laboratories. Suckling and Raventos were looking for a volatile nonexplosive anesthetic with low toxicity, rapid onset of action, high potency, and absence of respiratory irritation. In January 1953 they discovered halothane (Fluothane), which quickly became the inhalation anesthetic of choice, capturing and retaining the larger portion of the anesthetic market despite the introduction of such competitive products as methoxyflurane (Penthrane). Halothane, distributed in the United States by Wyeth-Ayerst, is recommended for induction and maintenance of general anesthesia. It has occasional adverse effects, which include hepatic dysfunction, hypotension, cardiac arrhythmias, and even cardiac arrest. Halothane has also been found to have teratogenic, embryotoxic, and fetotoxic properties in animals, so attempts to develop even safer anesthetics were not abandoned. In 1990 another volatile anesthetic was introduced—desflurane (Suprane), marketed in the United States by Ohmeda. It is a precisely controllable anesthetic with rapid onset and short duration of action and is particularly desirable for short office procedures. It does not impair kidney or liver functions but does depress respiration.

ANALGESICS

To control severe pain, physicians still depend largely on opioids, a group of drugs that are opium- or morphine-like in their properties. The word *opium* was derived from the Greek name for juice, because opium is obtained from the juice of the poppy, *Papaver somniferum*. The first undisputed reference to poppy juice can be found in the writings of Theophrastus, a Greek philosopher in the third century B.C. Paracelsus is credited with the reintroduction of opium in Europe in the sixteenth century. Originally opium was used to treat dysentery rather than as an analgesic. The behavioral effects of opium were particularly appreciated in the Orient, where opium smoking became popular in the seventeenth century.

Morphine and Related Drugs

The first of at least twenty alkaloids present in opium was isolated in 1806 and named morphine, after Morpheus, the Greek god of dreams. Its structure was described in 1925 by John Gulland and Robert Robinson. Many derivatives of morphine have been synthesized during the last fifty years, but none has replaced morphine. Extensive efforts have been undertaken to find a nonaddictive morphine derivative that retains its analgesic properties. In 1929 the U.S. government initiated a search for such a drug at the Universities of Virginia and of Michigan. Researchers spent ten years and millions of dollars but were unable to find the desired compound. The best compound derived from this project was methyldihydromorphinone, which was three times more potent than morphine and less likely than morphine to produce drowsiness or nausea, but which was still addictive. It was not marketed because it was difficult to manufacture.

At Hoffmann–La Roche in Switzerland another more potent derivative of morphine, levorphanolol, was discovered and marketed under the trade name of Dromorane. Its stereoisomer is devoid of analgesic and addictive properties but is highly effective as a cough suppressant; its generic name is dextromethorphan. Currently available over the counter in combination with antihistamines and other drugs, it is listed in the *Physicians' Desk Reference* as an ingredient in twenty-seven preparations.

In addition to being addictive, morphine and many related agents have other side effects as well. Of considerable concern is the respiratory depressant property of morphine. For that reason morphine and related drugs are contraindicated for patients with bronchial asthma and for use in childbirth, where such nonopioid analgesics as meperidine (Demerol) are preferred. Because morphine and related drugs also produce constipation, they are used, albeit infrequently, to treat diarrhea. They affect intestinal motility at doses lower than required for the control of pain.

In 1898, while looking for an analgesic that did not have respiratory depressant activity, Heinrich Dreser of Bayer discovered heroin; it was originally marketed under the false assumption that it was not addictive. Attempts to find a nonaddictive analgesic among morphine derivatives continued, and in 1942 John Weijard and Alan E. Erickson from Merck described nalorphine as an effective antagonist of the respiratory depressant effects of morphine. Its analgesic properties in humans were demonstrated in 1954 by Louis Lasagna and Henry Beecher of Harvard. Unfortunately it produced hallucinations and was therefore not useful clinically. In 1953 Albert Pohland and his coworkers at Lilly synthesized propoxyphene (Darvon), a compound that is structurally different from opioids but that binds to the same receptors. As an analgesic, propoxyphene is only half as potent as codeine, a methyl derivative of morphine, and it has no significant antipyretic or anti-inflammatory activities. It is less addictive than codeine, but the difference in their addicting properties is not significant if the drugs are compared at the equipotent analgesic doses. Propoxyphene is recommended for the treatment of moderate pain that cannot be adequately controlled by aspirin.

In 1964 Sydney Archer and his associates at Sterling Winthrop Research Institute in Rensselaer, New York, reported the synthesis of pentazocine (Talwin), a "nonaddictive" analgesic. It was less likely to produce addiction than codeine but was also less effective as an analgesic. Nevertheless it represented a step in the right direction. Pentazocine was the first agonist-antagonist among morphine derivatives: It mimicked the effects of morphine at some opioid receptors, while acting as an antagonist at a different subtype of the same receptors. This dual mechanism of action was probably responsible for inducing less psychological dependence in comparison with other opioids. Pentazocine has more rapid onset and shorter duration of action than morphine; it is used primarily for the treatment of moderate pain.

Since the 1960s no major breakthroughs have occurred in the field of analgesics. Mild-to-moderate pain is controlled with aspirin, acetaminophen, and various nonsteroidal anti-inflammatory drugs (see below). In addition to the above-mentioned opioids, hydromorphone (Dilaudid) was introduced to the American market by Knoll (a division of BASF). The drug has a higher relative potency (active at lower doses) by oral administration than morphine, but its side effects are similar. The major hazard is respiratory depression. A transdermal delivery system for the selective μ opioid agonist fentanyl was developed and marketed by Janssen (Duragesic). It is indicated in the management of chronic pain not controllable by nonsteroidal anti-inflammatory drugs. It also depresses respiration.

Among novel approaches to the control of intractable pain is the use of SNX-111, a peptide derived from marine snails and an antagonist of N-type calcium channels. It controls morphine-resistant pain but has to be administered directly into the brain (intrathecally). At the time of this writing SNX-111 has not yet been approved by the FDA.

ANALGESIC ANTAGONISTS

Nalorphine (see above) is only a partial antagonist of morphine. It antagonizes the respiratory depressant effect of morphine, but if administered alone, it can depress respiration as well. In the search for full antagonists devoid of any agonist properties, first naloxone (Narcan) and then naltrexone (Trexan) were synthesized. Naloxone, administered by injection, is indicated for reversal of postoperative depression caused by narcotics as well as in the treatment of narcotic overdose. Naltrexone is given orally to prevent the effects of narcotics and to help maintain a narcotic-free state in already detoxified drug abusers. In individuals using narcotics, naltrexone can precipitate a severe withdrawal reaction. Both antagonists are thought to act at the same opioid receptors as morphine and similar narcotics. There are at least three types of opioid receptors (μ, κ, and δ), with two or three subtypes for each type. Naloxone and naltrexone are not selective for any of these receptor subtypes, although the affinity of these drugs to μ receptors is higher than to other types.

More selective opioid antagonists are currently being developed. They may find clinical applications in a variety of disorders, including constipation, alcoholism, drug addiction, and immunological disorders. The importance of the search for selective antagonists is emphasized by the existence of endogenous peptides, enkephalins, and endorphins that bind preferentially to sigma opioid receptors and were found to abolish pain in animals.

ANTIANXIETY DRUGS

Anxiety is defined as an emotional state, a response to anticipation of imaginary or impending danger. It is a common symptom of many psychiatric disorders, including depression, personality disorders, and many phobias. Because these disorders cannot always be accurately diagnosed, psychiatrists are treating anxiety independently of the underlying disorder with antianxiety, or anxiolytic, drugs.

Muscle Relaxants

The first drug discovered capable of reducing anxiety symptoms was a muscle relaxant, mephenesin (Myanesin). Its discovery is a typical example of serendipity in drug research. When penicillin was introduced during World War II, the need to extend the spectrum of its antibacterial activity became obvious. William Bradley at the British Drug House in London decided to combine penicillin with an antiseptic, phenoxyethanol, which was known to kill bacteria resistant to penicillin. When phenoxyethanol was found to lack sufficient potency, Bradley decided to optimize its activity. He synthesized several glycerol ethers. While studying the toxicity of phenoxyethanol and its derivatives, Frank Berger, a pharmacologist, discovered that these compounds produced an unusual type of paralysis in animals. At doses lower than required for paralysis, these compounds caused relaxation, which Berger called "tranquilization." Bradley and Berger synthesized and tested more than one hundred compounds and selected mephenesin as the optimal choice. In the clinic mephenesin alleviated anxiety without producing anesthesia or loss of consciousness. The major drawback was the drug's short duration of action, which was sufficient for its use as a preanesthetic medication but not for the treatment of anxiety unrelated to anesthesia or surgery.

Berger subsequently moved to the United States and continued his work at Wallace Laboratories in New Jersey. In 1950 he and B. J. Ludwig synthesized and developed meprobamate (Miltown, Equanil), a longer-acting muscle relaxant and antianxiety agent. It rapidly became a large commercial success and is still sold today for short-term relief of symptoms of anxiety. Two other muscle relaxants with antianxiety properties were chlormezanone (Trancopal), developed at the Sterling-Winthrop Research Institute, and dantrolene (Dantrium), developed at Norwich Pharmaceutical Corporation. These two drugs never attained the commercial success enjoyed by meprobamate or later by the benzodiazepines.

Benzodiazepines

Leo Sternbach and Lowell Randall of Hoffmann–La Roche Laboratories in New Jersey are credited with the discovery of the most successful group of antianxiety agents, benzodiazepines. In 1954 Sternbach synthesized a series of tricyclic compounds related to chlorpromazine, a major tranquilizer. Randall, a pharmacologist, tested Sternbach's compounds for muscle relaxant, sedative, and anticonvulsant properties without any success. The last compound synthesized was not even submitted for pharmacological evaluation for more than a year. To everyone's surprise Randall found that this compound was superior to meprobamate as a muscle relaxant and antianxiety agent and that it also had tranquilizer activity. It turned out that Sternbach had used a different intermediate (a primary instead of a secondary amine) in the synthesis of this compound and unintentionally had created a completely different ring system (benzodiazepine instead of quinaxoline oxide). In 1958 a patent was filed for this compound, named chlordiazepoxide (Librium). Successful clinical studies on more that sixteen thousand patients followed before the FDA approved the drug in 1960.

Continuing their research of the series of new benzodiazepines, Sternbach and Randall discovered that certain structural features were not required to produce the desired activity and that some of the chlordiazepoxide analogs were more potent. They selected diazepam as a superior derivative, which was marketed in 1963 as Valium. Many more benzodiazepines followed. Another of Sternbach's compounds, clonazepam, was patented in 1963 and marketed as Klonopin. It has stronger anticonvulsant activity than some of the other benzodiazepines and is used primarily in the treatment of seizures. For the treatment of insomnia, Hoffmann–La Roche introduced flurazepam (Dalmane). Upjohn Company developed and marketed triazolam (Halcion) and alprazolam (Xanax). Wyeth-Ayerst developed oxazepam (Serax). It was estimated in 1993 that twenty-eight different benzodiazepines were on the world market.

Benzodiazepines differ from each other in potency and relative predominance of properties. Diazepam, for example, has stronger analgesic activity, while clonazepam is a stronger muscle relaxant. New benzodiazepines were introduced not for the initial indication, anxiety disorders, but for new indications, such as insomnia, epilepsy, and even depression (alprazolam), which were discovered mostly during clinical trials of initially developed compounds.

At the molecular level benzodiazepines appear to mimic the effects of gamma-aminobutyric acid (GABA), a chemical made by brain cells that transmits messages in the brain. Two types of GABA receptors are currently recognized: GABA-A and GABA-B. GABA-A receptors have two different interaction sites for drugs; benzodiazepines bind to only one of them. The antianxiety effect of benzodiazepines is thought to be associated with the stimulation of GABA-A receptors.

Buspirone

One of the recently developed nonsedative antianxiety agents is buspirone (BuSpar). It was discovered at Bristol-Myers Squibb Company and is marketed in the United States by two divisions of that company, Mead Johnson Pharmaceuticals and Princeton Pharmaceutical Products. Its main indication is in the treatment of anxiety symptoms, including tension, apprehension, and hyperactivity. Some of these symptoms are associated with depression. The clinical experience indicated that buspirone also has antidepressant properties and is useful in treating depression even when no anxiety symptoms are present.

The mechanism of action of buspirone involves its interaction with serotonin (5-HT_{1A} subtype) receptors. At these receptors it behaves as a mixed agonist-antagonist (for information on serotonin receptors, see "Serotinin [5H]

Agonists and Antagonists"). One of buspirone's drawbacks is its slow onset of action. It has to be taken for a few weeks before an effect is obtained. Several related compounds, such as gepirone and ipsapirone, were in clinical development but were abandoned by their sponsors (Bristol-Myers, Squibb, and Bayer) because of insufficient efficacy.

ANTIDEPRESSANT DRUGS

Drugs affecting the synthesis and metabolism of norepinephrine and serotonin have found their major use in psychiatry, where they have revolutionized the therapy of psychiatric diseases. The possibility of treating depression with drugs was first suggested at the 1957 meeting of the American Psychiatric Association in Syracuse, New York. George Crane of the Montefiore Hospital of New York City reported that iproniazid (Marsilid) improved the mood of several tuberculosis patients. This drug had been developed at the Hoffmann–La Roche Laboratories for the treatment of tuberculosis. Following Crane's observation Nathan Kline and his associates at Rockland State Hospital in Orangeburg, New York, successfully tested iproniazid in chronically depressed psychotic patients. This was the beginning of the chemical therapy of depression.

The antidepressant activity of iproniazid was attributed to the ability of the drug to inhibit an enzyme, monoamine oxidase (MAO), that facilitates the breakdown of monoamines, including norepinephrine and serotonin. The increased concentrations of these amines in the brain were associated with the antidepressant effect of iproniazid. Without waiting for FDA approval, many psychiatrists began to treat their depressed patients with iproniazid. Its liver toxicity was soon discovered, however, and iproniazid was eventually replaced by other MAO inhibitors—isocarboxazid (Marplan), phenelzine (Nardil), nialamide (Niamid), and others. All these drugs were related chemically, and all had some degree of liver toxicity. Their most disturbing side effect, however, was their potentially dangerous interaction with foods containing tyramine. The best-known example is the so-called "cheese effect": Patients receiving an MAO inhibitor who eat cheese can develop hypertension and stroke; cheese is rich in tyramine, a hypertensive amine, whose breakdown is blocked by MAO inhibitors.

Scientists soon began to search for a better antidepressant. The next breakthrough in this field was initiated by a nurse and a psychiatrist at a small psychiatric institution in Musterling on Bodensee, on the Swiss side of the lake. The Geigy pharmaceutical company conducted clinical studies with its potential antipsychotic drugs there. When one of Geigy's drugs was given to withdrawn schizophrenics, the nurse noticed that patients hallucinated more and became more difficult to handle. A careful analysis of their behavior suggested that the patients became less withdrawn. This observation prompted Geigy to evaluate this drug in the therapy of depression. The study was conducted by Raymond Kuhn. The results, reported in the *American Journal of Psychiatry* in

1958, were highly successful, and the drug became known as imipramine (Tofranil). Imipramine does not inhibit MAO but rather the uptake of norepinephrine and serotonin into their storage sites, and consequently increases their concentrations in the extracellular space in the brain.

The development of imipramine was followed by that of many other chemically and pharmacologically related drugs, known as "tricyclic" antidepressants; among these were amitriptyline (Elavil), desipramine (Norpramin), and nortriptyline (Aventyl). Their mechanism of action is similar, but some of the tricyclic antidepressants are more effective in inhibiting the uptake of norepinephrine, while others are relatively more specific in inhibiting the uptake of serotonin; some others block the uptake of still another neurotransmitter, dopamine. Their common side effects include cardiac toxicity (proarrhythmic activity), as well as dry mouth, constipation, and other consequences of anticholinergic action (blockade of acetylcholine).

The search for more effective and safer antidepressants led in 1972 to the discovery and development of fluoxetine (Prozac) by Bryan Molloy, David T. Wong, and Ray W. Fuller at Lilly Laboratories. Fluoxetine inhibits the uptake of neurotransmitters into their storage sites but is much more selective for serotonin than norepinephrine. It is a highly successful—and controversial—drug. It has been claimed to free many patients from depression and to greatly improve their quality of life. Many books have been written about fluoxetine, in which both patients and physicians shared their experience with the drug. Most of the books and articles were highly favorable, but others expressed concern. It was reported that fluoxetine may increase the rate of suicide. Other investigators state that all antidepressants increase the propensity toward suicidal thoughts. One possible explanation is that during the first weeks of treatment, an antidepressant drug boosts a patient's energy, making the patient more likely to act on impulse rather than remain immobilized or in a state of lethargy. The onset of the antidepressant action of fluoxetine is slow—two to three weeks. The drug may cause agitation and impair judgment. It tends to interact with many other drugs but is particularly dangerous in combination with MAO inhibitors.

Another antidepressant with a similar activity profile is sertraline (Zoloft). Developed at Pfizer Laboratories in Groton, Connecticut, by Kenneth Koe and his associates, it was launched in 1991 and rapidly gained a substantial portion of the antidepressant market. Like fluoxetine, sertraline primarily inhibits serotonin uptake, but unlike fluoxetine, sertraline is rapidly metabolized and excreted within twenty-four hours. The side-effect profile of sertraline also appears similar to that of fluoxetine. Other recently developed serotonin uptake inhibitors include paroxetine (Paxil), venlafaxine (Effexor), fluvoxamine (Luvox), and nefazadone (Serzone). The mechanism of action of these antidepressants is similar, although venlafaxine was claimed effective in patients resistant to fluoxetine.

A chemically and possibly pharmacologically different antidepressant is trazodone (Desyrel), which was developed in Italy in the late 1960s but did not become popular in the United States until the 1980s. A selective serotonin uptake inhibitor, trazodone was also found to act as an antagonist at 5-HT_{1a}, 5-HT_{1c}, and 5-HT_2 serotonin receptors (for information on these receptors, see "Serotinin [5-HT] Agonists and Antagonists"). At commonly used doses trazodone does not seem to affect the heart and circulation. It has sedative properties, however, and its major side effect is drowsiness; it is even used to treat insomnia.

ANTIPSYCHOTIC DRUGS

Although some plant extracts were used to modify human behavior in ancient Greece and Rome, the era of modern antipsychotic therapy did not begin until 1950 with the synthesis of chlorpromazine (Largactil, Thorazine) by Paul Charpentier at Rhône-Poulenc in France. Chlorpromazine was synthesized in the search for a better antihistamine. Its antipsychotic activity would probably never have been discovered if French surgeon and physiologist Henri Laborit had not included chlorpromazine in his "cocktail lytique," a drug combination he used as premedication in surgery to lower body temperature. Laborit noted that chlorpromazine augmented the effects of anesthetic drugs and thus suggested its use as a sleep-inducing agent.

The first use of chlorpromazine in psychiatric diseases is attributed to Jean Delay and his colleagues in France, who used chlorpromazine to treat patients with agitation, confusion, anxiety, depression, or schizophrenia and who described the results in 1952. Chlorpromazine was observed to make patients indifferent to external stimuli. The first detailed pharmacological analysis of chlorpromazine was conducted by Simone Courvoisier and his associates and published in 1953. These investigators were the first to observe that chlorpromazine inhibited conditioned avoidance behavior in rats. The animals were conditioned to climb a rope at the sound of a bell, but after treatment with chlorpromazine, they ignored the bell.

Rhône-Poulenc had no American organization at that time, so its management decided to license chlorpromazine to an American company for development in the United States. The drug was offered to Merck and to Pfizer, but neither company was interested. The management team of Smith Kline French Laboratories (now SmithKline Beecham) must be credited with making the correct decision to license and introduce chlorpromazine in the United States. Between 1952 and 1954 about five hundred publications on the clinical use of chlorpromazine appeared, primarily in France. The first clinical study on the American continent was conducted by Canadian investigators H. E. Lehmann and G. E. Hanrahan, who found chlorpromazine effective in controlling psychomotor excitement and the manic state; in 1954 they published their results. In 1955 Dexter Goldman reported in the *Journal of the American Medical Asso-*

ciation that chlorpromazine is effective in treating schizophrenic patients and stated that the drug created a sense of optimism rarely seen with any other treatment of psychotic states. Chlorpromazine was soon firmly established as the first antipsychotic drug. Its wide use led to an early recognition of its major side effects: tardive dyskinesia (distorted involuntary movements), sedation, hypotension, and occasional jaundice.

The discovery of the antipsychotic activity of chlorpromazine facilitated the establishment of laboratory tests for potential antipsychotic drugs and led to an intensive search for new drugs. Since 1955 many antipsychotic drugs have been discovered and developed, but all have the potential to produce tardive dyskinesia. The mechanism of action of chlorpromazine remained unknown until 1963, when Avid Carlsson in Sweden suggested that it and other antipsychotic drugs blocked dopamine receptors in the brain. There are five different known dopamine receptors (D_1 to D_5), and the antipsychotic activity of chlorpromazine and related phenothiazines is correlated with their potency in blocking D_2-receptors. Among related antipsychotic drugs the best known are thioridazine (Mellaril), mesoridazine (Serentil), fluphenazine (Prolixin), perphenazine (Trilafon), chlorprothixene (Taractan), and thiothixene (Navane). Their mechanism of action and side effects are qualitatively similar to those of chlorpromazine, but they differ in relative potency, duration of action, and the incidence or intensity of side effects.

A chemically different antipsychotic, haloperidol (Haldol) was discovered only a few years after chlorpromazine. Paul A. Janssen and his colleagues at Janssen Laboratories in Belgium (now part of Johnson and Johnson) synthesized and evaluated a series of compounds related to meperidine in an attempt to develop a better analgesic. One of their compounds was a poor analgesic but showed some chlorpromazine-like activity. With further synthetic efforts they developed a new derivative, which they called haloperidol, that maximized the chlorpromazine-like activity and lost the analgesic potency. In describing their discovery almost thirty years later, Janssen and J. P. Tollenaere conceded candidly that "hard work, an open mind for the unexpected, and serendipity eventually culminated in a structurally novel compound." Haloperidol is highly effective in treating acute psychoses, particularly in patients with overactivity, agitation, and mania. Although it is less sedating than chlorpromazine or other chemically related antipsychotics, haloperidol unfortunately has a greater tendency than chlorpromazine to produce dyskinesia. Like chlorpromazine, haloperidol is an antagonist primarily at the D_1- and D_2-receptors. It is conceivable, however, that its antipsychotic activity may not be related to its activity at the D_2-receptors. Haloperidol is a much more potent antipsychotic than is chlorpromazine, yet the difference in the potency of the two drugs at D_2-receptors is much less pronounced.

Another chemically novel antipsychotic is clozapine (Clozaril). Clozapine was discovered and patented in 1963 by a small Swiss company, Wander Chemical.

Its clinical efficacy in treatment-resistant schizophrenics was not reported until 1988 by J. M. Hane and associates. Unlike chlorpromazine or haloperidol, clozapine binds preferentially to D_4-dopamine receptors. It also binds to 5-HT_2, α_1-adrenoceptors, and histamine type 1 receptors and is therefore likely to affect more than one neurotransmitter pathway in the brain. Clozapine does not appear to cause dyskinesia but can produce agranulocytosis (decrease in mature white blood cells) and seizures, so its use is limited to patients resistant to other antipsychotics.

Clozapine is the first so-called "atypical" antipsychotic drug. It appears to be effective not only against "positive" symptoms of schizophrenia (delusions, hallucinations, and abnormal thought process), but also against "negative" symptoms (apathy, lack of drive, social withdrawal). Because of clozapine's side effects (agranulocytosis, seizures) it was not considered an ideal medication, and the search for new drugs was continued.

Highly successful is Lilly's olanzapine (Zyprexa). According to the May-June 1998 issue of *R & D Directions* (Engel Publishing Partners, West Trenton, New Jersey), the 1998 sales estimate for olanzapine is $1,325 million. In the initial clinical trials tardive dyskinesia in patients receiving olanzapine was not reported; its major side effect has been a fall in blood pressure in standing position (orthostatic hypotension), with a few patients (0.9%) developing seizures.

In 1997 the FDA approved a new antipsychotic, quetiapine (Seroquel). This drug was developed by Jeffrey Goldstein and his group at the American division of Zeneca in Wilmington, Delaware. It has been reported to interact with multiple receptors, including serotonin (5-HT_2) and dopamine (D_2) receptors. Laboratory studies indicated that quetiapine will be less likely to produce tardive dyskinesia than other antipsychotics. Clinical studies confirmed these expectations.

Currently another antipsychotic, also a 5-HT_2- and D_2-receptor blocker, ziprasidone (Zeldox), from Pfizer is awaiting approval by the FDA. It is likely to be marketed before the end of 1999 and will probably compete successfully with olanzapine and quetiapine.

ANTIPARKINSONIAN DRUGS

Although dopamine-receptor antagonists are potential antipsychotics, dopamine itself, as well as dopamine-receptor agonists, has been found useful in the therapy of shock, cardiac failure, and Parkinson's disease (characterized by tremor of resting muscles). When given intravenously, dopamine increases cardiac contractility and blood flow in the kidneys. These effects are attributed to the stimulation of D_1-receptors. The renal vasodilator effects of dopamine were described and extensively studied in the 1970s by Leon Goldberg of the University of Chicago.

In 1960 Oleg Hornykiewicz at the University of Vienna studied the autopsy results of patients with Parkinson's disease and proposed that the disease is associated with, or even caused by, depletion of dopamine from a small area in the brain, the basal ganglia. Because dopamine itself cannot enter the brain, Hornykiewicz decided to treat patients with a precursor of dopamine, dihydroxyphenylalanine, or dopa, which can enter the brain and form dopamine there. His initial clinical trial in twenty patients and a trial by André Barbeau in Montreal were successful, but subsequent trials by other investigators were not. Thus the use of dopa in the treatment of Parkinson's disease remained controversial until 1967, when George Cotzias at Brookhaven National Laboratories in Upton, New York, resumed clinical trials of dopa in patients with Parkinson's disease. Using very large oral doses of dopa, up to sixteen grams a day, he demonstrated consistent improvement of the general clinical condition in the majority of his patients. Soon thereafter Curt Porter at Merck demonstrated that only one of the dopa isomers, L-dopa (levodopa, Larodopa) was active; this isomer became the standard treatment for Parkinson's disease.

L-dopa was active at half the dose of dopa, so its optimal daily therapeutic dose did not exceed eight grams. Because L-dopa was known to be metabolized by an enzyme, dopa decarboxylase, Alfred Pletscher and his associates at Hoffmann–La Roche Laboratories in Switzerland synthesized an inhibitor of dopa decarboxylase, benzerazide, and found that L-dopa, if given simultaneously with benzerazide, is effective at lower doses. The combination product was marketed under the trade name of Madopar. Similar work was conducted independently at Merck Laboratories in West Point, Pennsylvania, by Victor Lotti, who found in 1971 that the L-isomer of carbidopa, another dopa decarboxylase inhibitor synthesized and patented in 1962, is capable of substantially reducing the therapeutic doses of L-dopa. The combination product of L-dopa with carbidopa, Sinemet, was marketed by Merck and since 1990 by DuPont Pharmaceuticals. Carbidopa reduced the total daily dose of L-dopa to one or two grams at most. A slow-release formulation, Sinemet CR, is also available.

Although L-dopa and its combination products have revolutionized the treatment of Parkinson's disease, they are not a cure, and tolerance to them is known to develop in some patients. The search for better drugs to treat Parkinson's disease continues. Attempts were made to use L-dopa with anticholinergic drugs, such as Merck's benztropine (Cogentin), with some success. Because dopamine is metabolized by monoamine oxidase as well as by dopa decarboxylase, monoamine oxidase inhibitors were combined with L-dopa, but this combination therapy led to a substantial increase in cardiovascular and other side effects.

In 1977 Walter Birkmayer and his associates described in *Lancet* a successful combination of L-dopa with selegiline (L-deprenyl, Eldepryl), an inhibitor of monoamine oxidase B (a different form of the same enzyme). The racemic selegiline, also known as deprenyl, was first discovered in 1964 at Chinoin, a Hungarian pharmaceutical company, and used as an antidepressant. Apparently if type B, and not type A, monoamine oxidase is inhibited, side effects are much less likely to occur. The patients may even eat cheese without developing hypertension or stroke. Selegiline inhibits the breakdown of dopamine in the important regions of the brain, so the dose of L-dopa in a combined therapy is further decreased.

Among other drugs used to treat Parkinson's disease are dopamine agonists, that is, drugs that are capable of mimicking the action of dopamine at its receptors. In July 1997 the FDA approved pramipexole (Mirapex), a dopamine agonist, developed jointly by Pharmacia & Upjohn and Boehringer Ingelheim. In September of the same year SmithKline Beecham received approval for ropinirole (Requip), also a dopamine agonist for the treatment of Parkinson's disease. Dopamine agonists can either be used as sole therapy or in combination with levodopa. The introduction of these two new drugs is likely to improve significantly the management of patients with Parkinson's disease. Different approaches were explored by other pharmaceutical companies. In January 1998 the FDA approved Hoffmann–La Roche's product, tolcapone (Tasmar), an inhibitor of catechol-O-methyl transferase (COMT), an enzyme responsible for the breakdown of levodopa. Tolcapone is expected to enhance the effects of levodopa when the two drugs are used together.

In the era of biotechnology many new approaches to the treatment of Parkinson's disease are being explored. They include transplants of fetal brain and the search for genetic factors involved in the pathogenesis of the disease.

Drugs Affecting the Autonomic Nervous System

The human nervous system is subdivided into the central (brain) and peripheral (nerves) nervous systems. The peripheral nerves are either voluntary (somatic) or involuntary (autonomic). Individuals cannot control the functions of the peripheral autonomic nervous system (e.g., cardiovascular or gastrointestinal functions). The autonomic nervous system is further subdivided into the sympathetic system, whose nerve fibers arise from the central portion of the spinal cord, and the parasympathetic system, whose fibers arise from the upper and lower portions of the spinal cord. The sympathetic nervous system is also called adrenergic and the parasympathetic cholinergic.

ADRENOCEPTOR AGONISTS

The main chemical mediators of nerve impulses in the sympathetic nervous system are norepinephrine (noradrenaline) or epinephrine (adrenaline); in the parasympathetic nervous system the mediator is acetylcholine. All three endogenous mediators are also used as drugs. Norepinephrine and epinephrine's receptors are adrenergic (the currently preferred term is adrenoceptors), while acetylcholine's receptors are cholinergic. Many drugs currently in use are either agonists or antagonists at the adrenergic or cholinergic receptors.

Because they are secreted by the adrenal medulla, the inner portion of the adrenal gland, epinephrine and norepinephrine mediate not only nerve impulses but also act as hormones. The presence of a pressor substance (one that elevates blood pressure) in the extract from the adrenal gland was first detected in 1894 by George Oliver and Edward Schäfer in England. Epinephrine was identified in 1897 in an extract from the adrenal medulla by the first American pharmacologist, John Jacob Abel, at Johns Hopkins University. A Japanese chemist, Jokichi Takamine, visited Abel's laboratory in 1900, studied his procedure for isolating epinephrine, and decided that it could be simplified. Takamine developed his own procedure, which he patented and licensed to Parke-Davis in 1901. Takamine is also credited with bringing Japanese cherry trees to Washington, D.C., as a gesture of good will. Parke-Davis is still selling epinephrine under the trade name Adrenalin.

The most common use of epinephrine is to relieve respiratory distress caused by bronchospasm. It is also used to provide relief from hypersensitivity reactions to drugs or allergens, to prolong the effects of local anesthetics, and to restore cardiac rhythm in cardiac arrest. Under the trade name Primatene, epinephrine is used as an inhalation aerosol in the treatment of acute attacks of bronchial asthma.

Closely related chemically to epinephrine is norepinephrine (Levophed), the major transmitter at the sympathetic nerve endings and a minor component in the adrenal medullary extract. Norepinephrine is currently used to elevate arterial pressure in various hypotensive states and as an adjunct in the treatment of cardiac arrest. In its chemical structure norepinephrine differs from epinephrine only in the absence of a methyl group on amino nitrogen. The prefix *nor* stands for the German *N ohne Radikal* (or "N without radical"). Norepinephrine was synthesized in 1904 by Friedrich Stolz, a chemist at Hoechst in Germany, long before it was recognized as a hormone and a neurotransmitter. The neurotransmitter function of norepinephrine was established in 1946 by Ulf von Euler at Karolinska Institute in Stockholm.

Many other chemically related synthetic substances were discovered and even used in medicine before it was realized that they act by mimicking the effects of natural neurotransmitters or hormones. Some of these substances were studied by Sir Henry Dale in England as early as 1910 and called "sympathomimetic amines." Of considerable historical interest was the discovery of another related amine, ephedrine, which is still widely used today, primarily in combination products, for the therapy of asthma. Ephedrine was isolated by a well-known Chinese-American pharmacologist, Ku Kuei Chen, from Mahuang, an herb used in Chinese medicine as early as 3000 B.C. After isolating ephedrine in 1923, Chen found to his surprise that it had already been isolated in 1897 by Nagajosi Nagai in Tokyo. The advantages of ephedrine over epinephrine are good oral activity and longer duration of action. Its disadvantage is tachyphylaxis (loss of activity with repeated administration).

Some of the related amines act indirectly by releasing epinephrine and norepinephrine from the adrenal medulla or from their storage sites at the nerve endings. An example of an indirectly acting amine is amphetamine (Adderall).

Of particular importance is its ability to enter the brain, where it exerts a powerful central stimulant effect, which may be followed by insomnia, fatigue, and even depression and psychosis. Amphetamine and similar drugs were widely used to maintain wakefulness or suppress appetite. Because of its severe effect on the brain and its addictive properties, amphetamine is currently indicated only in the therapy of narcolepsy (uncontrollable desire for sleep) and attention-deficit disorder.

In 1948 American pharmacologist Raymond Ahlquist postulated the existence of two types of adrenoceptors and named them α and β. According to him, some of the effects of the adrenergic neurotransmitters norepinephrine and epinephrine were mediated by α-adrenoceptors and others by β-adrenoceptors. Norepinephrine and epinephrine interact with both types of adrenoceptors, although in blood vessels norepinephrine stimulates predominantly α-adrenoceptors, whereas epinephrine activates β-adrenoceptors as well. In the heart muscle either of the two substances acts at the β-adrenoceptors.

A selective β-adrenoceptor agonist, isoproterenol (Aludrin, Isuprel), was discovered in 1940 by Heribert Konzett at Boehringer Ingelheim in Germany. It was considered the treatment of choice for acute asthmatic attacks because it was a good bronchodilator without the hypertensive effects of epinephrine. Isoproterenol was introduced in the United Kingdom and the United States in 1951 and was widely used as an antiasthmatic for the next twenty years. Because of its short duration of action it was used to treat acute asthmatic attacks rather than to prevent them. Isoproterenol is also a highly potent cardiac stimulant. In an aerosol formulation it can easily be overdosed and over the years has probably been responsible for the deaths of many asthmatic patients.

There was obviously a need for a more selective β-adrenoceptor agonist that would not stimulate the heart. The theoretical basis for such a development had existed since 1967 when A. C. Lands and his associates postulated the existence of two subtypes of β-adrenoceptors: β_1-receptors, which were cardiac receptors, responsible for the increase in heart rate and cardiac force; and β_2-adrenoceptors, which were claimed to mediate the vasodilator and bronchodilator effects of β-adrenoceptor agonists. The search for selective β_2-adrenoceptor agonists culminated at the British pharmaceutical company, Allen and Hanbury (now part of Glaxo), in the development of albuterol (salbutamol, Proventil, Ventolin) in 1967. Albuterol is longer acting than isoproterenol and is a considerably more potent bronchodilator than a cardiac stimulant. Within three years it largely displaced isoproterenol. In the United States albuterol is marketed by Schering-Plough.

The selectivity for β_2-adrenoceptors was further increased in another bronchodilator, terbutaline (Bricanyl, Brethine), which was originally developed at Astra in Sweden and patented in 1968; it is now marketed in the United States by Ciba-Geigy and Marion Merrell Dow. Despite a high degree of selectivity for β_2-adrenoceptors, terbutaline still has some cardiac stimulant effects. Occasional terbutaline-induced increases in the heart rate can be explained by the recently discovered presence of some β_2-adrenoceptors in the heart.

CHOLINERGIC AND ANTICHOLINERGIC DRUGS

Acetylcholine was first synthesized in 1867 by Adolf von Baeyer in Germany, but its first physiological effect, the ability to lower arterial pressure, was not discovered until 1906, by Reid Hunt at Johns Hopkins University. In 1914 Arthur Ewins, at Wellcome Laboratories in England, isolated acetylcholine from ergot extract, and Sir Henry Dale subsequently studied the substance extensively. Otto Loewi discovered that acetylcholine is a chemical transmitter of nerve impulses. Working in Graz, Austria, in 1921, Loewi perfused two frog hearts with physiological salt solution and electrically stimulated the vagus nerve of the first heart, which slowed the heart's contractions. When Loewi used the fluid from the first heart to perfuse the second heart, it slowed as well. Loewi concluded that, when stimulated, the vagus nerve releases a substance that he called "Vagusstoff." In 1926 he identified that substance as acetylcholine. In 1933 Walter Feldberg and Otto Krayer in London obtained evidence that acetylcholine is a chemical transmitter not only in frogs but also in mammals. In 1936 Loewi and Dale shared the Nobel Prize in physiology or medicine for the discovery of the chemical transmission of nerve impulses. Loewi escaped the Nazis in Austria and settled in New York, where he continued his research at New York University; he died in 1961.

Cholinergic Drugs

Drugs that mimic the action of acetylcholine are called cholinergic drugs. They can act either directly at the acetylcholine receptors or indirectly by enhancing the release of acetylcholine or preventing its breakdown. Acetylcholine acts at many sites in the body. Some of its effects resemble those of nicotine and are called nicotinic; others resemble those of muscarine, an alkaloid present in the poisonous mushroom *Amanita muscaria*, and are called muscarinic. The receptors that mediate the effects of acetylcholine are called accordingly nicotinic and muscarinic. Each receptor type has many subtypes, which makes for great diversity in the composition of the receptors.

The best-known cholinergic drugs are bethanechol (Urecholine) and pilocarpine (Salagen, Ocusert Pilo). Bethanechol is used to stimulate intestinal or urinary bladder functions, while pilocarpine is used primarily to reduce intraocular pressure.

Acetylcholine itself is very short acting and is therefore rarely used clinically. It is destroyed in the body by an enzyme called acetylcholinesterase. Drugs can inhibit this enzyme, which leads to accumulation of acetylcholine in the body and therefore to sustained stimulation of its receptors. The first discovered inhibitor of this enzyme was physostigmine (eserine, Antilirium), which is the active principle of the Calabar bean. Nigerians used to execute prisoners by forcing them to swallow a water extract of the bean. They either died or vomited; those who vomited were pronounced not guilty and released. This custom, reported in England in 1846 by William Daniell, an army surgeon, aroused the interest of Robert Christison, a toxicologist at the University of Edinburgh, who tried the Calabar bean extract himself and barely survived. One of his students, Thomas Fraser, isolated the active principle of the bean and called it "eserina." In 1864 J. Jobst and O. Hesse obtained the substance in pure form and named it physostigmine.

Physostigmine was used as a cholinergic agonist in gastrointestinal disorders and to lower intraocular pressure. Currently physostigmine has only one indication listed in the *Physicians' Desk Reference*, as an antagonist of cholinergic-blocking drugs. Other anticholinesterase drugs—neostigmine (Prostigmin), pyridostigmine (Mestinon), and edrophonium (Tensilon)—are used therapeutically either to treat myasthenia gravis, a disease characterized by skeletal muscle weakness and intestinal paralysis, or as an antagonist of curare, a neuromuscular blocking agent. Many inhibitors of acetylcholinesterase, including diisopropyl fluorophosphate (DFP), parathion, fenthion, malathion, and other phosphorus-containing compounds, were used as insecticides, and some are still used as such today. Some of the phosphorus-containing inhibitors of cholinesterase, such as tabun, sarin, and soman, are extremely toxic and were developed as nerve gases.

A new use for cholinesterase inhibitors emerged in recent years when it was discovered that Alzheimer's dementia was associated with a deficiency in the ability of certain nerve cells to produce acetylcholine. Even though this deficiency might not represent the cause of Alzheimer's disease, it appeared logical to try drugs capable of increasing acetylcholine levels in the brain. Many inhibitors of acetylcholinesterase were therefore evaluated in patients with Alzheimer's disease. In many instances their peripheral side effects prevented their use. Two such drugs were marketed: tacrine (Cognex), in 1994 by Warner-Lambert, and donepezil (Aricept) in 1996 by Pfizer and Eisai. These inhibitors of acetylcholinesterase appear to delay the progress of Alzheimer's dementia.

Anticholinergic Drugs

Inhibitors of acetylcholine were used in medicine long before that substance had been discovered. References to plants now known to contain anticholinergic alkaloids can be found in ancient Hindu, Greek, and Roman medical writings. Extracts from *Atropa belladonna* were used by the ladies of the Spanish court to beautify their eyes by dilating their pupils (therefore, belladonna). The name *Atropa* is derived from Atropos, the oldest of the Three Fates, who cuts the thread of life (*Atropa belladonna* was known to be very toxic). The active principle of the roots of *Atropa belladonna*, atropine, was isolated in 1831 by a German pharmacist with the last name of Mein and later in 1890 was studied extensively by Alfred Ladenburg at the University of Kiel. Atropine blocks muscarinic receptors nonselectively; it does not differentiate between receptor subtypes. The major pharmacological effects of atropine inhibit gastric and intestinal motility, gastric secretion, and cardiac slowing by stimulation of the vagus nerve. Atropine causes pupillary dilation and increased heart rate. The major therapeutic use is as an antispasmodic to relieve intestinal colics or diarrhea. Today atropine is used mostly in combination with other drugs (i.e., Donnatal). Atropine, particularly in toxic doses, can enter the brain and produce irritability and even hallucinations.

Another anticholinergic drug, scopolamine, enters the brain much better than atropine. It is used to disrupt recent memory and reduce salivation before anesthesia. Drugs that antagonize scopolamine in animals are considered candidates for possible memory-enhancing activity. A close derivative of scopolamine, methscopolamine (Pamine), lacks the central effects of scopolamine and is used in gastrointestinal diseases. Attempts to optimize the ability of atropine to inhibit gastric secretions led to the development of more specific inhibitors of gastric secretion—methantheline (Banthine) and propantheline (Pro-Banthine)—at Searle Laboratories. These drugs have higher nicotinic blocking activity than atropine and fewer side effects associated with the blockade of muscarinic receptors. They are used therapeutically to slow gastric emptying and were initially used as adjuncts in the therapy of gastric ulcers.

In 1947 dicyclomine (Bentyl) was synthesized and widely used as an antispasmodic in Europe. It relaxes intestinal smooth muscle, apparently not only by virtue of its anticholinergic activity but also by direct action on smooth muscle. Dicyclomine is used primarily in the therapy of so-called irritable bowel syndrome, a chronic intestinal disease characterized by abdominal pain and altered bowel habits.

One of the most effective drugs to control diarrhea is Lomotil, a combination product containing diphenoxylate and atropine. Diphenoxylate was synthesized and patented as an analgesic agent in 1959 by Janssen Laboratories (now a division of Johnson and Johnson) in Beerse, Belgium. Its antidiarrheal activity was discovered accidentally, when Gordon Van Arman, at that time a pharmacologist at Searle Laboratories, noticed a relative lack of feces in cages of rats treated with diphenoxylate. The drug is a controlled substance because it shares some pharmacological actions with opiates and other narcotics. At higher doses it produces respiratory depression and may be addictive. When combined with atropine, both drugs are given at sufficiently low doses so that side effects are kept at a minimum.

Antihistamines

Histamine, an amine present in all tissues of the human body, was first synthesized in 1907 by Adolf Windaus and his associates at the University of Göttingen. The physiological significance of histamine was not realized at first. In 1910 D. Ackerman at the University of Würzburg described the formation of histamine from an amino acid, histidine, by bacterial decomposition. Sir Henry

Dale in London introduced the name *histamine* after isolating it from ergot extract and demonstrating its properties as a uterine stimulant. Most investigators assumed at that time that histamine was present in various tissue extracts as a result of bacterial action and that it had no physiological function in the human body.

Charles Best from the University of Toronto, while visiting Dale in 1926 in London, isolated histamine from liver extracts under conditions that excluded bacterial decomposition and thus demonstrated that the vasodilator activity of the liver extract was caused by the presence of histamine. A major contribution to the understanding of the physiological role of histamine was made in London in 1927 when T. Lewis published a monograph titled *The Blood Vessels of Human Skin and Their Responses.* In this monograph he presented evidence that a "histamine-like" substance is released from the skin by injury or by foreign proteins (antigens) capable of producing immune responses. This substance was identified as histamine. The potential use of antihistamines in allergic disorders became an attractive target for the pharmaceutical industry.

ANTIHISTAMINES AS ANTIALLERGICS

The first antihistamine compounds were prepared in the early 1930s by Daniele Bovet and Ernest Fourneau at the Pasteur Institute. Their main reason for synthesizing antihistamines was to better define the role of histamine in the body. They demonstrated mild antihistaminic activity of piperoxan, an adrenergic-blocking agent previously discovered by Fourneau. Bovet's student Anne-Marie Staub continued to evaluate Fourneau's other compounds for antihistaminic activity, and in 1939 she published the first study on the structure-activity relationship of antihistamines. Unfortunately most of her compounds were too toxic for human use. The first clinically useful antihistamine was phenbenzamine (Antergan), synthesized by M.-M. Mosnier at Rhône-Poulenc and evaluated pharmacologically by B. N. Halpern in 1942. In 1943 George Rieveschl and Wilson Huber at the University of Cincinnati synthesized diphenhydramine (Benadryl) and assigned the patent rights to Parke-Davis.

Research on creating synthetic antihistamines accelerated after World War II. In 1946 Carl Djerassi and his associates at Ciba in New Jersey patented tripelennamine (Pyribenzamine). Paul Charpentier and his associates at Rhône-Poulenc synthesized a structurally different antihistamine, promethazine (Phenergan). It is chemically related to chlorpromazine and has central nervous system depressant and antiemetic properties. The main use of these antihistamines is in such allergic disorders as hay fever and skin eruptions. Because antihistamines were initially considered effective for the common cold, the market for them expanded rapidly. Their major side effect is sedation, so they are not recommended for patients who have to drive or remain alert. In a search for nonsedating antihistamines N. Sperber and his associates at American

Schering Corporation in 1951 synthesized and patented chlorpheniramine (Chlor-Trimeton). It was the most potent antihistamine at the time and produced less sedation than those previously available.

The sedative activity of antihistamines was also used successfully for therapeutic purposes. In 1956 UCB Pharmaceutical Corporation in Belgium developed hydroxyzine (Atarax, Vistaril) as a mild tranquilizer and licensed it to Pfizer. It is still sold today by the Roerig division of Pfizer for managing anxiety and emotional stress and appears to be particularly useful in rendering a disturbed patient amenable to psychotherapy.

The first nonsedating antihistamine was developed in 1973 by A. A. Carr and C. R. Kinsolving of Richardson-Merrell Company (now part of Hoechst). Their development of terfenadine (Seldane) was facilitated by the recognition of various types of histamine receptors. Three types have been found: H_1, H_2, and H_3 receptors. All the above-mentioned compounds block H_1-receptors as well as others, but terfenadine is highly selective for H_1-receptors. Another nonsedative and selective H_1-receptor antagonist is Schering-Plough's loratadine (Claritin), which is indicated for the treatment of seasonal allergies (rhinitis and urticaria). Still another nonsedative antihistamine is Rhône-Poulenc Rorer's ebastine (Kestine), which is awaiting approval in the United States and is already marketed in many European countries.

Antihistamines in Motion Sickness

In 1946 the usefulness of antihistamines in treating or preventing motion sickness was discovered accidentally. Most pharmaceutical research laboratories were trying to develop a nonsedative antihistamine. At Searle in Chicago the chemists decided to incorporate a mild stimulant component, a theophylline derivative, with diphenhydramine as a new salt. This new formulation, dimenhydrinate (Dramamine) still had sedative activity but showed an unexpected effectiveness in treating motion sickness. While Leslie Gay at Johns Hopkins University was evaluating dimenhydrinate as an antiallergic agent, one of the patients in the study reported that she did not experience her usual car sickness after taking the drug. This observation was confirmed in a few other patients who suffered from motion sickness and led in 1947 to a double-blind trial of the drug on a troop ship on the way to Europe. While 25 percent of the soldiers receiving the placebo suffered from seasickness, only 4 percent of those who took dimenhydrinate became sick. Drowsiness was the only significant side effect.

As dimenhydrinate rapidly became one of the best-selling antihistamines, most pharmaceutical companies reevaluated their antihistaminic drugs for their ability to prevent motion sickness. Pfizer licensed another antihistamine from UCB Pharmaceuticals in Belgium, meclizine, which was marketed as Bonine or Antivert. Its primary indication is the management of motion sickness, but like dimenhydrinate, it is also used to treat vertigo.

Most H_1-receptor antagonists appear to have some activity in the treatment or prevention of motion sickness. It is not clear, however, whether this effect is mediated by H_1 or another type of histamine receptor.

Antihistamines as Antiulcer Drugs

Because the antihistamines available in the early 1960s did not block the effects of histamine on gastric acid secretion, uterine contractions, or cardiac contractility, the existence of more than one type of receptor for histamine was suspected. In 1964 James Black and his associates at the Smith Kline and French Institute in England decided to search systematically for inhibitors of the stimulant effects of histamine on gastric secretion. It took eight years and the synthesis of more than seven hundred compounds before the first inhibitor, burimamide, was found. Burimamide was specific for gastric acid secretion and did not inhibit the effects of histamine on H_1-receptors at such sites as vascular or intestinal smooth muscle. The drug's activity was confirmed in humans, and gastric histamine receptors were named H_2-receptors.

Unfortunately burimamide was poorly absorbed orally, and Black continued to look for a better compound. A year later, in 1973, he announced the discovery of metiamide, an orally active derivative of burimamide. Two years later in clinical studies metiamide was found to produce neutropenia, a decrease in the white blood cell count, in two patients. The clinical trials were stopped, and Black and his associates had to continue to look for better and safer drugs. That same year, 1975, they announced the discovery of cimetidine (Tagamet), which was almost completely absorbed by oral administration, was highly effective in inhibiting histamine-induced secretions of gastric acid, and did not produce neutropenia. Cimetidine was introduced into clinical practice in 1976. It had taken twelve years to complete the project. For his work leading to the discovery of cimetidine and propranolol, James Black shared the 1988 Nobel Prize in physiology or medicine.

Cimetidine was first approved for the treatment of conditions associated with the hypersecretion of gastric acid and for short-term (no more than twelve weeks) treatment of duodenal or gastric ulcers. Its use was eventually expanded to longer-term therapy for ulcers and other gastrointestinal diseases. Cimetidine became the best-selling drug, earning more than $1 billion a year for Smith Kline and French (now SmithKline Beecham). The success of cimetidine led to the development of many other H_2-antagonists. The most successful among them was ranitidine (Zantac), which was developed by a team of scientists at Glaxo, led by Roy Brittain. Ranitidine is a more specific antagonist at H_2-receptors and has less severe side effects than cimetidine. Despite the recent discovery of the bacterial nature of gastric ulcer disease, H_2-antagonists will continue to be useful as an adjunct treatment for gastric or duodenal ulcers.

In 1987 another group of histamine receptors, H_3, was discovered. These receptors are present in the brain and at the endings of involuntary peripheral nerves. Substances that stimulate or block these receptors have been identified, but their clinical indications have not yet been clearly defined.

Another approach to the control of gastric secretion is to inhibit the enzyme called H^+, K^+ ATPase, or the "proton pump." This enzyme is the major mediator of gastric secretion in the stomach. In the early 1990s sustained inhibition of this enzyme was achieved with two drugs—omeprazole (Prilosec) and lanzoprazole (Prevacid). These drugs are activated by gastric acidity and are inactive in nonacidic medium. After ingesting one of these drugs, acid secretion does not resume until new enzyme molecules have formed. Prilosec was marketed by Astra Merck and became one of the best-selling drugs of 1997. It is indicated in the therapy of duodenal and gastric ulcers and of esophagitis.

Serotonin (5-HT) Agonists and Antagonists

Serotonin (5-hydroxytryptamine, 5-HT, enteramine) is an endogenous monoamine found in blood and in many other mammalian tissues, including the intestines and brain. It has many physiological functions in the human body, including stimulation of vascular or intestinal smooth muscle, inhibition of gastric secretion, and mediation of impulses in the brain and peripheral nerves. It is also a precursor of melanin, a dark pigment of the skin, hair, and various other tissues.

It was known for more than a century that blood contains a vasoconstrictor substance. In 1948 this substance was isolated, by Irvine Page and his colleagues at the Cleveland Clinic, and a year later identified as 5-hydroxytryptamine. Independently of the Cleveland Clinic group, V. Erspamer and his colleagues in Italy isolated a gut-stimulating factor from the intestinal wall, which they called enteramine. Although the early findings of Erspamer date back to the 1930s, he did not identify enteramine as 5-hydroxytryptamine until 1952. The role of serotonin as a neurotransmitter in the brain was established in the late 1950s by Bernard B. Brodie and Parkhurst A. Shore at the National Institutes of Health.

Serotonin then became a highly popular substance for scientific investigations. Many drugs were found to act by modulating the release or the uptake of serotonin into the tissues as well as by mimicking or blocking the effects of serotonin. The identification and classification of serotonergic receptors became a science of its own. By 1994 four classes, or types, of serotonergic receptors ($5\text{-}HT_1$ through $5\text{-}HT_4$) had been identified. Each of the four types has multiple subtypes, identified with capital letters as subscripts, for example, $5\text{-}HT_{1A}$, $5\text{-}HT_{1B}$, and so forth. Gene cloning work led to the tentative identification of three additional serotonergic receptors, the function of which remains to be determined. These receptors are identified with small letters—for example, $5\text{-}ht_5$.

The discovery of so many serotonergic receptors led to the development of selective agonists and antagonists for some of the receptor subtypes. These drugs found new therapeutic indications for which nonselective drugs are not useful. Because of the selectivity for only one receptor subtype, these drugs are expected to have fewer side effects than nonselective drugs.

SEROTONIN ANTAGONISTS AS ANTIALLERGIC AGENTS

Among the first drugs found to have a high affinity for serotonin receptors was cyproheptadine (Periactin), synthesized and patented in 1961 by Edward Engelhardt and his colleagues at Merck. Because this drug is also a potent antihistamine, it was used to treat allergies and pruritus (itching). It is also an effective appetite stimulant. Its use for this indication has not been approved in the United States, but cyproheptadine is widely used in South America for this purpose. The drug is no longer covered by patent and is no longer sold by Merck, but it is available as a generic drug from Mylan or Par. More recently it was found to have some selectivity as an antagonist at 5-HT_{2C} receptors.

SEROTONIN LIGANDS AS ANTIMIGRAINE DRUGS

Serotonin-receptor ligands have been used to treat migraine or vascular headache. The first antimigraine drug, the mechanism of action of which is likely to involve serotonin receptors, was ergotamine (Ergostat, Gynergen), one of the ergot alkaloids. Ergotamine is still used today for the symptomatic relief of pain in migraine. It was extracted in 1945 by Arthur Stoll and his associates at Sandoz in Basel; its structure was determined by the same group in 1951. Ergotamine interacts with many different receptors, including α-adrenoceptors and serotonergic (probably 5-HT_{1D}) receptors. But which pharmacological effect is primarily responsible for its usefulness in the treatment of migraine has not yet been established.

Another useful serotonin-receptor–based drug is methysergide (Sansert). Synthesized by Albert Hofmann and his colleagues at Sandoz, methysergide was patented in 1960 in the United Kingdom and in 1965 in the United States and marketed for the treatment of migraine. Its effectiveness was attributed to vasoconstriction caused by agonist activity at serotonin receptors. Methysergide, however, is not a very selective ligand at various subtypes of serotonin receptors; it is an agonist at 5-HT_{1A}, 5-HT_{1B}, and 5-HT_{1D} and a very potent antagonist at 5-HT_{2C} receptors. Its effectiveness in migraine was initially attributed to its serotonin antagonist effects, but it is now thought to involve agonist activity at 5-HT_{1D} receptors. The use of methysergide is limited by a serious side effect: Long-term use can lead to fibrotic thickening of membranes in the lung and abdominal cavity, as well as to the thickening of cardiac valves. The drug is therefore indicated only for the preventive therapy of otherwise uncontrollable vascular headache.

Methysergide is chemically related to serotonin; their structures contain an indole nucleus. It is even more closely related to lysergic acid and its diethylamide (LSD-25), which was synthesized much earlier (in 1938) in the same laboratory at Sandoz and by the same chemist, Albert Hofmann. The hallucinogenic or psychedelic activity of LSD-25 was not recognized, however, until 1943, when Hofmann decided to reexamine LSD-25. After a few hours of work he had inhaled enough of the substance to cause dizziness and restlessness. Fortunately he reached his home before hallucinations started. As he described it later, he fell into a peculiar state, characterized by exaggerated imagination, and when he closed his eyes, pictures of extraordinary plasticity and intense color seemed to surge toward him. In 1947 Arthur Stoll publicly announced that a few milligrams of LSD-25, taken by mouth, cause profound changes in human perception, similar to those experienced by schizophrenics. Attempts to use LSD-25 as an experimental tool to produce experimental schizophrenia in animals, however, were not successful. Doses of LSD-25, which caused hallucinations in humans, did not produce any detectable changes in the behavior of animals.

The mechanism of hallucinogenic action of LSD-25 has not been definitively established, although it appears to involve serotonergic receptors—LSD-25 binds with high affinity to 5-HT_{2A} and 5-HT_{2C} receptors in the brain. The relevance of this binding for the hallucinogenic effect of LSD-25 has been questioned, however, because closely related substances bind with the same affinity to the same receptors but are not hallucinogenic.

Methysergide's effectiveness in treating migraine aroused interest in the potential of serotonin ligands as antimigraine drugs. The next breakthrough development in this field was the discovery of sumatriptan (Imitrex) at Glaxo Laboratories in the United Kingdom. This drug was patented in 1983, but its pharmacological effects were not described until 1988. In the same year Alfred Doenicke and his colleagues reported the results of the first clinical trial of sumatriptan in thirty-four patients with migraine. Twenty-three showed dramatic improvement in the frequency and severity of the attacks, and the other eleven obtained some benefit as well.

The development of sumatriptan was greatly facilitated in 1985 when Pramod R. Saxena and colleagues in Rotterdam found that ergotamine produces selective vasoconstriction of arteriovenous anastomoses (vascular connections) in cats' heads. The closure of these anastomoses was allegedly mediated by a subtype of 5-HT_1 receptors and was proposed as the mechanism of the antimigraine action of ergotamine. Glaxo scientists used Saxena's technique and found that among their synthetic compounds sumatriptan had the best and most specific effect on arteriovenous anastomoses. Sumatriptan was subsequently shown to act as an agonist at 5-HT_{1D} receptors, while at higher concentrations it acted as an agonist also at 5-HT_{1A}, 5-HT_{1B}, and 5-HT_{2C} receptors.

After sumatriptan was developed, the search for new drugs for migraine greatly accelerated. In November 1997 the FDA approved Zenica's zolmitriptan, (Zomig), a selective 5$HT_{1B/1D}$ agonist. Two mechanisms of action were claimed for zolmitriptan: one in the brain and one peripheral, on the vascular system controlled by the trigeminal nerve. Glaxo Wellcome's naratriptan (Amerge) was approved in 1998 and Merck's selective 5-HT $_{1D}$ receptor agonist, rizatriptan (Maxalt), is awaiting approval. The application for Pfizer's eletriptan was scheduled to be filed with the FDA during the last quarter of 1998, while Pharmacia and Upjohn's almotriptan is in phase 3 clinical studies. It is not yet clear whether any of the "triptans" will have important clinical advantages over sumatriptan, but patients suffering from migraine will certainly have a choice of medication.

Hormones

Hormones are endogenous chemical substances produced in an organ or cells that are capable of regulating the function of the same or other organs or cells. Because under- or over-production of a hormone can lead to disease, hormones themselves or their antagonists can function as drugs. A classic example of a hormone used as a drug is insulin: A body's failure to produce insulin leads to diabetes; thus insulin is used as a drug to control diabetes. This section describes the major hormones as well as drugs than can either mimic or block their effects.

SEX HORMONES

The sexual development and characteristics of humans or animals are determined by sex hormones: Estrogens are primarily responsible for female characteristics and testosterone for male. Their levels affect features and behavior. These extremely powerful substances control almost every aspect of human life and have both beneficial and detrimental effects on many organs and their functions. During the last hundred years a considerable amount of knowledge has been accumulated on the nature and function of sex hormones. Some of these discoveries, as well as the use of sex hormones as drugs, are discussed here.

Testosterone

Many thousands of years ago primitive human societies knew that castrated males were unable to reproduce. The importance of the testes for reproductive capability in males was recognized, but the discovery that the testes are secretory glands is usually attributed to a German physiologist, A. Berthold.

The historical description of the discovery of sex hormones usually begins with the story of Charles Edouard Brown-Séquard, professor of medicine at the College de France, who in 1889, at the age of seventy-two, claimed that he was rejuvenated after injecting himself with an extract from guinea pig testicles. His claims created a sensation, and many elderly men followed his example. Brown-Séquard used water extracts of testes, so it is highly unlikely that his preparation contained any testosterone, which is not soluble in water. Many attempts to isolate the active principle from testes followed. An extract from animal testes, called Spermine, was marketed in Germany, but its effectiveness is doubtful because there were no methods for measuring its activity. Such a measure was not developed until 1929, when Fred C. Koch and his associates from the University of Chicago developed an assay in capons based on the stimulation of comb growth. With the help of this assay the isolation of male hormones became possible.

Adolf F. J. Butenandt in Göttingen, Germany, succeeded in 1931 in isolating androsterone, a metabolite of testosterone. To obtain fifteen milligrams of crystalline hormone, he used 15,000 liters of human urine. Soon thereafter Butenandt identified the chemical structure of testosterone, and Leopold Ruzicka and A. Wettstein in Switzerland synthesized it.

Testosterone is now considered to be the main androgen (male hormone) in the human body. It is metabolized to two other androgens—dihydrotestosterone and androsterone, which retain some of the male hormone activities and are present in the urine. It is important to realize that testosterone is also metabolized to estrogens and that the ovary and adrenal glands in females produce small amounts of testosterone.

The physiological effects of testosterone are not limited to virilization and masculinization; they have anabolic properties (that is, they increase muscle growth) and can retain sodium and water. The main therapeutic indication for testosterone and other androgens is hypogonadism (abnormally decreased gonadal function). They are also used as growth stimulants for growth retardation and are misused by athletes for stimulation of skeletal muscle growth. Androgens can stimulate erythropoiesis (red cell production) and are known to alleviate attacks of angioneurotic edema (sudden swelling of skin and mucous membranes). Many synthetic derivatives of testosterone are currently available. Some of them (such as testosterone enanthate, Delatestryl) are slowly degraded to testosterone and can be injected every two to four weeks; others (such as methyltestosterone, Metandren, Oreton Methyl) are active orally. Testosterone can now also be administered in the form of skin patches.

Estrogens

The existence of female sex hormones, known as estrogens, was suggested in 1900 by the experiments of Emil Knauer, a gynecologist in Vienna. He transplanted ovaries from mature animals into immature animals and found that their sexual maturity was accelerated. In 1912 Henri Iscovesco in Paris prepared ovarian extract, which was capable of producing sexual changes in castrated animals. The following year Ciba marketed an ovarian extract. A bioassay for ovarian extracts, developed in 1923 by Edgar Allen and Edward Doisy at Washington University in St. Louis, was based on changes in the appearance of cells in the vaginal lining of rodents that were correlated with the changes in the blood levels of female hormones.

The term *estrogens* for the female hormones was derived from the word *estrus*, defined as the recurrent period of sexual receptivity in female mammals. In humans the ovaries produce three major estrogens—estradiol, estrone, and estriol. Estrone is the oxidation product of estradiol, and estriol is the hydration product of estradiol. During pregnancy the placenta synthesizes large amounts of estrogens, which are excreted in urine. In 1927 Selman Aschheim and Bernardt Zondek in Berlin devised a pregnancy test based on the presence of estrogens in the urine. Doisy reported the isolation of crystalline estrone in 1929.

In 1932 James Cook and Charles Dodds in London were studying the structural requirement for estrogenic activity in an attempt to develop synthetic estrogens. These hormones belong to the chemical class of steroids; their chemical structure contains the typical steroid nucleus. Cook and Dodds discovered that the steroid nucleus is not required for estrogenic activity; chemicals of a completely different class, namely those containing a phenanthrene nucleus, also have estrogenic activity. Two years later they discovered that compounds even simpler than phenanthrenes—that is, compounds containing two benzene rings connected by a short carbon chain—have estrogenic activity. Dodds's further work in this field led to the discovery of another synthetic estrogen, diethylstilbestrol, which is still used today, as well as to the synthesis of two other compounds with estrogenic activity: dienostrol and hexoestrol.

In 1942 John Robson and Alexander Schonberg at the University of Edinburgh reported that a bromine-substituted triphenylethylene had estrogenic activity. In 1944 another related compound with estrogenic activity, chlorotrianisene, was discovered nearly simultaneously by Frederick Basford of ICI and Robert Shelton and Marcus Van Campen at William Merrell Company (now part of Hoechst) in Cincinnati, Ohio. The experience in the field of stilbenes led chemists at William Merrell to cholesterol-lowering compounds. In 1959 they patented a related compound, triparanol, as a hypocholesteremic agent. Triparanol unfortunately had to be withdrawn soon after its introduction because of its ability to produce cataracts.

Further synthetic work with estrogens led to the discovery of the ovulation-inducing agent, clomiphene (Clomid, Serophene), at William Merrell and to an antiestrogen, tamoxifen (Nolvadex), at ICI. Tamoxifen has no estrogenic activity but competes with estrogens for their receptors and can therefore antagonize the growth of estrogen-dependent breast tumors.

Estrogens have many therapeutic indications, but their current major use is as oral contraceptives (in combination with progesterone, a hormone produced by the corpora lutea in the ovary). The combination of the two hormones interfere with fertility by inhibiting ovulation. Other uses for estrogens include hormone replacement therapy in menopause and treatment of senile vaginitis, osteoporosis, and hirsutism. Estrogens were recently found to be beneficial in delaying the onset of Alzheimer's disease. Estradiol is currently available as Estrace, in tablets (generic) or as a transdermal formulation, Estraderm. Natural estrogens are present in the blood in conjugated form, either as sulfates or glucuronides. The conjugated estrogens can also be obtained commercially as Premarin.

During the last decade many new estrogen formulations and combinations were developed and marketed. Of particular interest are new delivery systems. In 1997 the FDA approved Parke-Davis's 17-β-estradiol skin patch under the trade name of FemPatch for the treatment of menopausal symptoms. This transdermal delivery system requires a substantially lower dose than oral formulations and is expected to produce fewer side effects. Rhône-Poulenc Rorer marketed Estalis Sequential in Europe; it provides a twenty-eight-day regimen (estrogen for fourteen days and estrogen-progesterone combination for another fourteen days) in the skin patch form. This formulation is awaiting approval in the United States. Raloxifene, the first of a new class of drugs called selective estrogen receptor modulators, was approved by the FDA and marketed in 1997 by Eli Lilly under the trade name of Evista for the prevention of osteoporosis in postmenopausal women. It is expected that raloxifene and similar drugs will act as estrogens in some but not other tissues.

Progestins

Around the turn of the century Ludwig Fraenkel at the University of Breslau (at that time in Germany) discovered that the corpora lutea, "yellow bodies," in the ovaries are needed for the maintenance of pregnancy. Their destruction in rabbits caused abortion. Various attempts were made to use corpora lutea extracts to treat menstrual disorders. An effective extract eluded scientists until 1931, when it was produced by George Corner and Willard Allen at the University of Rochester. In 1934 three groups of scientists independently isolated the active principle. They were Willard Allen and Oscar Wintersteiner at Columbia University, Adolf Butenandt and K. H. Slotta at Göttingen University, and A. Wettstein and Max Hartmann in Switzerland. The active principle was named progesterone because of its ability to maintain gestation. Because progesterone is inactive orally, it has to be administered by injection. It is also rapidly metabolized in the liver.

Many derivatives with progesterone-like activity (progestins) with longer duration of action, better oral absorption, or both, were synthesized during the 1950s. They include medroxyprogesterone acetate (Depo-Provera) and megestrol acetate (Megace). The major use of progestins is in contraception (see below), but there are also many therapeutic indications for these agents, including endometriosis, threatened abortion, endometrial carcinoma, and premenstrual tension. Progestins are also used with estrogens to control dysfunctional uterine bleeding.

Oral Contraceptives

The use of progestins, with or without estrogens, for oral contraception was based on the research of Gregory Pincus at the Worcester Institute for Experimental Biology in Worcester, Massachusetts. In 1950, in an attempt to develop a safe contraceptive, Pincus administered large oral doses of progesterone to female rabbits, assuming that, as in pregnancy, progesterone would prevent the release of ova and therefore act as a contraceptive. His assumption was correct. Despite the repeated matings of rabbits no pregnancies occurred.

Having heard of the Pincus experiments, John Rock, a gynecologist at Harvard, administered progesterone to infertile women to regulate uterine bleeding, which occurred regularly after withdrawal of progesterone. After repeated courses of progesterone therapy some of the patients were able to conceive, but oral absorption of progesterone was inadequate and erratic. Rock asked Pincus for a progestin with better oral absorption.

In 1953 Pincus invited pharmaceutical companies to submit samples of potential progestational agents for evaluation. He received nearly two hundred compounds and selected fifteen for clinical evaluation by Rock. Among the submitted steroids 19-norprogesterone derivatives were well absorbed orally and were highly potent. In 1954 Frank Colton from G. D. Searle in Chicago patented norethynodrel (a component of Brevicon and Norinyl), and Carl Djerassi and George Rosenkrantz from Syntex in Mexico City synthesized norethindrone (a component of Norethin and Estrostep).

Most of the currently available oral contraceptives are combination products of a progestin and an estrogen. The discovery that estrogens enhance the contraceptive activity of progestins was made during the first large-scale clinical trial of norethynodrel as a contraceptive in Puerto Rico. Breakthrough bleedings were reported in patients who received the purer batches of norethynodrel, but not in batches contaminated with trace amounts of an estrogen, mestranol. When mestranol was intentionally incorporated, the breakthrough bleedings did not occur. The combination product with these two steroids (Enovid) was evaluated in sixteen hundred women before it was approved by the FDA as a contraceptive in 1960.

Because side effects of oral contraceptives (thrombosis, thrombophlebitis, tumors) were thought to be caused by the estrogenic component, many of the subsequently developed contraceptives contained smaller amounts of estrogen. In Norinyl 1+50, the content of mestranol was reduced to half that in Enovid-E, and the norethynodrel was replaced by norethindrone. In the 1998 edition of the *Physicians' Desk Reference* twenty-one trade names for oral contraceptives are listed, most of them for various combinations of a progestin with an estrogen. Pure progestin contraceptives are also available as so-called "mini-pills" (Micronor, Ovrette). They are considered to be less effective contraceptives than preparations containing various combinations of progestins and estrogens. At large doses estrogens alone (such as diethylstilbestrol) are used as postcoital contraceptives (within seventy-two hours after intercourse, twice a day for five days).

The market for oral contraceptives rapidly expanded during the 1960s. It was estimated in 1975 that 55 million women worldwide were taking oral contraceptives. The rate of market expansion diminished somewhat after several side effects became apparent; of particular concern was a reported increase in morbidity and mortality from cardiovascular diseases in women taking oral contraceptives.

In an attempt to improve the safety of contraceptives the pharmaceutical industry introduced new combinations of estrogens and progestins. According to the May 1998 issue of *R & D Directions*, eighteen new contraceptive products were in various stages of development. In 1997 the FDA approved two such combinations: Wyeth-Ayerst Alesse (levonoprogestrel and ethinyl estradiol) and Organon's Mircette (desogestrel and ethinyl estradiol). Mircette is the first oral contraceptive with a shortened hormone-free interval and reduced estrogen content. It was approved by the FDA in April 1998.

INSULIN

Insulin is a protein hormone, produced by so-called β cells in the islets of Langerhans of the pancreas. The islets were named for their discoverer, Paul Langerhans, a nineteenth-century German pathologist. The first observation that led eventually to the discovery of insulin was made in 1889 by Josef von Mering and Oscar Minkowski at the University of Strassburg. They found that removing the pancreas from a dog rapidly leads to the development of diabetes.

In 1908 Georg Zuelzer, a physician in Berlin, found that daily injections of pancreatic extract can prevent an increase in urinary sugar excretion in dogs after removal of the pancreas. Zuelzer also administered his extract to eight diabetic patients and observed an improvement in their disease in all cases. Zuelzer's work was supported by the German Schering corporation, but the company was reluctant to develop the extract commercially. In 1912 Zuelzer signed a contract with another German company, Hoechst, for the commercial development of his pancreatic extract. The extract was never marketed, however, apparently because Hoechst encountered difficulties in manufacturing an extract with reliable activity.

In 1916 Nicolas Paulesco, a Rumanian physiologist, prepared an aqueous pancreatic extract that was effective in diabetic dogs. In 1920 he claimed successful isolation of an antidiabetic hormone, which he called pancreine. He was unable to produce enough material for clinical use, however. The actual discovery of insulin is attributed to Frederick Banting and Charles Best of Toronto. Banting was an orthopedic surgeon who became interested in the reported activity of the pancreatic extract. He approached John McLeod, a well-known expert in diabetes at the University of Toronto, who offered Banting research facilities, advice on how to proceed with the investigations, and the assistance of Best, a biochemistry student. Banting and Best started their project in May 1921 and by December had isolated enough active material to inject themselves. They encountered no untoward effects, and in January 1922 they tested the extract in a young diabetic boy. Unfortunately the extract produced considerable local irritation, and it was only after further purification that a positive effect was obtained. They called their material insulin.

The manufacturing facilities at Connaught Laboratories, which the university provided to Banting and Best were of critical importance for further development of insulin. The university established the Insulin Committee and helped Banting and Best to obtain a patent for isolation of insulin. To make insulin rapidly available for diabetics, the committee decided to provide Eli Lilly Company in Indianapolis with all available information on insulin manufacturing. Within a year Eli Lilly was producing enough insulin to satisfy the demand in North America. In 1925 the Nobel Prize for physiology or medicine was awarded to Banting and McLeod. Banting shared his award with Best, and McLeod shared his with James B. Collip of the University of Edmonton, who had made a major contribution to the isolation procedure used by Banting and Best.

Insulin initially manufactured by Eli Lilly was not completely pure. The first crystalline insulin was obtained by John Abel at Johns Hopkins University in 1926. To his surprise it was shorter acting than that manufactured by Lilly. The explanation was provided by H. C. Hagedorn of Nordisc Laboratories in Copenhagen, who demonstrated that the addition of a small protein, protamine, to the crystalline insulin prolonged its activity. A longer-acting insulin formulation containing protamine was marketed in 1936. The chemical structure of insulin was elucidated in 1955 by Frederick Sanger of Cambridge, who in 1958 received the Nobel Prize in chemistry for this accomplishment. Sanger also discovered structural differences in the insulins from various species. This discovery led eventually to the production of human insulin (Humulin, Novolin). Human insulin is currently synthesized by a non–disease-producing strain of *E. coli* that has been altered by the addition of the gene for human insulin production. The tendency to produce antibodies to insulin appears to be lower with human insulin than with either beef or pork insulin, previously the most common sources of the hormone. In the United States insulin is available from two companies: Eli Lilly and Novo Nordisc. Both companies marketed a total of forty-seven formulations of insulin that differ in their origin (beef, pork, human) or additives (such as zinc or protamine), which determine its duration of action.

Among recently marketed insulin products of particular interest is Eli Lilly's Humalog. It is a rapidly acting analog of human insulin, insulin lispro, produced by a nonpathogenic laboratory strain of *E. coli* bacteria genetically altered by the addition of a gene for insulin lispro. Awaiting FDA approval is another new Lilly product called Humalog Mixtures. Hoechst Marion Roussel is planning to market their recombinant human insulin under the trade name of Insuman. The ongoing search for more convenient insulin formulations led Pfizer to the development of an insulin inhalation product that was successfully tested clinically in the spring of 1998, but is not likely to be marketed before the year 2000.

During the last twenty years considerable progress has been made in basic research dealing with the mechanism of action of insulin, its formation, and interaction with the specific receptors. Insulin is now considered a member of a large family of related peptides, known as insulin-like growth factors (IGFs). Unlike insulin, IGFs are produced by most tissues of the body. They interact with specific cell receptors, which are similar but not identical to the insulin receptors. Their physiological effects are similar to that of insulin: enhancement of glucose transport into many types of cells. The binding of insulin to its receptors is thought to lead to formation of so-called second messengers that mediate the effect of insulin on cellular metabolism. Insulin-like growth factor receptors mediate not only metabolic but also the long-term mitogenic effects of insulin or IGFs. Of considerable interest is a more recent discovery of insulin-like growth factor binding proteins (IGFBPs). These proteins are of high molecular weight and serve as IGF carrier molecules. Six such proteins were purified from several human sources, their genes were localized, and their structures identified. The discovery of IGFs and IGFBPs is likely to lead to new drugs for the treatment of diabetes.

ANABOLIC STEROIDS

Male hormones such as testosterone have anabolic (muscle-building) as well as androgenic (masculinizing) properties. In 1948 a project was initiated at G. D. Searle Laboratories, which was designed to separate these two properties and to create a pure muscle-building steroid. Searle's biologists, Francis Saunders and Victor Drill, used castrated rats as the test objects. An ideal anabolic agent would be expected to increase the weight of levator ani muscle without increasing the weight of prostate glands. The project continued for seven years without success until they discovered norethandrolone, which was as anabolic as testosterone but had only one-sixteenth its androgenic potency.

Efforts to develop an even more selective anabolic steroid continued at Searle and other laboratories. In 1959 scientists at Organon in the Netherlands discovered ethylestranol, while in 1962 stanozolol (Winstrol) was patented by Sterling-Winthrop. In 1964 Searle's scientists succeeded in replacing norethandrolone with a more specific anabolic steroid, oxandrolone (Oxanorin). Despite the higher anabolic specificity none of the currently available anabolic steroids is completely free of either androgenic properties or of sodium- and water-retaining properties. In short-term treatment most of the weight gain in patients receiving anabolic steroids is caused by salt and water retention. Although anabolic steroids are often used to improve athletic performance, there is no scientific evidence that these drugs are effective for this purpose.

ANTI-INFLAMMATORY CORTICOSTEROIDS

Since Brown-Séquard demonstrated the importance of the adrenal gland for maintaining life in 1856, many investigators have attempted to isolate a hormone that could keep adrenalectomized animals alive. In 1930 Wilbur Swingle and his associates at Princeton University isolated a potent lipid-soluble substance from the adrenal cortex that maintained life in adrenalectomized animals. It was thought to be a single substance and was named cortin.

Subsequent research has established that cortin contained numerous steroids, which are referred to as adrenal corticosteroids. Among the many investigators who tried to isolate and identify pure steroids from the adrenal cortex was Edward Kendall of the Mayo Clinic, who isolated cortisol and corticosterone. He expected that adrenal cortical hormones could be used to treat adrenal insufficiency, shock, and burns. In 1943 Tadeus Reichstein and his colleagues in Basel isolated twenty-six steroids from the adrenal cortex and determined their chemical structures. The first observation that adrenal corticosteroids could be useful in the treatment of rheumatoid arthritis was made by Kendall's colleague at the Mayo Clinic, Philip Hench, who observed that patients with the disease had no arthritic symptoms when they were ill with jaundice. Arthritic pain was also known to disappear during pregnancy. He suspected, therefore, that adrenal hormones released during stress have anti-inflammatory activity. In 1948 Hench decided to try one of Kendall's adrenal corticosteroids in the treatment of a young woman with severe rheumatoid arthritis. Kendall asked Lewis Sarett at Merck for a supply of cortisone (Cortone), which Sarett had synthesized. After three injections the patient improved dramatically. Hench reported his findings in 1949, which prompted two other pharmaceutical companies to develop new methods for producing adrenal corticosteroids. Searle devised a semisynthetic method to produce hydrocortisone (Hydrocortone) using bovine adrenal glands, and Upjohn learned to produce the same drug using a mold, *Rhizopus nigricans*. In 1950 Reichstein, Kendall, and Hench shared the Nobel Prize in physiology or medicine for the discovery of adrenal corticosteroids and their antiarthritic effects.

Cortisone is still used today for a variety of indications, but primarily for treating rheumatic and allergic disorders. Soon after cortisone and hydrocortisone were introduced, their side effects and limitations became apparent. Of particular concern were salt and water retention as well as the development of gastric ulcers. Attempts to separate the anti-inflammatory activity from the salt- and water-retaining properties of corticosteroids were partially successful in the laboratories of Schering-Plough, where Arthur Nobile obtained two new steroids by fermentation of cortisone: prednisone (Sterapred) and prednisolone (Prelone). Either of the two compounds is about five times more potent than hydrocortisone as an anti-inflammatory agent, but equipotent in respect to its salt- and water-retaining properties. At therapeutic doses prednisone or prednisolone is therefore less likely than hydrocortisone to produce edema. During the first year after its introduction, in 1955, the sales of prednisone reached $20 million, a record figure at that time. Prednisone is converted in the body to prednisolone, and because the rate of this conversion can vary among patients, prednisolone became, at least for a few years, the corticosteroid of choice.

In further attempts to reduce the salt- and water-retaining properties of corticosteroids, Seymour Bernstein and his colleagues at Lederle Laboratories synthesized triamcinolone (Aristocort, Nasacort). Triamcinolone was particularly successful in the treatment of psoriasis and allergic dermatitis because of superior absorption of its acetonide formulation through the skin, while nausea tended to limit the usefulness of its oral formulation. Similar efforts at Merck led to the synthesis and introduction of dexamethasone (Decadron) in 1958. Dexamethasone is about six times more potent than prednisolone as an anti-inflammatory agent and causes significantly less water and salt retention. Similar to dexamethasone is betamethasone (Beta-Val), which was also synthesized at Merck and patented in 1962. Other pharmaceutical companies besides Merck, including Lederle, Upjohn, and Schering-Plough, were synthesizing and developing corticosteroids, so patents for the same compounds were occasionally granted in different countries to different companies, and complicated cross-licensing arrangements had to be worked out.

Among the many anti-inflammatory corticosteroids developed during the 1950s and 1960s, two more compounds should be mentioned. They are methylprednisolone and beclomethasone (Beclovent, Vanceril, Vancenase). Methylprednisolone was patented in 1959 by Upjohn as an anti-inflammatory corticosteroid, but its indications were expanded to other uses. In addition to its use in rheumatoid and allergic diseases, methylprednisolone is now approved for use in multiple sclerosis. More recently methylprednisolone was found to benefit patients with spinal cord injury. Beclomethasone was patented in 1962 by Merck as an anti-inflammatory corticosteroid but is currently marketed by Schering-Plough, Inc., for oral inhalation therapy of asthma and as a nasal spray for the therapy of rhinitis. Either of the two drugs can serve as an example of how the discovery of new indications expands the use of a drug and prolongs its life after expiration of initial patents.

The cellular mechanism of action of the adrenal corticosteroids has been the subject of intensive research during the last thirty years. Recent findings indicate that there are two types of receptors for corticosteroids in cell membranes: mineralocorticoid receptors (MRs) and glucocorticoid receptors (GRs). The MRs mediate the effects of steroids on salt and water retention, while GRs mediate their effects on deposition of liver glycogen as well as their anti-inflammatory effects. The finding that MRs and GRs are different supports the claims that some corticosteroids can have anti-inflammatory activity without salt and water retention. Activation of GRs in the cell membrane increases the formation of a large protein, lipocortin-1, which reduces the formation of pro-inflammatory lipids. Of particular interest are the recent findings that MRs and

GRs are also present in the brain and can modify behavior. These findings are likely to expand the therapeutic indications for the corticosteroids even further.

OTHER STEROIDS

Among many formulations of testosterone marketed during the last decade, the most novel approach is represented by Testoderm, a controlled transdermal delivery system for testoterone marketed by Alza. Its indication is testosterone deficiency in males. Daily application of Testoderm approximates the natural pattern of blood testosterone.

Among other steroids recently introduced into clinical medicine is finasteride (Proscar), a synthetic azasteroid derivative that inhibits an enzyme, steroid 5-α-reductase, which catalyzes the formation of 5-α-dihydrotestosterone (DHT) from testosterone. Growth and development of the prostate gland is stimulated by DHT. Finasteride was discovered in a specifically designed screening program for inhibitors of 5-α-reductase initiated by Roy Vagelos, now retired chief executive officer of Merck. In an interview with *Harvard Business Review* (November–December 1994) Vagelos described the events leading to the discovery of finasteride. According to him, he got the idea from an article by a group from Cornell who described patients with a congenital deficiency of steroid 5-α-reductase. These patients had underdeveloped prostate glands. Vagelos, together with Merck chemists, assumed that inhibition of steroid 5-α-reductase by a drug could control the benign prostatic hyperplasia (BPH) in elderly men that is often associated with a decrease in urinary flow. This expectation was confirmed by clinical studies, although only half of the patients treated with finasteride for twelve months showed improvement in urine flow and other symptoms. Inhibition of 5-α-reductase with the consequent increase in plasma testosterone concentrations was also expected to improve hair growth, and so finasteride was marketed in 1997 for the therapy of male baldness under the trade name Propecia.

Nonsteroidal Anti-inflammatory Drugs and Immunosuppressant Agents

Nonsteroidal anti-inflammatory drugs, or NSAIDs, belong to the most widely used group of drugs. They include aspirin, ibuprofen, ketoprofen, naproxen, and many other drugs used to treat not only inflammatory diseases but also minor headaches and pain of any origin. Their indications include fever, the common cold, and arthritis. Although NSAIDs belong to different chemical classes, many of them share a common mechanism of action: inhibition of production of endogenous chemicals called prostaglandins.

Either anti-inflammatory steroids or immunosuppressant agents may be used to alleviate arthritic pain when NSAIDs fail. Immunosuppressants block responses to foreign proteins, and their major use is in preventing rejection of transplanted organs.

NONSTEROIDAL ANTI-INFLAMMATORY DRUGS

Aspirin

Analgesics are drugs that reduce or abolish pain; antipyretics reduce fever, and anti-inflammatory drugs reduce inflammation associated with arthritis or other diseases. Acetylsalicylic acid, better known as aspirin, is one of the drugs that exhibits all three activities. It is the most commonly used analgesic and antipyretic drug today. Aspirin belongs to the chemical class of salicylates.

The first salicylate used in medicine was salicylic acid, used initially as a disinfectant. Its antipyretic activity was discovered in the early 1870s, when Carl Buss of St. Gallen, Switzerland, administered salicylic acid to typhoid patients for "internal cleansing" and found that it reduced fever without affecting the disease process. After that discovery salicylic acid became a standard fever-reducing (antipyretic) drug and has also been successfully used as an analgesic in the treatment of arthritis. Patients treated with salicylic acid, however, objected to the drug's unpleasant taste, which caused them to vomit. Felix Hoffman, a chemist at Bayer laboratories in Wuppertal, Germany, decided to give his father, who suffered from arthritis and was taking salicylic acid, a closely related but better-tasting substance, acetylsalicylic acid, instead. The "trial" was a success, and after further confirmation in a few German clinics, acetylsalicylic acid was introduced as Aspirin for the treatment of pain, fever, and inflammation. Over the years this term became generic. In terms of the total number of tablets sold, aspirin is now the best-selling drug in the world, even though it is not free of side effects.

The most common adverse effect of aspirin is gastrointestinal irritation, usually associated with the appearance of blood in feces. The bleeding is caused by petechial hemorrhages (small, pinpoint-sized bleedings), which also occur in rats receiving long-term aspirin therapy at doses comparable to those used in human therapy. The hemorrhages can lead to ulcers, particularly when associated with stress or when aspirin is coadministered with other ulcerogenic drugs.

Aspirin was used extensively for more than a hundred years before its mechanism of action was established. The famous British pharmacologist John Vane received the Nobel Prize and was knighted for his work with prostaglandins, which included discovery that aspirin and other NSAIDs act by inhibiting production of pro-inflammatory prostaglandins (fatty acids first discovered by Ulf von Euler in seminal fluid and later found to be present in most tissues). By inhibiting an enzyme called cyclooxygenase, aspirin prevents the formation of pro-inflammatory and tissue-protective prostaglandins from their precursor, arachidonic acid, an essential fatty acid normally present in most organs. Inhibition of the formation of pro-inflammatory compounds is thought to be responsible for the anti-inflammatory activity of aspirin, while inhibition of the formation of prostaglandin-like compounds that have a tissue protective activity is likely to explain the gastrointestinal side effects. A more specific inhibitor of the formation of pro-inflammatory prostaglandins could therefore be expected to have fewer side effects than aspirin; such compounds are currently under development.

The current *Physicians' Desk Reference* (1998) lists twenty-two preparations containing aspirin. Many of them are combination products; others are formulations designed to reduce aspirin-induced gastric irritation by buffering either with calcium carbonate, magnesium oxide, and magnesium carbonate (Bufferin) or with calcium carbonate, magnesium, and aluminum hydroxides (Ascriptin). An enteric-coated formulation of aspirin is also available (Ecotrin). Many other popular analgesic formulations, such as Anacin, Easprin, and Alka-Seltzer, contain aspirin.

Other Nonsteroidal Analgesic and Anti-inflammatory Drugs

In the 1870s physicians began to associate analgesic with antipyretic and anti-inflammatory properties of the drugs. Franz Stricker in Berlin found that salicylic acid had anti-inflammatory and antirheumatic activities in addition to its antipyretic properties. In the 1880s phenazone (antipyrine) and aminopyrine (amidopyrine or aminophenazone) were introduced as analgesics; they were also effective as anti-inflammatory and antirheumatic drugs. These drugs were used for many years, alone or in various combinations, despite serious, initially unrecognized side effects. These included a blood disorder called agranulocytosis, a disease characterized by a marked reduction in the white blood cell count; increased susceptibility to bacterial infections; and ulcerations of mucous membranes and skin.

Another antipyretic and analgesic drug discovered in the 1880s was acetanilide. It was discovered accidentally by A. Cahn and P. Hepp at the University of Strassburg, who were asked by their professor to treat a patient with naphthalene as an internal antiseptic. They requested naphthalene from the pharmacy but received acetanilide instead. While treating their patient with acetanilide, they discovered its antipyretic activity. Because naphthalene does not affect body temperature, they suspected that they may have received the wrong drug—a suspicion that proved correct. Acetanilide was subsequently marketed by Hoechst and was widely used despite its side effects, which included methemoglobinemia. This condition is characterized by formation of methemoglobin, a modified form of oxyhemoglobin, which firmly binds oxygen and therefore interferes with normal respiration. Bayer chemists synthesized various derivatives of acetanilide. The 4-ethoxy derivative was found to be less toxic than acetanilide and was marketed under the trade name of Phenacetin.

Another derivative, paracetamol (acetaminophen), was also synthesized at Bayer and tested clinically by Josef von Mering, who incorrectly claimed that the drug produced methemoglobinemia. This claim was successfully challenged in 1947 by David Lester from Yale, who demonstrated the relative safety of paracetamol, which was marketed in 1953 by Sterling-Winthrop under the

Sir John Robert Vane (1927–) shared the 1982 Nobel Prize in physiology or medicine with Sun K. Bergström and Bengt I. Samuelsson for major contributions to the knowledge of pharmacology of prostaglandins, including discovery of the mechanism of action of aspirin. Sir John taught pharmacology at Oxford and Yale, as well as at the Royal College of Surgeons in London and directed research at the Wellcome Research Laboratories in the United Kingdom. He established the William Harvey Research Institute in London, which he currently chairs. Courtesy Sir John Robert Vane.

trade name Panadol and by McNeil as Tylenol. The *Merck Index* lists thirty-five different trade names for paracetamol. The U.S. Pharmacopoeia adapted the generic name of acetaminophen for this drug; currently seventy-four versions of acetaminophen, alone or in various analgesic formulations, are available on the American market. The main advantage of acetaminophen over aspirin is the absence of gastrointestinal bleeding and ulceration. It is less effective than aspirin, however, in arthritic disorders, dysmenorrhea, and other conditions requiring high anti-inflammatory efficacy.

Another group of analgesics was studied after World War II. Because aminopyrine was poorly soluble and poorly absorbed following oral administration, the Swiss company Geigy developed a soluble complex of aminopyrine with its acidic analog, phenylbutazone, under the trade name of Irgapyrine. Soon after the compound was introduced in 1949, it was discovered that phenylbutazone was primarily responsible for the therapeutic effects of Irgapyrine. These findings led to the marketing of phenylbutazone alone, without aminopyrine, as Butazolidin.

The pharmacological effects of phenylbutazone were first described in 1951. In animal studies phenylbutazone compares favorably with other nonsteroidal anti-inflammatory drugs, such as aspirin or ibuprofen. Its main mechanism of action—inhibition of cyclooxygenase—is similar to that of aspirin. In addition phenylbutazone inhibits the migration of white blood cells, increases the excretion of uric acid, and blocks the release of enzymes that destroy the integrity of the cell. These properties probably contribute to the beneficial effects of phenylbutazone.

Phenylbutazone is beneficial in many rheumatic diseases, including acute rheumatoid arthritis and degenerative joint disease of the hip and knee. Like other NSAIDs it provides symptomatic relief without curing the disease. Unfortunately, like aminopyrine, phenylbutazone can produce agranulocytosis; it can also lead to aplastic anemia, a deficiency of blood cell formation. Because of the risks phenylbutazone is no longer available in the United States.

While Geigy was working with phenylbutazone and related agents, Merck tried a different approach. Because serotonin mediates or facilitates the inflammatory reaction, Merck scientists reasoned that its antagonists should have anti-inflammatory properties. Working on that assumption, Tsung-Ying Shen synthesized indomethacin, which Merck marketed under the trade name of Indocin and which became one of the most successful NSAIDs. Indomethacin's development was greatly facilitated by an experimental technique that involved injecting an irritant (carrageenan) into a rat's paw to produce swelling and then measuring this swelling by immersing the paw in mercury. This technique was originally developed by Gordon Van Arman at Wyeth in Radnor, Pennsylvania. While visiting Van Arman's laboratory, Charles Winter, Merck's pharmacologist, was impressed with the technique and used it at his laboratory at Merck. After testing many hundreds of compounds in rats, Winter identified indomethacin as the best candidate for clinical development. The original hypothesis that indomethacin acts by blocking the pro-inflammatory effects of serotonin proved to be wrong. Like aspirin, indomethacin inhibits cyclooxygenase, and its anti-inflammatory properties probably depend on the inhibition of this enzyme.

Indomethacin has been shown clinically effective in the treatment of various forms of arthritis, including rheumatoid and osteoarthritis, in acute painful shoulder syndrome (bursitis or tendinitis), and in acute attacks of gouty arthritis. Its major side effects are gastrointestinal. Single or multiple ulcerations of the esophagus, stomach, duodenum, or small or large intestines have been reported in patients receiving indomethacin. Despite this side effect indomethacin was highly successful. Merck scientists and management correctly identified the medical need and then met that need with indomethacin. It was clear to Merck management, however, as well as to the management of other pharmaceutical companies, that indomethacin could be further improved and that similar drugs with fewer side effects could be developed.

The continuation of this project at Merck led in the early 1960s to the discovery of sulindac (Clinoril) by Van Arman, who replaced Winter at Merck. Sulindac is a pro-drug—most of its pharmacological activity depends on the formation of its sulfide metabolite. The incidence of gastrointestinal side effects is lower with sulindac than with indomethacin, possibly because of its pro-drug nature. At Pfizer, Edward H. Wiseman and his associates developed piroxicam (Feldene), which is completely absorbed in the gastrointestinal tract and has a very long half-life.

The researchers at Boots Pharmaceuticals in the United Kingdom concentrated their efforts on a project designed to develop a safer and more potent aspirin. This project led in 1964 to the development of ibuprofen, marketed by Boots in the United States as Rufen and by Upjohn as Motrin. Ibuprofen is a derivative of propionic acid and is chemically unrelated to indomethacin. At doses used for antiarthritic therapy, it produces fewer gastrointestinal side effects than aspirin. Its relative safety allowed the 200-milligram, nonprescription tablets known as Advil to be introduced. Use of ibuprofen was extended to include treatment for pain after surgery, childbirth, and injuries, as well as in dysmenorrhea, which created a highly impressive market.

The development of ibuprofen is an excellent example of how a slight improvement in the side-effect profile can serve a medical need and provide substantial profits for the industry. Since ibuprofen was introduced, many other chemically related drugs (phenylpropionic acid derivatives) were synthesized and developed. Among them are naproxen (Naprosyn) and ketoprofen (Orudis), which are still unfortunately not free from gastrointestinal side effects. Their mechanism of action appears similar to that of aspirin and indomethacin—inhibition of cyclooxygenase and consequently of prostaglandin synthesis.

The anti-inflammatory products marketed in the 1980s and early 1990s were slow-release formulations of older NSAIDs, their analogs, or combination products. In 1997 the FDA approved G. D. Searle's Arthrotec, a combination product of diclofenac (Voltaren) and misoprostol, a synthetic prostaglandin-like compound. Misoprostol was expected to antagonize the gastrointestinal side effects of diclofenac. The extended-release formulation of etodolac (Lodine XL) was developed by Wyeth-Ayerst and approved by the FDA in February 1998 for the treatment of osteoarthritis and rheumatoid arthritis.

Among the drugs currently undergoing clinical trials are potential breakthrough products—cyclooxygenase-2 (COX-2) inhibitors. In the early 1990s it was recognized that one of the enzymes involved in the formation of prostaglandins, cyclooxygenase, had two isoforms (COX-1 and COX-2). Gastric damage, the major side effect of NSAIDs, is caused by inhibition of COX-1 only; therefore drugs that inhibit COX-2 primarily or exclusively are expected to have anti-inflammatory activity with little or no gastrointestinal side effects.

Aspirin and indomethacin are primarily COX-1 inhibitors. Their ratio of COX-1 over COX-2 inhibitory activities is approximately 50, while diclofenac and naproxen inhibit both isoforms equally (ratio = 1). Etodolac has a slightly higher selectivity for COX-2. During the 1990s many pharmaceutical companies identified potential drugs that inhibit COX-2 selectively or exclusively. Two of them, celecoxib (Celebrex) from G. D. Searle and rofecoxib (Vioxx) from Merck, were approved in 1999.

In 1996 the ad hoc committee of the American College of Rheumatology published guidelines for the treatment of rheumatoid arthritis. According to the committee's recommendations, all patients with active rheumatoid arthritis despite treatment with NSAIDs are candidates for additional therapy with disease-modifying antirheumatic drugs. These drugs include methotrexate, hydroxychloroquine (Plaquenil), sulfasalazine (Azulfidine), azathioprine (Imuran), and microemulsion of cyclosporine (Neoral). The committee's recommendation was based on the recognition that these drugs have the potential to prevent or reduce joint damage and that they are not as toxic as originally thought.

Another potentially attractive approach to the treatment of rheumatoid arthritis is the use of "biological response modifiers." These drugs include antagonists of mediators of inflammation, tumor necrosis factor and interleukin-1. Immunex in collaboration with Wyeth-Ayerst is developing Enbrel, an inhibitor of tumor necrosis factor.

IMMUNOSUPPRESSANTS

The first drug described to have immunosuppressant activity was mercaptopurine (Purinethol), an antileukemic agent originally discovered in 1952 by George Hitchings and Gertrude Elion at Wellcome Research Laboratories. In the late 1950s William Dameshek and Robert Schwartz at Tufts University in Boston were looking for drugs capable of depressing immune response after bone-marrow transplantation in patients with leukemia. They tested many existing drugs for their ability to depress production of antibodies to human serum albumin in rabbits; mercaptopurine was the most effective. Roy Calne at Harvard Medical School tested mercaptopurine in dogs undergoing kidney transplantation and found that transplanted kidneys functioned in mercaptopurine-treated animals longer than in the control dogs. In an attempt to optimize the activity of mercaptopurine further, Hitchings and his associates at Burroughs Wellcome screened many compounds related to mercaptopurine. They selected azathioprine (Imuran). Calne introduced the drug to transplant surgeons, and it was shown to be effective in suppressing rejection of the transplanted organs.

Other laboratories joined the search for better immunosuppressants. In 1972 Jean Borel at Sandoz discovered the immunosuppressant activity of cyclosporin A (Sandimmune). In 1978 Calne demonstrated its effectiveness in patients undergoing bone-marrow transplants.

In 1984 scientists from Fujisawa Pharmaceutical Corporation in Japan isolated a new immunosuppressant, known as tacrolimus (FK-506, Prograf) from a fermentation broth of *Streptomyces tsukubaensis*. The first clinical trials were conducted in patients with liver transplants. Subsequent studies demonstrated clinical efficacy in preventing rejection in a broad range of allografts. Tacrolimus has been approved by the FDA. Many experimental studies in various laboratories are now attempting to develop new immunosuppressants with similar efficacy but lower toxicity than tacrolimus. Among the new promising drugs is DSG (15-deoxysperqualin), which appears to interfere with antigen process-

Gertrude Belle Elion (1918–1999) shared the 1988 Nobel Prize in physiology or medicine with George Hitchings and Sir James Black. She made major contributions in the fields of antileukemics (6-mercaptopurine), immunosuppressants (azathioprine), and hypouricemics (allopurinol). All of her research work was conducted at Wellcome Research Laboratories, Burroughs Wellcome Company, initially in Tuckahoe, New York, and later in Research Triangle Park, North Carolina. She described her career in 1993 in the Annual Review of Pharmacology and Toxicology. *Courtesy Gertrude Elion.*

ing and to affect B- and T-lymphocytes directly; it is also less toxic than previously available immunosuppressants. DSG is in advanced clinical evaluation.

In 1997 the FDA approved daclizumab (Zenapax), a highly selective immunosuppressant from Hoffmann–La Roche. It is the first drug capable of blocking only those immune cells that are ready to attack the transplanted organ without compromising the whole immune system. In February 1998 the FDA approved mycophenolate mofetil (Cellcept), another immunosuppressant from Hoffmann–La Roche. It inhibits production of an enzyme called inosine monophosphate dehydrogenase (IMPDH) and therefore the synthesis of guanosine nucleotides. According to the May 1998 issue of *R&D Directions*, there are twenty-six other immunosuppressants in various stages of development by twenty-four different pharmaceutical or biotechnology companies.

Drugs for Metabolic Diseases

Metabolism is defined as the sum of all the processes by which living matter is built or broken down. Metabolic diseases are abnormalities in metabolism caused by the interaction of genetic and environmental factors. The genetic factor is usually an inherited mutation of a gene, whereas environmental factors include diet and stress. Diabetes mellitus and gout are examples of metabolic diseases. The available drugs do not cure metabolic diseases, but only alleviate the symptoms and consequences of the diseases. They prolong life expectancy and improve the quality of life.

ANTIDIABETIC SULFONAMIDES

One of the series of drugs derived from the antibacterial sulfonamides are sulfonylureas, orally active hypoglycemic agents. Their discovery and development can be traced to M. Janbon and his associates in Montpellier, France, who in 1942 treated a typhoid patient with sulfamidothiadiazol, an antibacterial sulfonamide, and noticed "hypoglycemic accidents," a lowering of blood sugar after drug treatment. August Loubatiéres, also from Montpellier, subsequently administered the same type of sulfonamide to a pancreatectomized animal, found no hypoglycemic activity, and concluded that the drug acts in normal animals by stimulating the pancreas to release insulin. Another antibacterial sulfonamide, carbutamide, was subsequently found to lower blood sugar in patients. The U.S. patent for carbutamide as a hypoglycemic was issued in 1959 and assigned to Boehringer Mannheim.

Soon thereafter tolbutamide (Orinase, Oramide) was discovered at Hoechst in Germany. It had no antibacterial activity and was less toxic than carbutamide. Upjohn developed tolbutamide in the United States. Its duration of action is about six hours, so it has to be taken two or three times a day. In 1960 a U.S. patent was issued to Pfizer for another sulfonylurea antidiabetic, chlorpropamide (Diabenese). It has longer duration of action than tolbutamide and can be administered once a day. The second-generation sulfonylureas with longer duration of action, glipizide (Glucotrol) and glyburide (Micronase), were introduced in the early 1980s. Because of their high relative potency their therapeutic doses are lower than those of tolbutamide or chlorpropamide. Their absolute potency, or efficacy, is not any different from that of the two other drugs, however, and their pharmacological profile is very similar in many respects.

In 1961 a cooperative clinical trial was initiated in twelve medical centers to determine whether control of blood sugar reduces the cardiovascular mortality rate of diabetic patients. The study, completed eight years later, concluded that the incidence of deaths from cardiovascular diseases is higher in patients receiving oral antidiabetic drugs. The American Diabetes Association initially supported this conclusion but later withdrew its support after many deficiencies in the study were discovered. The current recommendations state that the sulfonylureas should be used only in patients with non–insulin-dependent diabetes mellitus (NIDDM), whose disease cannot be controlled by diet and who are unable or unwilling to take insulin.

There are many contradictory studies on the cellular or molecular mechanisms of action of sulfonylureas. Most studies agree that they act on the so-called β-cells of the pancreas to facilitate insulin release. The most recent hypothesis involves potassium channels, special proteins in the cell membrane, that are blocked by sulfonylureas. This blockade reduces the outward potassium current in the pancreatic β-cells. Because these channels regulate potassium currents in other organs, such as the heart or vascular smooth muscle, it is unlikely that the effects of sulfonylureas are limited only to the pancreas.

OTHER ANTIDIABETIC AGENTS

In 1995 metformin was marketed in the United States by Bristol-Myers Squibb under the trade name Glucophage. Metformin is an older drug first patented in 1948 and introduced in Europe as an antidiabetic in 1957. Its introduction in the United States was delayed because it tends to cause lactic acidosis, particularly in patients with renal disease. The approval of metformin was based on the assumption that its benefits outweigh the risks. It is indicated in patients with NIDDM as an adjunct to the diet and is contraindicated in patients with renal disease or acidosis.

In 1996, twenty years after its discovery, Bayer's acarbose (Precose) was approved by the FDA and marketed in the United States. Acarbose acts by delaying absorption of carbohydrates in the gastrointestinal tract. Consequently, it prevents the peak increases of glucose blood levels after meals. Acarbose is indicated in patients with NIDDM as an adjunct to the diet or as a combination therapy with sulfonylureas. Its major side effect is flatulence.

In 1997 the FDA approved the first drug to treat insulin resistance, troglitazone (Rezulin), codeveloped by Sankyo in Japan and Parke-Davis in the United States, as well as repaglinide (Prandin), marketed in the United States jointly by NovoNordisk and Schering-Plough. Troglitazone acts by improving target cell response to insulin, while repaglinide acts by stimulating the secretion of insulin. Many new products for the therapy of diabetes and particularly for the prevention of diabetic complications are currently in development.

ANTIGOUT DRUGS

The oldest antigout drug, which is still used today, is colchicine. It was isolated in 1820 from the root of the herb meadow saffron by Joseph Pelletier and Joseph Caventou at the Superior Pharmacy School in Paris. The herb was used by Arabian and Byzantine physicians to treat gout and rheumatism.

The discovery of another antigout drug, a uricosuric (drugs that cause an increase in uric acid excretion), was an unexpected bonus from another project at Merck. In the early 1940s penicillin was extremely scarce and badly needed by the Armed Forces. The need was so urgent that urine from patients treated with penicillin was collected and the excreted penicillin recovered. Karl Beyer and his associates at Merck undertook a project designed to find a substance that would delay the excretion of penicillin and, it was hoped, prolong the duration of its action. In cooperation with James A. Shanon of New York City, Beyer had established a modern renal function laboratory at Merck and was well prepared to study renal excretion of drugs.

The first question the Merck researchers asked was whether penicillin is secreted by the kidney. The first few experiments provided positive results: The excretion rate of penicillin was higher than could be expected from simple renal filtration of the drug. The next question was whether another substance secreted by the kidneys, para-amino hippurate (PAH), could slow the excretion rate of penicillin. The response to the second question was also positive, but extremely high amounts of PAH were necessary to affect the excretion rate significantly. A team of Merck chemists under the leadership of James M. Sprague decided to try to optimize PAH and find a related compound that would inhibit penicillin secretion. These efforts led to probenecid (Benemid), which was clinically effective in delaying penicillin secretion. By 1950, when Merck was ready to market probenecid, penicillin was no longer scarce, so the new drug was no longer needed. During their studies Beyer and his associates found that probenecid increases the excretion of uric acid, apparently by blocking its tubular reabsorption. Because uric acids crystals are deposited in joints and blood uric acid levels are increased in patients with gout, it was logical to try probenecid in gout. This was accomplished successfully. Probenecid is now indicated not only as an adjuvant to penicillin and other antibiotics, to prolong their action, but also to treat hyperuricemia (elevated uric acid levels) associated with gout. It is also available in combination with colchicine as Col-Benemid for treatment of chronic gout.

Another drug used in the therapy of gout is allopurinol (Zyloric, Zyloprim), a drug developed in 1957 by Hitchings and Elion at Burroughs Wellcome in an attempt to improve the therapy of leukemia. They decided to develop an inhibitor of an enzyme, xanthine oxidase, which was known to degrade mercaptopurine, an antileukemic drug. By inhibiting the enzyme they were hoping to slow the metabolism of mercaptopurine and therefore prolong its effects. They discovered that allopurinol inhibits xanthine oxidase, but no therapeutic benefits in leukemia were obtained when the two drugs were used together in the clinic. Xanthine oxidase was also known to inhibit formation of uric acid. In 1962 Wayne Rundles at Duke University tested allopurinol in patients with gout and found that the drug lowered blood uric acid levels and reduced the severity of symptoms of gout. In 1966 Burroughs Wellcome introduced allopurinol to prevent acute attacks of gout.

BONE METABOLISM REGULATORS

Bone metabolism regulators include estrogens to treat osteoporosis as well as the recently developed phosphonates for osteoporosis and some of the other bone deformities (e.g., Paget's disease). Estrogens as drug therapy were discussed earlier. One of their effects is the prevention of bone resorption. They are thought to be more effective when used before any significant loss of bone mass. Raloxifene (Evista), a new estrogen receptor modulator, is also useful in osteoporosis allegedly because it acts as estrogen in some tissues and as antiestrogen in others. SmithKline Beecham is currently evaluating idoxifene, another estrogen receptor modulator, also for the treatment of osteoporosis. In 1997 alendronate (Fosamax), Merck's aminobiphosphonate, was approved by the FDA for the prevention of osteoporosis. It acts by inhibiting the activity of osteoclasts, bone cells involved in bone resorption. Sanofi's biphosphonate, tiludronate (Skelid), was approved six months earlier for the treatment of Paget's disease because clinical studies with tiludronate in osteoporosis were not yet completed. In March 1998 the FDA approved Procter & Gamble's phosphonate, residronate (Actonel), also for the treatment of Paget's disease.

ANTI-OBESITY DRUGS

This group of drugs is also known as appetite suppressants or "diet pills." The term *anti-obesity* is preferable because drugs that control obesity without suppressing appetite may eventually become available. Anti-obesity drugs have a potentially huge market in the United States: Americans spend nearly $30 billion annually on diet foods and programs. In the 1950s and 1960s amphetamine was used in the treatment of obesity. After amphetamine's potential for abuse was recognized, phentermine (Fastin) or fenfluramine (Pondimin) was prescribed instead. Between 1973 and 1996 the FDA did not approve any new anti-obesity drugs. In 1992 a report was published in *Clinical Pharmacology and Therapeutics* that a combination of fenfluramine and phentermine, known as fen-phen, produced a sustained weight loss in a small group of obese patients. The use of this combination was promoted through various diet centers. In 1996, after lengthy debate, the FDA Advisory Committee recommended the approval of dexfenfluramine (Redux), an isomer of fenfluramine developed by Interneuron and Wyeth-Ayerst. There was concern that the drug infrequently caused primary pulmonary hypertension and a post-marketing study of its safety was requested by the FDA. In 1996 the results of the International Primary Pulmonary Hypertension Study were reported by Hayden and associates in the *New England Journal of Medicine*. According to this study fenfluramines greatly increase the risk for primary pulmonary hypertension. The manufacturers of fenfluramines informed physicians about this risk in a customary "Dear Dr." letter. The use of fenfluramines declined drastically, but they were not withdrawn until September 1997 when reports about the serious and unusual heart valve problems among users of fen-phen were published and the FDA asked Interneuron and Wyeth to withdraw fenfluramines. Heart valve abnormalities were found in 30 percent of fen-phen users and were attributed to fenfluramine. Phentermine is thought to enhance fenfluramine's effects but not to produce heart valve damage on its own. Milo Gibaldi, in the March 1998 issue of *Pharmaceutical News*, called the story of diet pills an "American fiasco." In spite of fenfluramine's failure, new anti-obesity drugs are being developed. In November 1997 the FDA approved sibutramine (Meridia), an inhibitor of serotonin and norepinephrine uptake, from Knoll Pharmaceutical

Company. Awaiting approval is Hoffmann–La Roche's orlistat (Xenical), an inhibitor of fat absorption from the gastrointestinal tract.

Cancer Chemotherapeutics

A diagnosis of cancer was, and still is, viewed by many as a death sentence. Fortunately this is no longer necessarily the case. If diagnosed early, some types of cancer can be controlled by drugs. The drugs are also used as adjuncts to surgery to prevent metastasis formation and to prolong life or improve the chances of a complete cure. During the last fifty years many anticancer drugs have been developed; they are usually referred to as cancer chemotherapeutics.

Inhibition of tumor growth by chemicals or tissue extracts was first observed in experimental animals in the late 1930s. Richard Lewisohn, at Mount Sinai Hospital in New York City, noted that primary tumors are rarely found in the spleen, which led him to suspect that the spleen contains a substance with antitumor activity. Lewisohn decided to treat mice carrying transplanted tumors with spleen or liver extracts and found that the extracts caused tumors to regress in almost half of the animals. Because vitamins of the B group are found in liver, Lewisohn tested yeast extracts, which were also rich in B vitamins. He found that yeast and even barley extracts were capable of causing regression of spontaneous sarcoma 180 in mice. The validity of Lewisohn's results was questioned, and scientists at Memorial Hospital in New York City could not confirm his findings. To continue his work, Lewisohn looked for industrial support and found a partner in Lederle Laboratories.

FOLIC ACID ANTAGONISTS

Among the many substances Lewisohn tested were concentrates of folic acid. Two folic acid derivatives, diopterin and teropterin, were synthesized at Lederle Laboratories and submitted to Sidney Farber at the Harvard Medical School for clinical evaluation. In 1947 Farber and his associates reported temporary improvement in some patients receiving teropterin, but in patients with leukemia, folic acid derivatives appeared to accelerate disease progression. Farber reasoned that if folic acid–like substances accelerate the progression of leukemia, folic acid antagonists might have an opposite effect. Antagonists of folic acid and other growth factors were already known as antimetabolites and were available at Lederle and other companies.

Among folic acid antagonists made available to Farber were pteroyl glutamic acid and aminopterin, synthesized at Lederle's parent company, American Cyanamid. Farber initially obtained sustained remissions in ten of the sixteen children with leukemia that he treated with aminopterin. These findings were confirmed by further studies. But aminopterin had a pronounced effect on white blood cells, reducing the counts even to below-normal levels and thereby exposing the patients to infections. Aminopterin was soon replaced by methotrexate, which appeared to be safer. In 1952 George Hitchings and Gertrude

George Herbert Hitchings (1905–1997) shared the 1988 Nobel Prize in physiology or medicine with Gertrude B. Elion and Sir James Black. He spent most of his professional life at Wellcome Research Laboratories and made major contributions to the discovery and development of antileukemic, immunosuppressant, and antiviral drugs. He described his accomplishments in 1992 in the Annual Review of Pharmacology and Toxicology. *Courtesy Gertrude Elion.*

Elion developed mercaptopurine (Purinethol), which caused year-long remissions of leukemia. In 1957 Charles Heidelberger from the University of Wisconsin, in collaboration with Robert Duschinsky and Robert Schnitzer from Hoffmann–La Roche, discovered fluorouracil (Efudex, Fluoroplex), which caused regression of transplanted tumors in mice. Fluorouracil was found to have clinical use in treating gastrointestinal and breast tumors.

ALKYLATING AGENTS

Alkylating agents represent another class of cancer chemotherapeutics. Their discovery is a by-product of a search for chemical warfare agents. In 1942 Yale pharmacologists Louis Goodman and Alfred Gilman studied the mechanism of the action of nitrogen mustard and found that it primarily affected rapidly dividing cells, such as blood cells and gastrointestinal lining cells. Because cancer cells also divide rapidly, they decided to test nitrogen mustard and similar compounds in mice with transplanted lymphoma. This work led to the discovery of mechlorethamine (Mustargen), which is still used today to treat Hodgkin's disease, leukemia, and even brain tumors. In the United Kingdom similar research was conducted by Alexander Haddow and his associates at the Chester Beatty Institute. In 1948 they discovered that aromatic nitrogen mustards are effective cytotoxic agents and in 1952 identified chlorambucil (Leukeran) as a potential anticancer drug. Chlorambucil initially appeared to be less toxic than mechlorethamine and was widely used in Hodgkin's disease, lymphomas, leukemia, and ovarian cancer.

The discovery of chlorambucil stimulated many other laboratories to conduct an intensive search for new alkylating agents. In 1956 Herbert Arnold and his associates at the Asta-Werke in Germany developed cyclophosphamide (Cytoxan), a pro-drug that is metabolized in the liver to active alkylating substances. Cyclophosphamide is used in the treatment of many types of cancer, including leukemias, lymphomas, ovarian tumors, and breast carcinomas.

ANTIBIOTICS AS ANTICANCER DRUGS

The reorganization of the Chemical Warfare Service after World War II led to the discovery not only of alkylating agents but also of other types of anticancer drugs. At the Sloan-Kettering Institute in New York City, Cornelius Roads hired scientists from the Chemical Warfare Service to screen chemicals and culture filtrates from molds for anticancer activity in rats with transplanted tumors. This program led to the discovery of the anticancer activity of certain antibiotics, including actinomycin. Unfortunately none of the antibiotics discovered by the Sloan-Kettering group was safe enough for clinical use. In 1955 the Cancer Chemotherapy National Service Center in Bethesda, Maryland, took over and greatly expanded the Sloan-Kettering cancer screening program. In 1949 Hans Brockmann of the University of Göttingen isolated actinomycin C (cactinomycin), and Christian Hackmann at Bayer demonstrated its antitumor effect in mice. It was found useful in treating Hodgkin's disease and was marketed as Sanamycin.

In 1953 Selman Waksman, at Rutgers University, isolated actinomycin D (dactinomycin, Cosmigen), which was found to be effective in treating kidney tumors (Wilms' tumor), bone tumors (Ewing's tumor), and Hodgkin's disease. Hamao Umezawa, at the Tokyo Institute for Microbiological Chemistry, organized a large screening program for antitumor antibiotics and in 1956 discovered mitomycin (Mutamycin), which is used in combination with other cancer chemotherapeutic agents to treat stomach and pancreas carcinomas. In 1962 Umezawa discovered bleomycin (Blenoxane), which is currently used to treat head, neck, and testicular carcinomas. Aurelio Di Marco at Farmitalia in Milan discovered daunorubicin (Cerubidine) in 1962 and doxorubicin (Adriamycin PFS) in 1967. Both drugs are used in leukemia, and doxorubicin is also used to treat certain solid tumors, such as breast and ovarian carcinomas.

During the last decade some new synthetic derivatives of antibiotics were introduced as cancer chemotherapeutics. In 1997 Pharmacia & Upjohn marketed a synthetic derivative of daunorubicin, idarubicin (Idamycin), for the treatment of acute leukemia in adults. Its use is associated with severe side effects, including suppression of bone-marrow activity, heart failure, loss of hair, nausea, and vomiting. Another synthetic compound chemically related to daunorubicin, mitoxantrone (Novantrone), was marketed by Immunex. Its indications and side effects are similar to those of idarubicin.

ANTIMITOTIC DRUGS

Some of the currently used cancer chemotherapeutics are derived from plants. Among them are alkaloids from periwinkle (*Vinca rosea Linn.*): vinblastine (Velban) and vincristine (Oncovin). Periwinkle extract has been used in folk medicine for centuries. In 1949 an endocrinologist from the University of Western Ontario, Ralph Noble, heard about its use in diabetes and decided to test the effect of the extract on the blood sugar of rats. The extract had no effect when taken orally, but when Noble administered it intravenously, the rats died from infections. Noble found that the white cell count in the blood of these rats was dramatically reduced. In collaboration with Charles Beer, a chemist, Noble isolated an active principle from the periwinkle extract, an alkaloid later named vinblastine.

Noble reported his findings in 1958 at a meeting of the New York Academy of Science. At the same meeting Gordon Svoboda from Lilly stated that he had also found that periwinkle extract reduces the white blood cell count and that it is effective in mice with experimental leukemia. In 1961 Svoboda reported isolation of another active alkaloid, vincristine, from periwinkle extract. Both alkaloids are currently available for the therapy of leukemia, Hodgkin's disease, and other malignancies. They are thought to act by blocking cell division (mitosis) and are therefore called mitotic poisons.

Another mitotic poison of plant origin is podophyllotoxin, which was isolated from the roots of the mayapple (*Podophyllum peltatum*). The roots were used by Native Americans as a purgative and an anthelmintic. Podoxyphilline, another derivative of podophyllinic acid, used topically to remove warts, is highly irritating and toxic. Because warts were at one time thought to be skin tumors, podoxyphilline was assumed to have antitumor activity and attempts were made to separate its irritant from its antitumor effects using a series of synthetic analogs. Hartmann F. Stähelin, from Sandoz Laboratories, developed a podophyllotoxin derivative, etoposide (VePesid), in 1970, which had antitumor activity but was less toxic than podoxyphilline. To everybody's surprise etoposide acted by a different mechanism than podophyllotoxin or *Vinca* alkaloids; it was found to inhibit DNA synthesis. Etoposide is currently indicated in the treatment of otherwise refractory testicular tumors as well as in small-cell lung cancer.

During the 1970s and 1980s the search for new cytotoxic agents with higher specificity for cancer cells continued. Among more recently developed drugs of interest are gemcitabine (Gemzar) from Lilly and capecitabine (Xeloda) from Hoffmann–La Roche. Gemcitabine exhibits cell phase specificity; it kills primarily the cells undergoing DNA synthesis and is indicated in the treatment of pancreatic tumors (adenocarcinoma). Its side effects include suppression of bone-marrow activity, hair loss, nausea and vomiting, and elevation of serum transaminase levels. Capecitabine was claimed to kill cancer cells selectively. It is considerd to be the first "tumor-activated" anticancer drug. It was approved in April 1998 by the FDA and is recommended for the treatment of metastatic breast cancer in patients whose tumors are resistant to paclitaxel and anthracycline.

SEX HORMONES

The pioneering studies on the use of sex hormones in the therapy of tumors were done in the late 1930s by Charles Huggins, professor of surgery at the University of Chicago. He noticed that castration caused regression of malignant prostatic tumors in dogs and concluded that these tumors must be hormone dependent. He decided to try an estrogen, stilbestrol (diethylstilbestrol), to produce the chemical equivalent of castration and obtained regression of prostatic tumors in dogs. In 1941 he undertook a clinical study with stilbestrol in patients with advanced prostatic cancer and again observed regression of tumors. Subsequent clinical studies revealed that stilbestrol is effective in treating breast cancer in about 40 percent of postmenopausal women. For these discoveries Huggins shared the 1966 Nobel Prize in physiology or medicine with Francis Rous. Subsequently, however, stilbestrol by itself was found to be carcinogenic and to produce abnormalities in the genital tract of daughters of women treated with the drug.

Many synthetic estrogens were developed and clinically tested. In 1964 ICI patented tamoxifen (Nolvadex). Two ICI scientists, Michael Harper and Arthur Walpole, had discovered that tamoxifen was a racemic mixture of two isomers, only one of which had estrogenic activity, while the other had antiestrogenic properties by virtue of its ability to block estrogen receptors. Harper and Walpole suggested that tamoxifen should be tried in the therapy of mammalian tumors. The first clinical study, done in 1971, demonstrated both the efficacy and the safety of tamoxifen in comparison with previously used estrogens. The drug is now widely used to treat metastatic cancer in women and is effective in delaying the recurrence of breast cancer after surgery. It appears that women with breast cancer, whose tumors are estrogen-receptor positive, respond better to tamoxifen.

Tamoxifen is, however, not a pure estrogen antagonist. It has some estrogenic properties. In a few patients it was reported to produce vision disturbances, including cataract. There was also concern that tamoxifen may have been responsible for the increased rate of endometrial cancer in women treated with the drug for two years or longer. Since the effectiveness of tamoxifen in the treatment of breast cancer was attributed to its ability to bind to estrogen receptor, further research was directed toward development of pure estrogen antagonists or inhibitors of estrogen formation. These efforts led to the development of an estrogen antagonist, toremifine (Fareston), by Schering-Plough and Orion and to its approval in 1997 by the FDA. Toremifine is used to treat metastatic breast cancer in postmenopausal women. This drug was the first new antiestrogen treatment to become available in the United States in nineteen years. Three months later the FDA approved another innovation in breast cancer therapy, letrozole (Femara), a Novartis product. This drug is an inhibitor of an enzyme called aromatose, which is required for estrogen synthesis.

MISCELLANEOUS CANCER CHEMOTHERAPEUTICS

Among miscellaneous compounds used today in cancer chemotherapy, cisplatin (Platinol) is of considerable interest. Its discovery was serendipitous. In 1964 Barnet Rosenberg and his associates at Michigan State University observed that electric current interfered with the cell division of *E. coli* bacteria in suspension. They discovered that the effect was caused by platinum electrodes, which formed cytotoxic platinum salts. Various platinum-containing compounds were tested, and many were found to block cell division. Rosenberg reported his findings in 1969, and in the early 1970s cisplatin, a chemical containing platinum, was successfully tested in the clinic. It has now became established in treating metastatic testicular and ovarian tumors and bladder cancer.

Probably the most exciting anticancer drugs developed during the last two decades are paclitaxel (Taxol) and docetaxel (Taxotere). Paclitaxel was isolated from the bark of the Pacific yew. The use of yew tree products in medicine dates back to the Celts. In this century the development of paclitaxel dates back to the late 1950s, when the National Cancer Institute screening program surveyed many natural products and selected extracts from Pacific yew bark for testing. The active principle (paclitaxel) was identified in 1971, but its initial clinical testing led to hypersensitivity reactions. The phase I evaluation (safety studies in normal volunteers) was not completed until 1988. After successful phase II and III studies paclitaxel was approved for the therapy of ovarian tumors. Its supply was initially severely limited because of the small amounts present in the yew bark and the slow growth of the trees. Many approaches to improve the paclitaxel supply have been successful. A precursor of paclitaxel was extracted from yew needles, making a semisynthetic process feasible. A fungus capable of producing paclitaxel has been found, and paclitaxel can now be completely synthesized.

Because paclitaxel is poorly soluble in water, it was used clinically in a solution containing modified castor oil and alcohol. The hypersensitivity reactions observed initially in the clinic were attributed to the solvent, and a new formulation of paclitaxel in liposomes (phospholipid suspensions) was developed. Neutropenia (low white blood cell count) was found to be the limiting factor in the clinical use of paclitaxel. If paclitaxel is administered by a very slow infusion (over three hours), neutropenia is much less of a problem. The mechanism of action of paclitaxel is unique. It interferes with mitosis, but by a completely different mechanism than other previously known antimitotic agents such as vincristine. Paclitaxel and its derivative, docetaxel, were found to inhibit cell division by promoting the formation of very stable but nonfunctional microtubules inside the cells. The discovery of the mechanism of action of paclitaxel enhanced scientific interest in the drug. Docetaxel is slightly more potent than paclitaxel, but otherwise has a similar mechanism of action and similar side effects. Both drugs are now called taxoids. They were recently shown to be effective not only in ovarian but also in lung, breast, head, and neck cancers.

The search for new cancer chemotherapeutics was greatly intensified in the 1990s. This was partially because of the growth of the biotechnology industry and greater involvement of academic scientists in the search for new therapeutic approaches. The May 1998 issue of *R&D Directions* lists 387 cancer or cancer-related products introduced or in development in 1997 and 1998 worldwide. Many of these products are new formulations or delivery systems for old drugs. But among them is a remarkable number of innovative approaches to cancer chemotherapy. Recently marketed products include those that act by enhancing immune response rather than by killing rapidly multiplying cells. One of them is the recombinant interferon α-2b (Intron A) from Schering-Plough, which was approved in October 1997 and is indicated in various types of tumors, including melanoma, lymphomas, and Kaposi's sarcoma. Another is aldesleukin (Proleukin) from Chiron. It was produced by recombinant DNA technology and has the biological activity of human interleukin-2. In other words it acts as a growth factor and stimulates the growth and function of T and B cells. Aldesleukin is used to treat adults with metastatic renal cell carcinoma. Another new approach to cancer chemotherapy is the use of monoclonal antibodies as drugs. Awaiting approval is Genentech's trastuzumab (Herceptin), a monoclonal antibody to a growth factor receptor that is present in excessive amounts on the surface of cancer cells.

Sanofi is marketing an innovative photodynamic cancer treatment system that combines the light-activating drug porfimer (Photofrin) with laser systems for selective destruction of cancer cells. Merck marketed asparaginase (Elspar) for use as an adjunct therapy with other cancer chemotherapeutic agents to induce remission of leukemia in children. This product is based on the discovery that unlike normal cells cancer cells cannot synthesize asparagine. Asparaginase reduces the availability of asparagine to cancer cells, while normal cells survive by synthesizing asparigine for their own use.

Among many potential products still in the laboratory or in phase I clinical evaluation are angiogenesis (blood vessel generation) inhibitors: angiostatin and endostatin. These two substances and one of their advocates, Judah Folkman of Harvard University in Boston, received considerable publicity in the press and television in early 1998. The concept that tumor growth depends on blood supply and growth of new blood vessels is not new, but Folkman and his colleagues have shown that natural substances (angiostatin and endostatin) that antagonize the growth of blood vessels can regress tumors in mice. Angiostatin is a fragment of plasminogen usually found in tumor-bearing animals, while endostatin is a fragment of collagen produced by some tumors. Both compounds can prevent the formation of metastases as well as reduce the size of the original tumors. Moreover, tumors do not develop resistance to endostatin; therefore repeated rounds of therapy are just as effective as the first round. Both compounds are being developed by EntreMed, but angiostatin is already licensed to Bristol-Myers Squibb.

Lifestyle Drugs

On 11 May 1998 *Business Week*'s cover story was "The New Era of Lifestyle Drugs." The story was about Pfizer's sildenafil (Viagra) and its use not only to treat impotence but also to improve male sexual performance. The article, written by Joseph Weber and Amy Barnett with Michael Mandel and Jeff Lederman, went beyond sildenafil's appeal and Pfizer's profits. It introduced the term *lifestyle drugs* and analyzed the potential importance of these drugs for the pharmaceutical industry and even for the entire economy. The authors suggested that lifestyle drugs could double the pharmaceutical industry's size within the next five years. This projection is probably overly optimistic, but certainly conceivable.

Pfizer was not the first company to market a drug for the treatment of impotence. Pharmacia & Upjohn and Schwarz Pharma were marketing aprostadil (prostaglandin E) under the trade names of Caverject and Edex, respectively, for the same purpose, but by intracavernous rather than oral administration. Aprostadil is also available as a urethral suppository under the trade name Muse from Vivax. The main advantages of sildenafil over aprostadil are its effectiveness and safety by oral administration. Further, the promotion of sildenafil is supported by extensive and successful clinical trials as well as by a considerable financial investment. Pfizer tested sildenafil in over 4,500 male patients with erectile dysfunction. No such trials were ever conducted with generic drugs like yohimbine (Yocon, Yohimex, Aphrodyne, Erex), which were suspected but never proved to have aphrodisiac properties.

Sildenafil was also not the first drug to be used to enhance life experience rather than to treat diseases. Oral contraceptives, discussed in the section on hormones, can also be called lifestyle drugs. Their economic impact was and still is substantial. The misuse of anabolic steroids by athletes for muscle-building purposes also represents lifestyle modification. In this volume, under antihypertensive drugs, we discussed a vasodilator, minoxidil (Rogaine), marketed as a hair growth stimulant. While hair loss can represent a symptom of a disease or a side effect of drug therapy, baldness is often inherited and is not viewed as a disease by the medical profession or insurance companies. At the end of 1997 the FDA approved finasteride (Propecia) for the treatment of male baldness. It is a new formulation of the same drug that was sold under the trade name Proscar for the treatment of benign prostatic hypertrophy. It is an inhibitor of an enzyme (5-α-reductase) that converts testosterone to 5α-dihydrotestosterone, a very potent androgenic steroid, that enhances the prostate enlargement and inhibits hair growth in males. Finasteride is expected to be a "blockbuster" product with $700 million in sales three to five years after launch.

In addition to their use as lifestyle drugs, sildenafil and minoxidil have a similar history of discovery. Neither drug was synthesized for its current use. Sildenafil was in development as an antianginal drug, while minoxidil was first used as an antihypertensive. Erection and excessive hair growth, respectively, were viewed originally as side effects and clinicians rather than medicinal chemists or pharmacologists should be credited with the discovery of their usefulness for the current indications. Our understanding of the mechanism of their "therapeutic" actions is also far from complete. Selective increase in blood flow in particular areas does not tell us much about molecular mechanisms involved, and opening of potassium channels (KATP) by minoxidil may have nothing to do with hair growth stimulation.

Some drugs used in dermatology can be classified as lifestyle drugs as well. Among them is tretinoin (Renova) recommended for the treatment of fine skin wrinkling and roughness. Wrinkles are also treated by local injections of minute amounts of botulinum toxin.

The future of lifestyle drugs appears to be bright. Even ten years ago most large pharmaceutical companies would not have initiated research projects for drugs to stimulate erectile function. According to industrial sources, researchers at Merck & Co., Bristol-Myers Squibb, Schering-Plough, and Abbott are currently involved in such projects. Within the next few years we can expect many new lifestyle drugs. Pfizer will certainly be imitated with further improvements over sildenafil (more rapid onset of action, higher efficacy, stimulation of male or female libido, and so forth). The ultimate lifestyle drugs will be designed to delay aging. Research on aging has now progressed to the stage that development of anti-aging drugs is not only possible but likely within the next decade or two.

Sources

J. D. Adams; K. P. Flora; B. R. Goldspiel; J. W. Wilson; S. G. Arbuck; R. Finley. "Taxol: A History of Pharmaceutical Development and Current Pharmaceutical Concerns." *Journal of the National Cancer Institute, Monographs* 15 (1992), 141–147.

T. Akera; T. M. Brody. "The Role of (Na^+, K^+)-activated ATPase in the Inotropic Action of Digitalis." *Pharmacological Reviews* 29 (1977), 187–210.

A. W. Alberts; J. S. MacDonald; A. E. Till; J. A. Tobert. "Lovastatin." *Cardiovascular Drug Review* 7 (1989), 89–109.

J. C. Allen; A. Schwartz. "A Possible Biochemical Explanation for the Sensitivity of the Rat to Cardiac Glycosides." *Journal of Pharmacology and Experimental Therapeutics* 168 (1969), 42–46.

K. H. Beyer. *Discovery, Development and Delivery of New Drugs.* New York: SP Medical & Scientific Books, 1978.

J. T. Bigger; B. F. Hoffman. "Antiarrhythmic Drugs." In *The Pharmacological Basis of Therapeutics.* Edited by L. S. Goodman and A. Gilman. Sixth edition. New York: MacMillan, 1980.

R. J. Bing. *Cardiology: The Evolution of the Science and the Art.* Philadelphia: Harwood Academic Publishers, 1992.

J. S. K. Boyd. "Tetanus in the African and European Theaters of War, 1939–1945." *Lancet* 1 (1946), 113–119.

H. Braun. "Bedeutung des Adrenalins für die Lokalanästhesie." *Langenbeck's Archiv für Klinische Chirurgie* 69 (1903), 150.

M. Brodie; W. H. Park. "Active Immunization against Poliomyelitis." *American Journal of Public Health* 26 (1936), 119–125.

T. L. Brunton. "On the Use of Nitrite of Amyl in Angina Pectoris." *Lancet* 2 (1867), 97–98.

T. C. Butler. "The Introduction of Chloral Hydrate into Medical Practice." *Bulletin of the History of Medicine* 44 (1970), 168–178.

Q.-L. Choo; G. Kuo; A. J. Weiner; L. R. Overby; D. W. Bradley; M. Houghton. "Isolation of a cDNA Clone Derived from a Blood Borne Non-A, Non-B Viral Hepatitis Genome." *Science* 244 (1989), 359–362.

I. Creese; C. M. Fraser, editors. *Dopamine Receptors.* New York: Alan R. Liss, 1987.

P. Descomby. "L'anatoxine tetanique." *Journal of the Canadian Royal Society of Biology* 91 (1924), 239–241.

A. Einhorn. "Über neue Arzneimittel." *Justus Liebig's Annalen der Chemie* 371 (1910), 125–131.

J. F. Enders; T. H. Weller; F. C. Robbin. "Cultivation of the Lansing Strain of Poliomyelitis Virus in Cultures of Various Human Embryonic Tissues." *Science* 109 (1949), 85–87.

G. B. Elion. "The Quest for a Cure." *Annual Review of Pharmacology and Toxicology* 33 (1993), 1–23.

U. S. von Euler. *Noradrenaline: Chemistry, Physiology, Pharmacology and Clinical Aspects.* Springfield, Ill.: Charles C. Thomas, 1956.

E. Fischer; J. von Mering. "Über eine neue Klasse von Schlafmitteln." *Therapie der Gegenwart* 44 (1903), 97–101.

A. Fleckenstein. *Calcium Antagonism in Heart and Smooth Muscle.* New York: John Wiley & Sons, 1983.

T. Francis Jr.; R. F. Korns; R. B. Voight; M. Boisen; F. M. Hemphile; J. A. Napier; E. Tolchisky. "An Evaluation of 1954 Poliomyelitis Vaccine Trials (Summary Report)." *American Journal of Public Health* 45 (1955), 1–50.

A. G. Gilman; L. S. Goodman; T. W. Rall; F. Murad, editors. *The Pharmacological Basis of Therapeutics.* Seventh edition. New York: Macmillan, 1985.

A. T. Glenny; B. E. Hopkins. "Diphtheria Toxoid as an Immunizing Agent." *British Journal of Experimental Pathology* 4 (1923), 283–288.

H. Gold; McK. Cattell. "Mechanism of Digitalis Action in Abolishing Heart Failure." *Archives of Internal Medicine* 65 (1940), 263–278.

J. G. Hardman; L. E. Limbird; P. B. Molinoff; R. W. Ruddon; A. G. Gilman, editors. *The Pharmacological Basis of Therapeutics.* Ninth edition. New York: McGraw-Hill, 1996.

F. G. Hayden; A. D. Osterhaus; J. J. Treano; et al. "Efficacy and Safety of Neuraminidase Inhibitor Zanamavir in the Treatment of Influenzavirus Infections." *New England Journal of Medicine* 337 (1997), 874–880.

A. Heffter. "Über die Einwirkung von Chloral auf Glucose." *Chemische Berichte* 22 (1889), 1050–1051.

M. R. Hilleman; E. B. Buynak; R. E. Weibel; J. Stokes. "Life-Attenuated Mumps-Virus Vaccine." *New England Journal of Medicine* 278 (1968), 227–232.

M. R. Hilleman et al. "Development and Evaluation of the Moraten Measles Virus Vaccine." *Journal of the American Medical Association* 206 (1968), 587–590.

G. H. Hitchings. "Antagonists of Nucleic Acid Derivatives as Medicinal Agents." *Annual Review of Pharmacology and Toxicology* 32 (1992), 1–6.

S. Hosoda; H. Kasamuki; K. Miyata; M. Endo; K. Hirosawa. "Results of a Clinical Investigation of Nifedipine in Angina Pectoris with Special Reference to Its Therapeutic Efficacy in

Attacks at Rest." In *First International Nifedipine (Adalat) Symposium.* Edited by K. Hashimoto, E. Kimura, T. Kobayashi. Tokyo: University of Tokyo Press, 1975, pp. 185–189.

H. A. Howe. "Duration of Immunity in Chimpanzees Vaccinated with Formalin Inactivated Poliomyelitis Virus." *Federation Proceedings* 11 (1952), 471–472.

International Collaborative Study Group. "Reduction of Infarct Size with the Early Use of Timolol in Acute Myocardial Infarction." *New England Journal of Medicine* 310 (1984), 9–15.

E. Jenner. *An Inquiry into the Causes and Effects of the Variolae Vaccinae.* London: Low, 1798.

S. L. Katz et al. "Studies on an Attenuated Measles Virus Vaccine. VIII. General Summary and Evaluation of Results of Vaccine." *New England Journal of Medicine* 263 (1960), 180–184.

C. Koller. "Über die Verwendung des Cocains zur Anästhesierung am Auge." *Wiener Medizinische Wochenschrift* 34 (1884), 1276–1278, 1309–1311.

J. A. Kolmer. "Vaccination against Acute Anterior Poliomyelitis." *American Journal of Public Health* 26 (1936), 126–135.

H. Koprowski; G. A. Jervis; T. W. Norton. "Immune Responses in Human Volunteers upon Oral Administration of a Rodent-adapted Strain of Poliomyelitis Virus." *American Journal of Hygiene* 55 (1952), 108–126.

J. B. Kostis. "Angiotensin Converting Enzyme Inhibitors in Hypertension." *Cardiovascular Drug Reviews* 7 (1989), 173–176.

J. C. Krantz Jr.; C. J. Carr; G. Lu; F. K. Bell. "Anaesthetic Action of Trifluoroethyl Vinyl Ether." *Journal of Pharmacology and Experimental Therapeutics* 108 (1953), 488–495.

J. G. Kuhn. "Pharmacology and Pharmacokinetics of Paclitaxel." *Annals of Pharmacotherapy* 28 (Suppl. 5) (1994), 15–17.

M. M. Levine. "Vaccines and Vaccination in the Historical Perspectives." In *New Generation Vaccines.* Edited by G. C. Woodrow, M. M. Levine. New York/Basel: Marcel Dekker, 1990, pp. 3–29.

J. T. Lie. "Recognizing Coronary Artery Disease. Selected Historical Vignettes from the Period of William Harvey (1578–1657) to Adam Hammer (1818–1878)." *Mayo Clinic Proceedings* 53 (1978), 811–817.

N. Löfgren; P. Lundqvist. "Studies on Local Anesthetics II." *Svensk Kemisk Tidskrift* 58 (1946), 206–217.

A. P. Long; P. E. Sartwell. "Tetanus in U.S. Army in World War II." *Bulletin of the U.S. Army Medical Department* 7 (1947), 371–385.

G. Mabuchi; H. Kishida; K. Suzuki. "Clinical Effects of Nifedipine on Variant Form of Angina Pectoris." In *First International Nifedipine (Adalat) Symposium.* Edited by K. Hashimoto, E. Kimura, T. Kobayashi. Tokyo: University of Tokyo Press, 1975, pp. 177–184.

D. V. Madore et al. "Safety and Immunologic Response to *Haemophilus influenzae* Type b Oligosaccharide—CRM197 Conjugate Vaccine in 1- to 6-Month-Old Infants." *Pediatrics* 85 (1989), 331–337.

F. Markwardt. "Hirudin: The Promising Antithrombotic." *Cardiovascular Drug Reviews* 10 (1992), 211–232.

A. Maseri; O. Parodi; S. Severi; A. Pesola. "Transient Transmural Reduction of Myocardial Blood Flow, Demonstrated by Thallium-201 Scintigraphy, as a Cause of Variant Angina." *Circulation* 54 (1976), 280–288.

R. A. Maxwell; S. B. Eckhardt. *Drug Discovery: A Casebook and Analysis.* Clifton, N.J.: Humana Press, 1990.

V. J. Mc Auliffe; R. H. Purcell; J. L. Gerin. "Type B Hepatitis: A Review of Current Prospects for a Safe and Effective Vaccine." *Reviews of Infectious Diseases* 2 (1980), 470–492.

I. M. Morgan. "Immunization of Monkeys with Formalin-Inactivated Poliomyelitis Viruses." *American Journal of Hygiene* 48 (1948), 394–406.

E. A. Mortimer. "Pertussis Vaccine." In *Vaccines.* Edited by S. A. Plotkin, E. A. Mortimer. Philadelphia: W. B. Saunders, 1988, pp. 74–97.

R. M. Moskowitz; P. A. Piccini; G. Nacarelli; R. Zelis. "Nifedipine Therapy for Stable Angina Pectoris: Preliminary Results of Effects on Angina Frequency and Treadmill Exercise Response." *American Journal of Cardiology* 44 (1979), 811–816.

W. Murrell. "Nitroglycerine as a Remedy for Angina Pectoris." *Lancet* 1 (1879), 80–81, 113, 151, 223.

R. L. Noble; C. T. Beer; J. H. Cutts. "Further Biological Activities of Vincaleukoblastine—an Alkaloid Isolated from *Vinca rosea* (L.)." *Biochemical Pharmacology* 1 (1958), 347–348.

Norwegian Multicenter Study Group. "Timolol-Induced Reduction in Mortality and Reinfarction in Patients Surviving Acute Myocardial Infarction." *New England Journal of Medicine* 304 (1981), 801–807.

L. Pasteur. "Methode pour prevenir la rage apres morsure." *Comptes Rendus de l'Academie des Sciences (Paris)* 92 (1881), 1378–1383.

H. Peltola. "*Haemophilus influenzae* Type B Capsular Polysaccharide Vaccine in Children: A Double-Blind Field Study of 100,000 Vaccinees 3 Months to 5 Years of Age in Finland." *Pediatrics* 60 (1977), 730–737.

D. Pennica; W. E. Holmes; W. J. Kohr; et al. "Cloning and Expression of Human Tissue-Type Plasminogen Activator cDNA in *E. coli*." *Nature* 301 (1983), 214–221.

S. J. Peroutka, editor. *Serotonin Receptor Subtypes: Basic and Clinical Aspects.* New York: John Wiley & Sons, 1991.

Physicians Desk Reference. Forty-eighth and fifty-third editions. Montvale, N.J.: Medical Economics, 1994 and 1999.

R. F. Pitts. *The Physiological Basis of Diuretic Therapy.* Springfield, Ill.: Charles C. Thomas, 1959.

S. L. Plotkin; S. A. Plotkin. "A Short History of Vaccination." In *Vaccines*. Edited by S. A. Plotkin, E. A. Mortimer. Philadelphia: W. B. Saunders, 1988, pp. 1–7.

J. Raventos. "The Action of Fluothane, a New Volatile Anaesthetic." *Journal of Pharmacology and Experimental Therapeutics* 111 (1956), 394–409.

C. G. Ray. "Measles (Rubeola)." In *Harrison's Principles of Internal Medicine.* Edited by E. Braunwald, K. J. Isselbacher, R. G. Petersdorf, J. D. Wilson, J. B. Martin, A. S. Fauci. Eleventh edition. New York: McGraw-Hill, 1987, pp. 682–683; "Mumps," pp. 709–712; "Rubella," pp. 684–686.

B. W. Richardson. "On a New and Ready Mode of Producing Local Anesthesia." *Medical Times Gazette* 1 (1886), 115–117.

A. A. Rubin, editor. *New Drugs: Discovery and Development.* New York/Basel: Marcel Dekker, 1978.

A. B. Sabin; W. A. Hennessen; J. Winsser. "Studies on Variants of Poliomyelitis Virus. I. Experimental Segregation and Properties of Avirulent Variants of Three Immunological Types." *Journal of Experimental Medicine* 99 (1954), 551–576.

O. Schmiedeberg. "On the Pharmacologic Action and Therapeutic Application of Some Ethereal Salts of Carbamic Acid. *Practitioner* 35 (1985), 275–280, 328–332.

A. J. F. Schwartz. "Preliminary Tests of a Highly Attenuated Measles Vaccine." *American Journal of Diseases of Children* 103 (1962), 386–389.

A. Scriabine; C. S. Sweet, editors. *New Antihypertensive Drugs.* New York: Spectrum Publications, 1976.

E. D. Shapiro; J. D. Clemens. "A Controlled Evaluation of the Protective Efficacy of Pneumococcal Vaccine for Patients at High Risk of Serous Pneumococcal Infections." *Annals of Internal Medicine* 101 (1984), 325–330.

W. Sneader. *Drug Discovery: The Evolution of Modern Medicines.* New York: John Wiley & Sons, 1985.

C. W. Suckling. "Some Chemical and Physical Factors in the Development of Fluothane." *British Journal of Anaesthesia* 29 (1957), 466–472.

D. L. Tabern; E. H. Volwiler. "Sulfur Containing Barbiturate Hypnotics." *Journal of the American Chemical Society* 57 (1935), 1961–1963.

J. H. Triebwasser; R. E. Harris; R. E. Bryant; E. R. Rhodes. "Varicella Pneumonia in Adults: Report of Seven Cases and a Review of Literature." *Medicine* 46 (1967), 400–423.

F. Van de Werf; P. A. Ludbrook; S. R. Bergmann; et al. "Coronary Thrombolysis with Tissue-Type Plasminogen Activator in Patients with Evolving Myocardial Infarction." *New England Journal of Medicine* 310 (1984), 609–613.

W. Vater; G. Kroneberg; F. Hoffmeister; H. Kaller; K. Meng; A. Oberdorf; W. Puls; K. Schlossmann; K. Stoepel. "Zur Pharmakologie von 4-(2'Nitrophenyl)-2,6-Dimethyl-1,4-Dihydropyridin-3,5-Dicarbonsäure Dimethylester (Nifedipin, BAY a 1040)." *Arzneimittelforschung* 22 (1972), 1–14.

E. M. Vaughan-Williams. "Classification of Antidysrhythmic Drugs." *Pharmacology and Therapeutics* 1 (1975), 115–138.

J. A. Vedin; C. E. Wilhelmson. "Beta Receptor Blocking Agents in the Second Prevention of Coronary Heart Disease." *Annual Reviews of Pharmacology and Toxicology* 23 (1983), 29–44.

E. H. Volwiler; D. L. Tabern. "5,5-Substituted Barbituric Acids." *Journal of the American Chemical Society* 52 (1930), 1676–1679.

A. J. Wagstaff; H. M. Bryson. "Foscarnet: A Reappraisal of Its Antiviral Activity, Pharmacokinetic Properties and Therapeutic Use in Immunocompromised Patients with Viral Infections." *Drugs* 48 (1994), 8–199.

P. E. Wallemacq; R. Reding. "FK 506 (Tacrolimus), a Novel Immunosuppressant in Organ Transplantation: Clinical, Biomedical and Analytical Aspects." *Clinical Chemistry* 39 (1993), 2219–2228.

M. Weatherall. *In Search of a Cure: A History of Pharmaceutical Discovery.* Oxford: Oxford University Press, 1990.

W. Withering. "An Account of Foxglove and Some of Its Medicinal Uses: With Practical Remarks on Dropsy and Other Diseases." London: C. G. J. J. Robinson, 1785. Reprinted in *Medical Classics* 2 (1937), 305–443.

Chapter Three

The Role of Biotechnology in Drug Development

ALEXANDER SCRIABINE

According to *Webster's Third New International Dictionary* the term *biotechnology* is defined as "the aspect of technology concerned with the application of biological and engineering data to problems relating to the mutual adjustment of man and the machine." The term *biotechnology* was first used in 1917 by Karl Ereky, a Hungarian scientist, to describe new agricultural technology that was expected to save mankind from starvation. But the meaning of the word gradually changed during the century. Since the early 1980s the term *biotechnology* has been applied to an industry that uses new biological technology to produce drugs or diagnostics previously unobtainable either in pure form or in sufficient quantities. The genes that can generate the potential products are often discovered, cloned, and introduced into bacteria, yeast, or mammalian cells capable of producing the desired products in large quantities.

Most biotechnology companies were started by scientists who became entrepreneurs or by scientists in a partnership with entrepreneurs. Not all of these companies are interested in the development of pharmaceuticals or diagnostics; some develop agricultural products, such as insecticides and genetically modified plants, while others offer services only to the pharmaceutical industry. The major goal of therapy-oriented biotechnology companies, however, is to discover and develop new treatments for diseases for which treatment is lacking or inadequate.

Unlike large pharmaceutical companies, research and development in the biotechnology industry is primarily financed not by the sale of products, but by venture bankers, public stock offerings, or individual investors. According to one industry survey, there were 1,308 biotechnology companies in the United States in 1996; 260 of them were public.

Table 1. Examples of Disease-Oriented Specialization of Biotechnology Companies

Major disease groups	Selected companies
Central nervous system diseases	Anergen, Athena Neurosciences, Cambridge Neurosciences, Cephalon, Interneuron, Neurex, Neurogen, Regeneron, Scios Nova, Synaptic
Cardiovascular diseases	Biogen, Centocor, CoCensys, Cor, Corvas, Genentech, Gensia
Inflammatory diseases	BioCryst, Biomatrix, Icos, Signal, Xoma
Cancer	Cantab, Cytogen, Idex, Imclone, Ingenix, Medarex, NeoRx, Oncogen Sciences, Seragen
Viral infections, including human immunodeficiency virus	Aviron, Agouron, Antivirals, Biogen, Hybridon, Virus Research Institute
Bacterial and fungal infections	Bactex, Biomedix, Helix, IntraBiotics, Magainin
Immunological disorders	Anergen, Autoimmune, Immunex, Immunlogic, Immunogen, Immunomedics, Immunotherapeutics, Medimmune, T Cell Sciences, Tanox Biosystems

More biotechnology companies have been formed in the United States than in the rest of the world. The American industry grew rapidly in the 1980s and 1990s, in part because Western European governments severely restricted genetic research in Europe. Forced to cross the Atlantic to keep up with their American competitors, the major European pharmaceutical companies established their own biotechnological research facilities in the United States or provided funds to small American companies, or both.

The biotechnology industry represents an unprecedented concentration of brainpower devoted to improving health. The companies either have in-house research laboratories or have contracted with academic scientists to work on their projects. Many famous biologists serve on the scientific advisory boards of biotechnology companies or have established collaborative research projects with them. The biotechnology industry has thus become an important source of research funds for academic institutions. At the same time academic institutions have licensed discoveries to biotechnology companies and others to develop those discoveries for market. According to the 1995 licensing survey of the Association of University Technology Managers, Inc., some 1,100 new companies were formed between 1980 and 1994 on the basis of licenses granted by academic institutions, including 175 new companies in 1994 alone. Among the patents licensed from academic institutions, those on recombinant DNA technology form the basis for many new drug developments.

Most large American and European pharmaceutical companies have entered into partnerships with biotechnology companies. In discovering and developing new drugs, large companies often find it more cost-effective to use innovative technologies developed by small biotechnology companies than to introduce new technology in their own facilities. Because the development of new drugs is extremely expensive and risky, the pharmaceutical industry is willing to share the costs and risks of drug research as well as the potential profits with biotechnology companies and their shareholders or investors. These partnerships can range from research collaborations to strategic alliances with equity participation by the larger company. Fortunately, most biotechnology companies specialize in the development of their own technologies, so duplication in research effort is limited; usually only clusters of six to twelve companies work in the same niche. The companies often specialize in one or more disease areas, accumulating expertise in evaluating and developing potential drugs for the treatment of a certain disease or groups of diseases (Table 1). Other companies specialize in a type of product rather than in a disease (Table 2). Many biotechnology companies are not expected to survive as single entities, but those that do are likely to be the ones that develop revolutionary products for the treatment of many diseases.

Molecular Biology and Biotechnology since World War II

Biotechnology is based to a large extent on molecular biology, defined in *Dorland's Medical Dictionary* as "the study of molecular structures and events underlying biological processes, including the relation between genes and the functional characteristics they determine." Molecular biology uses the techniques of genetics and morphology as well as physical and structural chemistry to elucidate the structure and function of biological molecules and their mechanisms

Table 2. Examples of the Specialization of the Biotechnology Industry by Type of Product

Major types of products	Selected companies
Drug delivery systems, gene delivery systems	Alkermes, Alza, Cygnus, Emisphere Technologies, Genevec, Inhale, Liposome Company, Megabios, Theratech
Monoclonal antibodies	Cell Genesis, Corvas, Cytogen, IDEC, Medarex, MedImmune, Protein Design Labs, Seragen
Growth factors, hormones, protein fractions, antibodies	Amgen, Biogen, Chiron, Genentech, Genetics Institute, Genzyme, Immunex, Prizm
Vaccines	Aviron, Biogen, Chiron, Genentech, Immune Response, MicroGenSys, Medimmune, Oravax, Viagene, Vical
Cell therapy, organ transplants, stem cells	Alexion, BioSurface, Amgen, CellPro, Cytotherapeutics, DNX, Genzyme, Neocrin, Progenitor, Systemix
Gene therapy, genomic sequencing	Ariad, Avigen, Canji, Gene Medicine, Genetic Therapy, Genzyme, Human Genome Sciences, Incyte, Somatix, Systemix, Targeted Genetics, Viagene, Vical
Receptor technology, rational drug design	Affymax, Agouron, Ariad, Immunex, Nexagen, Signal

Table 3. Major Discoveries in Genetics, Molecular Biology, and Biotechnology in the Twentieth Century

Discovery	Discoverer(s)	Country	Year
Effect of genes on metabolic pathways	Sir Alfred B. Garrod	United Kingdom	1909
Nature and function of genes	Max Delbrück	United States	Mid-1930s
DNA transfers the properties of one organism to another	O. T. Avery C. MacLeod M. McCarty	United States	1944
Identification of DNA as the hereditary material	A. D. Hershey M. Chase	United States	1952
Elucidation of DNA structure	James Watson Francis Crick	United Kingdom	1953
Amino acid sequence of genes	Vernon M. Ingram	United States	1957
Genetic code	Severo Ochoa Marshal Nirenberg	United States	1963
Monoclonal antibodies	Cesar Milstein Niels Kaj Jerne George Köhler	United Kingdom	1975

Data from G. J. Ingle, *Life and Disease* (New York/London: Basic Books, 1963); and S. L. Eck and J. M. Wilson, "Gene-based Therapy." In *Pharmacological Basis of Therapeutics*, edited by J. G. Hardman and L. F. Limbird (New York: McGraw-Hill, 1996), pp. 77–101.

for carrying and transmitting biological information. After World War II the field of molecular biology grew rapidly, primarily in the United States, the United Kingdom, and France, where booming economies were capable of supporting expensive basic research. Generous support of basic science by the U.S. government in the 1950s and 1960s was an important factor contributing to the development of molecular biology as a science. As the biotechnology industry has developed, it has come to rely not only on the discoveries of molecular biologists but on those made by all biological scientists.

During the past twenty years scientists began to realize that genetic factors are involved in many more diseases than previously anticipated. Genetic defects were identified in patients with cystic fibrosis and Alzheimer's, Parkinson's, and many other diseases. Researchers therefore thought that the techniques of molecular biology would be useful not only in diagnosing but eventually in curing these diseases. One promising approach is the use of genes as therapeutics, an approach known as gene repair or gene replacement therapy. It is considered theoretically possible to synthesize healthy human genes and use them to replace defective genes. That is the major goal of a few biotechnology companies that specialize in gene therapy.

The usefulness of molecular biology techniques in manufacturing human hormones, such as growth hormone, or factors capable of stimulating some of the normal body functions, such as interferons and erythropoietin, was demonstrated during the past decade. Transfection of genes capable of inducing the production of these hormones or factors into microorganisms or cell lines is the technique that is primarily responsible for the success of the biotechnology industry. (The basic scientific discoveries that led to rapid progress in molecular biology are listed in Table 3.)

Because genes control the production of proteins in the cells and because many important hormones or growth factors are proteins, the early goal of biotechnology was to develop proteins as drugs. Now many biotechnology companies focus on small synthetic compounds that mimic the effects of proteins. The major problem with pursuing development of protein drugs is their delivery, because most proteins are not absorbed from the gastrointestinal tract and are therefore orally inactive. Attempts to solve this delivery problem led to the formation of biotechnology companies that specialize in the development of alternative delivery systems. The cost of manufacturing large amounts of proteins was also a problem for some biotechnology companies that were not prepared to invest hundreds of millions of dollars in manufacturing facilities.

In 1975 Cesar Milstein, George Köhler, and Niels Kaj Jerne of the British Medical Research Council's Molecular Biology Laboratory in Cambridge found a way to mass-produce monoclonal antibodies, homogenous proteins that can be used as potential drugs. They fused B-cells from the human body's immune system with mouse cancer cells to produce hybridomas, which in turn produce monoclonal antibodies. One way of delivering protein-based drugs to their site of action is to use mammalian cells capable of producing these proteins in situ. With advances in transplantation technology and the availability of immunosuppressant drugs, cell transplantation therapy for treating Parkinson's disease or diabetes has become a realistic new therapeutic approach.

Another important target for disease intervention or modulation are drug receptors. These animal or human proteins are capable of specific interactions with drugs, hormones, or neurotransmitters (endogenous chemicals capable of transmitting signals from nerves to other tissues). Paul Ehrlich, who received the Nobel Prize in medicine in 1908, suggested the existence of receptors in the late 1800s. Not until the 1950s, however, did advances in protein chemistry make identification of these proteins possible. With the advances in protein chemistry and molecular biology in the 1960s and 1970s, many receptors were isolated, identified, and even cloned. During the 1980s genes for many receptors were isolated as well, so it is now possible to control a specific receptor population using gene therapy. With the advance of drug receptor technology it became theoretically possible to design small molecules to fit a particular receptor. Among the problems with this approach is the multiplicity of drug-receptor subtypes and the many, still unknown aspects of drug-receptor interaction. Rational drug design is currently still in the exploratory stage, although Miguel Ondetti and David Cushman at Squibb successfully used it in the 1970s to synthesize captopril, the first angiotensin-converting enzyme inhibitor effective in treating hypertension or heart failure.

Since scientists learned how to introduce or replace new genes in animals in the laboratory, the development of gene therapy has been a goal of numerous biotechnology companies as well as some of the major pharmaceutical companies, including Novartis, Rhône Poulenc Rorer, and SmithKline Beecham.

Interactions between Biotechnology and Large Pharmaceutical Companies

At first most biotechnology companies intended not only to introduce and develop a particular technology, but also to manufacture and sell the products of their research. A few companies—Amgen, Biogen, Chiron, Genentech, and Genzyme—achieved this goal. Most biotechnology companies, however, lacked the financial and human resources even to develop their products to the point where they could apply for approval from the Food and Drug Administration (FDA). Established pharmaceutical companies were the logical partners for the biotechnology industry. The relations between large pharmaceutical and biotechnology companies varied from complete acquisitions, or buyouts, to consortia, partnerships, joint ventures, alliances, and research collaboration on certain projects. Because "breakthrough" products are few and far between, large pharmaceutical companies strive to acquire access to new technologies and rights to potential products. At the same time biotechnology companies look primarily for cash, but also for clinical trial resources, management support, and, ultimately, of course, a share in potential profits.

Larger pharmaceutical companies have gained control of several biotechnology companies. Among the major acquisitions are those of Genentech by Hoffmann–La Roche, Chiron by Ciba-Geigy (now Novartis), Genetics Institute by American Home Products, Sphinx Biotechnologies by Lilly, and Affymax by Glaxo. Licensing to pharmaceutical companies by biotechnology companies of the rights to certain products or technologies is more common, however, as are the numerous strategic alliances between biotechnology and pharmaceutical companies. These alliances can involve collaboration on certain research projects or even an exchange of technology for a product. A good example of technology acquisition is the agreement in May 1993 in which SmithKline Beecham paid Human Genome Sciences $125 million in return for the rights to most of the company's genomic technology. The two companies also collaborated to establish a human gene database; several large companies, including Schering-Plough, agreed to pay between $30 million and $55 million to access this database.

During 1993 and 1994 biotechnology companies experienced increasing difficulties in raising sufficient funds from venture bankers or public stock offerings. The search for partners in the pharmaceutical industry was intensified, and the resulting deals saved many biotechnology companies from bankruptcy while expanding the technological capabilities of large pharmaceutical companies. The biotechnology companies lost some of the freedom they enjoyed to manage and direct their own research and had to orient their efforts toward potential breakthrough products for the larger markets desired by major pharmaceutical companies.

Relations between Biotechnology Companies

In addition to various deals between biotechnology and larger pharmaceutical companies an increasing number of partnerships have been formed in recent years between two or more biotechnology companies. In 1994 the total value of deals made between two biotechnology companies was higher than those between biotechnology and larger pharmaceutical companies, in part because some biotechnology companies accumulated sufficient capital to acquire smaller companies. Chiron's acquisition of Cetus and Amgen's takeover of Synergen are examples. These acquisitions often involved stock rather than cash transactions; such was the case with the acquisition of Creagen by Neurex and Genica Pharmaceuticals by Athena Neurosciences.

A new trend in the biotechnology industry is the creation of consortia of more than two companies. These consortia can be as complicated as RPR Gencell, a creation of Rhône-Poulenc Rorer that involves a partnership with fifteen biotechnology companies and academic institutions, with a goal of achieving a dominant position in gene and cell therapies. Pfizer created another gene therapy consortium, appropriately named Pfizergen. It represents the alliance of Pfizer, Incyte Pharmaceuticals, Immusol, and Myco Pharmaceuticals in the United States and AEA Technology and Oxford Asymmetry in the United Kingdom.

The consolidation of the biotechnology industry by acquisitions and mergers is likely to continue, as is the formation of consortia, and the relationships between the pharmaceutical and biotechnology industries will probably become even more complicated with fractions of some companies owned by different partners.

Major Biotechnology Companies and Their Products: Problems and Perspectives

Eight well-established biotechnology companies that have discovered or developed their own products—Amgen, Biogen, Centocor, Chiron, Genentech, Genetics Institute, Genzyme, and Immunex—are discussed below. Table 4 provides an overview of the companies, showing their recent financial information and major products. These companies were generally well financed in comparison with most other biotechnology firms. Nonetheless, each had to enter into alliances or partnerships with major pharmaceutical or other biotechnology companies to obtain additional funds for product development, to expand technological capabilities, or to complete late-stage clinical trials. The partnerships ended for four of these companies in their partial or complete takeover by their pharmaceutical partner.

Table 4. Major Biotechnology Companies: Their Revenues, Income (Loss), Major Marketed Products, and Drugs in the Pipeline

Company	Year Established	Total Revenue*		Net Income (loss)*		Major Marketed Products (number)	Drugs in the Pipeline (number)
		1994	1996	1994	1996		
Amgen	1980	1,647	2,240	320	680	Epogen Neupogen	Stemgen KGF GDNF Interleukin-1ra R-568
Biogen	1978	140	277	20	41	Avonex Intron A Hepatitis B vaccines	LFA3TIP CVT124 Gelsolin CD40L antibody
Centocor	1979	67	135	(127)	(13)	Rheopro Myoscint Panorex	Tumor markers Infliximab ATTRACT
Chiron	1981	454	1,313	18	55	Betaseron Proleukin (interleukin-2) Vaccines (10)	Gene therapy Myotrophin Regranex TNF-antibody Vaccines (12)
Genentech	1976	795	969	124	118	Activase Actimmune Protropin Pulmozyme Rogeron-A Nutropin	Rituximab Anti-Her2 NGF Second-generation tissue plasminogen activator
Genetics Institute	1980	131	†	(19)	†	Recombinate Epogin Recormon Leucomax Factor VIII	Plasminogen activator (RhNPA) Interleukin-3 and -6
Genzyme (General Division)	1981	311	511	32	(31)	Ceredase Cerazyme Carticel Epicel	Seprafilm Sepracoat Thyrogen Sepragel Antithrombin Gene therapy
Immunex	1981	144	151	(33)	(54)	Leukine Novantone Thioplex Leucovorin Amicar	Enbrel Flt-3 ligand Interleukin-4 receptor

Data from company reports.

* In millions of U.S. dollars.

† Since Genetics Institute is now completely owned by AMP, financial data for the institute alone are no longer available.

AMGEN

Today many consider Amgen (the name was derived from Applied Molecular Genetics) as the brightest star of the biotechnology industry, with total revenues of $2.4 billion in 1997, nearly 5,400 employees, and a net income of $644 million. Venture bankers William K. Bowes and Sam Wohlstadter founded Amgen in 1980 with an initial investment of $50,000. Scientific guidance was provided primarily by Amgen's first president and CEO, George Rathman, who has a doctorate in physical chemistry from Princeton University and came to Amgen from the Diagnostics Division of Abbott. The company initially raised $19 million in private financing, and in 1983 a public offering raised an additional $43 million. Amgen's success was largely attributed to Rathman's scientific guidance, generous initial financing, and successful litigation against the Genetics Institute, which had claimed the rights to Amgen's major product, erythropoietin, or Epogen, which is used to treat anemias. Amgen's patent claims include methods for manufacturing and purifying erythropoietin as well as for cloning of its gene. The substance itself was in the public domain.

The existence of erythropoietin was suspected as early as 1906, and its activities were the subject of an extensive review in 1972 by James Fisher of Tulane University. One of Amgen's consultants, Eugene Goldwasser of the University of Chicago, provided Amgen with a small sample of purified erythropoietin. The sequence of the erythropoietin gene and its synthesis by microorganisms proved patentable and turned out to be highly profitable for Amgen.

Another profitable Amgen product is granulocyte-colony stimulating factor, or filgrastim, marketed under the trade name Neupogen. It is a protein containing 175 amino acids, manufactured by recombinant DNA technology. The existence of granulocyte-stimulating factors, like that of erythropoietin, was known for many years, but Amgen succeeded in applying modern biotechnological methods to its manufacture. This is accomplished by inserting the human gene for filgrastim into bacteria (*Escherichia coli*). The major clinical indication for filgrastim is to restore the white blood cell (neutrophil) count, a therapy used to treat patients with congenital neutropenia (low white blood cell count) and cancer patients receiving chemotherapy, which lowers the white cell count.

Amgen currently has a dozen potential therapeutics in development, although the company discontinued work on MGDF (pegylated megakaryote growth and development factor) to restore platelet count. Amgen's acquisition of Synergen and of the obesity gene from Rockefeller University should permit the company to broaden its scope into central nervous system diseases and obesity.

BIOGEN

Biogen was founded in 1978 by a group of scientists and entrepreneurs. The founding scientists included leaders in genetic engineering: Phillip Sharp,

Charles Weissmann, Kenneth Murray, Peter-Hans Hofschneider, and Brian Hartley. The research was initially directed by Walter Gilbert, a Nobel laureate, whose scientific guidance established the company's reputation. During its first few years Biogen farmed its research projects out to academic laboratories. And like most biotechnology companies Biogen desperately needed cash. The company was saved by one of its founders, Moshe Alafi, who convinced W. H. Conzen, Schering-Plough's CEO, to invest $8 million in Biogen. In return Schering-Plough secured the rights for three Biogen products, including alpha-interferon, which was cloned for Biogen by Shigezaku Nagata and Charles Weissmann in 1980. Schering-Plough spent another $170 million for the large-scale production of alpha-interferon, and the investment paid handsomely. In 1993 Schering-Plough's sales of alpha-interferon 2 (Intron A) totaled $600 million.

Alpha-interferon 2 is approved for the treatment of hairy cell leukemia, Kaposi's sarcoma, hepatitis B and C, and venereal warts. The mechanism of action is not precisely known, but all interferons bind to specific receptors on the cell surface and initiate many intracellular events, including enzyme induction, suppression of cell proliferation, and inhibition of viral replication.

Biogen derived its income initially from contract research for large pharmaceutical companies, but gradually replaced that income with licensing fees for their products: alpha-, beta-, and gamma-interferons; hepatitis B vaccine; and hepatitis diagnostics. Biogen is known for the strength of its patents, which cover technology for mass-producing interferons and other endogenous proteins and hormones by bacteria or various cell lines.

In 1993 and 1994 Biogen drastically revised its research programs to concentrate on the most promising and potentially profitable products. The current programs emphasize new uses for the interferons, including the treatment of multiple sclerosis and discovery of new immunomodulators and anticancer agents. In 1997 Biogen had the fifth largest total revenue ($434 million) among the publicly owned American biotechnology companies. It is on its way to becoming a fully integrated pharmaceutical company. In 1997 Biogen's product sales ($249 million) finally exceeded royalties from licensing ($172 million).

CENTOCOR

The story of Centocor is that of a typical biotechnology company, containing all of the typical elements: rapid success and dramatic failure, aggressiveness and overconfidence of some of the executives, dangers of protracted lawsuits, overextended debt, reliance on overly optimistic market research, inexperience in dealing with the FDA, slow acceptance of products by the medical establishment, aggressive pricing of potential products, volatility of the stock market, and excessive dependence on one product.

Established in 1979, Centocor by 1986 was already listed among the top ten biotechnology companies in the United States, and its market value was the fifth largest. Between 1986 and 1992 the company raised $572.4 million from partners, as secondary equity financing, and from convertible debts. In early 1992 Centocor entered into a historic deal with Lilly. In anticipation of the FDA's approval of Centocor's monoclonal antibody, Centoxin, Lilly obtained the marketing rights for it and an option on another monoclonal antibody, RheoPro.

Centoxin had been recommended for approval by an FDA advisory committee in September 1991; but final approval never materialized, and Centocor lost $1.5 billion in market capitalization within days. Failure to present convincing clinical data to the FDA and an interim analysis of the data that introduced a bias were the reasons underlying the denial. Centocor was subsequently hit with numerous lawsuits, including a class-action suit from its shareholders alleging that the company management omitted certain facts and made false and misleading statements about Centoxin. The company's venture capital partners voiced similar complaints. Some of the suits were settled; others are still in litigation. Centocor was also charged with patent infringement by Xoma, a suit that Centocor eventually lost.

Centocor's failure to win FDA approval for Centoxin forced the company to downsize. It reduced its work force from 1,600 to 535, primarily through cutbacks in its sales force, and postponed its plans to become a fully integrated pharmaceutical company. Centocor is likely to depend on its corporate partners for some time for the development, marketing, and distribution of its products. In addition to Lilly, its partners include Tanabe Seiyaku Company, Glaxo-Wellcome, Abbott, Roche, and Boehringer Mannheim.

In early 1995 Centocor's business began to improve. The company won FDA approval for abciximab (RheoPro), a monoclonal antibody that binds selectively to specific glycoprotein receptors (GPIIb and IIIa) located on the platelet surface and prevents fibrinogen from interacting with these receptors. As a result platelet aggregation is inhibited. The major indication for abciximab is to prevent the coronary artery from closing after angioplasty. In accordance with the previously granted option, Lilly markets and distributes abciximab. In January 1995 another Centocor product, an antibody for postoperative treatment of colorectal cancer, was approved in Germany and marketed by Burroughs Wellcome (now Glaxo Wellcome) under the trade name of Panorex. Soon thereafter Centocor received FDA approval for two diagnostic products: Captia, for detection of antibodies to *Treponema pallidum*, an organism that causes syphilis; and CA 125II, a second-generation cancer assay, for detection of ovarian cancer.

In 1997 the FDA approved expanded labeling for RheoPro, allowing the use of the drug in a broader range of patients undergoing percutaneous transluminal coronary angioplasty. This approval significantly increased

RheoPro's sales ($254 million in 1997). In February 1998 Centocor announced the proposed acquisition from Boehringer-Ingelheim of the recombinant reteplase (Retavase) for the treatment of heart attacks. Another Centocor product, anti-TNF (tumor necrosis factor) monoclonal antibody, Avakine, has been found to be effective in the treatment of Crohn's disease (inflammatory disease of the lower intestinal tract) and to benefit patients with rheumatoid arthritis.

CHIRON

Chiron was started in 1981 by William J. Rutter, chairman of biochemistry at the University of California at San Francisco (UCSF); Edward E. Penhoet, the current president and CEO; and Pablo Valenzuela. Rutter was initially interested in the application of recombinant DNA technology to the production of vaccines. His efforts were supported by research projects from Merck (development of hepatitis B vaccine), Ciba-Geigy (insulin growth factor), and Nova-Nordisk (single-chain insulin in yeast). The company also established joint ventures with Ciba-Geigy (known as the Biocine Company) and with Johnson and Johnson (called Ortho-Chiron).

Chiron raised $20 million through a public stock offering in 1983 and obtained an additional $115 million in 1991. Its establishment as a major biotechnology company was greatly facilitated by its acquisition of Cetus. In 1993 the FDA approved the Cetus beta-interferon (Betaseron) for treating muscle paralysis of patients with multiple sclerosis. In the same year Chiron won approval for another product, interleukin-2 (Proleukin), for treating renal carcinoma in its early stage. Sales of Proleukin reached $41 million in 1994.

Chiron entered a research collaboration with Janssen Pharmaceutica NV, a subsidiary of Johnson and Johnson, on "peptoids," which are small peptide analogs for treating cancer and infectious diseases. The company also established a collaborative research project with Searle on a tissue factor to control coagulation in microsurgery. Still another research project was initiated with Procept to develop new immunomodulatory compounds. Ciba-Geigy made a major investment in Chiron in 1994, acquiring a 49.9 percent interest for $2.1 billion.

In 1996 Chiron was the world's second largest biotechnology company, with 7,000 employees, $1.3 billion in total revenue, and $55 million in net income. Its products—diagnostics, therapeutics for the treatment of cancer and infectious diseases, and vaccines—were marketed in ninety-seven countries. In addition to its well-known products for treating renal cell carcinoma and multiple sclerosis, Chiron marketed ten vaccines in a joint venture with Behring and had twelve vaccines in development. The applications for two new therapeutic products—Myotrophin, an insulin-like growth factor, and Regranex, a platelet-derived growth factor for the treatment of diabetic foot ulcers—were under review by the FDA. Chiron was exceptionally skilled at combining acquired business with strategic planning. Correct business decisions made at the right time and the willingness of management to take risks were the major factors that contributed to the company's success.

In 1997 the FDA cleared Regranex, but not Myotrophin. A revised application for approval of Myotrophin was subsequently submitted. In addition the FDA approved two more of Chiron's products: a rabies vaccine, Rab-Avert, and ACS Centaur, an immunoassay system.

GENENTECH

Genentech, one of the stars of the biotechnology industry, was founded in 1976 by Robert Swanson, an investment banker, and Herbert Boyer, a professor at UCSF. The company's first scientific achievement was the synthesis of the gene for human somatostatin. Somatostatin is a polypeptide present in hypothalamus and many other tissues, which inhibits growth hormone release, among other effects. Major research contributions were made at Genentech by three young scientists Swanson hired in 1978: David Goedell, Axel Ullrich, and Peter Seeburg. Their research efforts at Genentech culminated in the cloning of human insulin in 1978 and of human growth hormone genes in 1979. These achievements were noticed by Eli Lilly, the major manufacturer of porcine and bovine insulin, which provided substantial financial and technical help to Genentech. This collaboration led to the marketing of the first genetically engineered drug, human insulin (Humulin), by Lilly in 1982.

Genentech's scientists used gene cloning technology in the development of its first major product, tissue plasminogen activator (tPA, recombinant alteplase, Activase), which the FDA approved in 1987 for dissolving blood clots in patients with acute myocardial infarction. Activase is a human peptide, a protease, that contains 527 amino acids. Its existence has been known since the 1940s, but its production in pure form and in large amounts was not feasible without techniques of modern molecular biology. In 1974 plasminogen activator was isolated from the blood vessels of human cadavers; in 1979 it was found in human uterine tissue. The usefulness of tPA as a thrombolytic agent in humans was first demonstrated in 1983 using material produced in vitro by human melanoma cells.

The putative advantage of Activase is that it activates plasminogen in blood clots and to a lesser extent circulation, while previously available thrombolytic agents, such as streptokinase and urokinase, do not differentiate between tissue and plasma plasminogen. Activase was therefore expected to produce less bleeding than previously available drugs and to consume less plasminogen. After initial favorable clinical reports, some investigators claimed that Activase was not superior to streptokinase, bleeding was still of concern, circulating levels of plasminogen and fibrinogen decreased in some patients, its onset of action was slower than desired (sixty to ninety minutes), and its metabolism in humans was exceptionally fast (in vivo half-life was four minutes). Moreover the price was very high ($2,200 for a single dose).

In 1988 and 1989 the sales of Activase did not reach the expected levels, and the company had to downsize. Genentech was resuscitated by G. Kirk Raab, who took over as CEO in 1990 and persuaded the board to invest $55 million in a new clinical study of Activase that involved 41,000 cardiac patients. In 1993 the study demonstrated the superiority of Activase over streptokinase, and Genentech soon increased its share of the thrombolytic market. Raab took another gamble in 1991 when he built a $37 million plant to produce alphadornase (Pulmozyme), an enzyme that selectively cleaves DNA and is useful in the therapy of cystic fibrosis, two years before Pulmozyme was approved by the FDA. The gamble was successful; Pulmozyme became another major product of the company.

Raab was also instrumental in arranging the sale of the majority interest in Genentech (60 percent) to Roche Holdings Ltd., the parent company of Hoffmann-La Roche, for $2.1 billion. The sale provided an infusion of needed cash, and Roche agreed to let Genentech operate independently. The deal ended Raab's career with Genentech, however. On 10 July 1995 Genentech's board of directors fired Raab and replaced him with research director Arthur D. Levinson. The official reason was conflict of interest in Raab's dealing with Roche Holdings Ltd.

In addition to Activase and Pulmozyme, Genentech manufactured human growth hormone formulations (Protropin, Nutropin, and Nutropin AQ) to promote growth in children and interferon gamma-1b (Actimmune) for treating chronic granulomatous disease, an inherited deficiency of the immune system. Genentech's total revenues grew from $9 million in 1980, to $968.6 million in 1996, to over $1 billion in 1997. The company began with 166 employees in 1980, but by 1996 the number had increased to 3,071 and in 1997 was 3,242. Unlike many other biotechnology companies Genentech never reported a net loss. Its net income in 1996 was $118 million and in 1997 was $129 million.

In 1997 Genentech received FDA approval to market rituximab (Rituxan) for the treatment of non-Hodgkin's lymphoma. The company also completed phase 3 clinical studies (large-scale efficacy trials) with trastuzumab (Herceptin), an antibody to growth factor receptor known as Her2, for the treatment of breast cancer. During the same year Genentech established numerous collaborative projects with other companies, including a project with Novartis that involved a phase 3 clinical trial of anti-IgE (immunoglobulin E antibody) for the therapy of allergic asthma; a phase 3 clinical trial, with Alkermes, of sustained-release human growth factor (hgH); and development, with Alteon, of pimagedine for the treatment of kidney disease in diabetic patients.

The establishment of strategic alliances with Sumitomo Pharmaceutical Company and Pharmacia and Upjohn contributed to the increase of royalties income (from $215 million in 1996 to $241 million in 1997).

GENETICS INSTITUTE

The Genetics Institute was founded in 1980 to develop and commercialize protein-based therapeutics. Its formation was initiated by Derek Bok, the former president of Harvard, who was interested in starting a company to exploit new technologies developed at the university. He approached two molecular biologists at Harvard, Mark Ptashne and Tom Maniatis, who in turn located a group of young scientists from various disciplines. Together they would develop product ideas for licensing to other companies. At first Harvard's administration assisted Ptashne and Maniatis in finding business consultants and in obtaining financing, but in September 1980 the Harvard faculty voted against any further participation of the university in a commercial venture. Left to their own devices, Ptashne and Maniatis decided to raise the capital from private investors. They approached Benno Schmidt, at that time with J. H. Whitney and Company; William Paley, founder of CBS; and others to invest in the company and obtained $6 million within a few months.

First located in the Boston Lying-In Hospital, the scientists moved to their current home in Cambridge, Massachusetts, in 1985. The Genetics Institute went public in 1986 as the ninth largest biotechnology company in the United States. In 1992 American Home Products Corporation acquired a 66 percent stake in the Genetics Institute for $300 million. The biotechnology company continued to operate as an independent unit, and its shares were publicly traded, until 31 December 1996, when American Home Products acquired the outstanding shares and made the Genetics Institute a wholly owned subsidiary.

Some of the products developed by the Genetics Institute are sold by other pharmaceutical companies. These products include rhEPO, a protein to regulate the production of red blood cells. It is marketed as Recormon by Boehringer Mannheim in Germany and as Epogin by Chugai in Japan. Granulocyte macrophage-colony stimulating factor (rhGM-CSF) is being sold by Novartis and Schering-Plough in Europe and Latin America under the trade name of Leukomax. This product protects the immune system in cancer patients undergoing chemotherapy.

In September 1996 the Genetics Institute announced a collaborative program with Genentech and Chiron designed to accelerate the discovery of protein-based drugs. The program, named DiscoverEase, is capable of quickly identifying fragments of genes for therapeutically important hormones. The partners will have access to the Genetic Institute's protein library and will be able to license and develop them, while the institute will retain co-marketing rights. This agreement represents the first partnership of that type among biotechnology companies.

In February 1997 the FDA approved recombinant coagulation factor IX (Benefix) for treatment of hemophilia B. The factor is structurally and functionally similar to the endogenous factor IX, but because it is genetically engineered, it is free from any infectious agents. This product represents a breakthrough development and a major milestone in the history of the Genetics Institute.

In 1997 the company had three major new products in development. RhIL-11, a protein to enhance platelet production, was being tested in cancer patients undergoing chemotherapy; rhIL-12, a protein to stimulate the activity of T- and natural-killer cells, was expected to increase immunity in patients with cancer or infections; and rhBMP-2, a protein to induce bone growth, was intended for use in local bone repair.

GENZYME

Genzyme is different from other biotechnology companies in several respects. It was founded by Henry Blair, who soon after acquired Whatman Biochemicals with venture capital support and merged it with Genzyme. In 1982 Genzyme acquired a British fine chemicals company—Koch Light Laboratories. Genzyme's stated mission is to apply enzyme technologies to the development of therapeutics, but the company management, particularly Henri Termeer, who became president in 1983 and CEO in 1985, has emphasized the need to provide continuous cash flow by selling diagnostics and fine chemicals and to grow by acquisitions. In 1986 Genzyme went public and raised $28.2 million; in 1991, in one of the largest public offerings ever for a biotechnology company, it raised $143 million.

Genzyme grew rapidly through mergers and acquisitions as well as by establishing new divisions in the United States and abroad. In 1989 Genzyme merged with Integrated Genetics of Framingham, Massachusetts. During the next five years Genzyme acquired five other companies and opened a research and development center in the United Kingdom. In 1994 Genzyme acquired BioSurface Technology of Cambridge, Massachusetts, and created a tissue repair division to develop, manufacture, and market products for treating and preventing tissue damage. The major product of that division is Carticel, cultured cartilage cells for transplantation into damaged knees. The division also markets Epicel skin grafts for the therapy of skin ulcers and has many other products for tissue growth promotion and wound debridement in development. Another division of the company, Genzyme Transgenic, was established to develop and market transgenic animals—rats and mice with genetic disease caused by mutated genes.

The major part of the company, Genzyme General, had $378.6 million in total revenues in 1995, with a net income of $43.7 million. Genzyme Tissue Repair Division had $5 million in total revenues and a net loss of $22 million. In 1996 Genzyme acquired Deknatel Snowden Pencer Inc. (DSP), a manufacturer and distributor of surgical devices, sutures, and fluids, with $100 million in annual sales. In 1996 total revenues for Genzyme General, including those of DSP, were $511 million, with a loss of $30.5 million. In 1997 the total revenue was $597 million, with a net profit of $96 million.

The Genzyme General division became a complex organization, marketing not only specialty therapeutics but also diagnostic products and services, surgical products, bulk pharmaceuticals, fine chemicals, and dietary supplements. Genzyme's most successful specialty therapeutic was aglucerase (Ceredase), an enzyme for treating Gaucher's disease (a lipid-storage disease affecting the liver, spleen, lymph nodes, and many other organs). Aglucerase, produced from human placental tissue, was marketed in 1991. Because of difficulties in obtaining sufficient quantities of placental tissue, Genzyme developed a recombinant analog of aglucerase, imiglucerase (Cerezyme), produced by mammalian cell culture. Imiglucerase was approved by the FDA and came on the market in 1994. The combined sales of both enzymes reached $260 million in 1996.

In 1997 Genzyme formed a joint venture with Geltex Pharmaceuticals to market RenaGel, a phosphate binder to control phosphate levels in patients undergoing kidney dialysis. By the end of the same year the new drug application (NDA) for RenaGel was submitted to the FDA. The phase 3 clinical trials with the recombinant human thyroid-stimulating hormone (Thyrogen) were successfully completed in 1997, and the NDA was filed. Among the new products Genzyme has under development are antithrombin III (AT-III) produced by genetically modified animals. Clinical studies with AT-III were started in 1998. It is expected to be useful in patients with AT-III deficiency associated with liver disease, shock, burns, and hip and knee replacements as well as in patients having organ transplants.

A bioresorbable membrane that separates tissues during the healing process (Seprafilm) was launched by Genzyme Tissue Repair Division in 1996 and approved in Japan in 1997. It considerably reduces scar formation after surgery. Genzyme General Division remains the leading provider of genetic diagnostic services. Four new genetic procedures, including tests for inherited colorectal cancer and cystic fibrosis, are in development.

IMMUNEX

Immunex was founded in 1981 in Seattle, Washington, by Steven Gillis, Christopher S. Henney, and Stephen A. Duzan. The company set out to discover and develop new immunotherapeutics, which it licensed to others to manufacture. In 1987, however, the company management decided to manufacture and market its own drugs. Four years later, in 1991, the FDA approved what has been the major Immunex product, sargramostim (Leukine), a granulocyte macrophage colony–stimulating factor that supports the survival and differentiation of hematopoietic progenitor cells. It is a glycoprotein consisting of 127 amino acids that is used primarily in patients undergoing bone-marrow transplantation to improve the survival of the transplants. To promote sargramostim and other potential products, the company hired marketing and sales staff. To support expansion, Immunex signed a merger agreement in 1993 with American Cyanamid's Lederle Oncology Division. American Cyanamid received 53.5 percent of the shares of the company, which continued operating under the Immunex name. In 1994 Lederle Laboratories was taken over by American Home Products, which now holds the majority interest in Immunex.

In November 1996 the FDA approved mitoxantrone (Novantrone), a synthetic anticancer drug for treating acute nonlymphocytic leukemia; in June 1997 it was approved for treating advanced prostate cancer. In addition to sargramostim and mitoxantrone, Immunex was also selling leucovorin, methotrexate, and thiotepa (Thioplex) for cancer chemotherapy and aminocaproic acid (Amicar), a fibrinolysis inhibitor, to treat bleeding emergencies. The company inherited all four products from Lederle's oncology divisions after its takeover by American Home Products.

In 1997 Immunex had four potential new drugs under clinical investigation: tumor necrosis factor receptor (Enbrel) for the treatment of rheumatoid arthritis; Flt-3 ligand to improve the recovery of patients with blood cell or bone-marrow transplants (Flt-3 is a receptor present on the surface of primitive bone-marrow cells); interleukin-4 (IL-4) receptor for the treatment of asthma and allergic rhinitis; and CD40 ligand for the treatment of B-cell lymphoma and epithelial cancers. Enbrel's effectiveness in the treatment of rheumatoid arthritis was confirmed in phase 3 clinical trials. Immunex also had four potential products in preclinical development, including interleukin-15 (IL-15) for the treatment of infectious diseases or radiotherapy-induced mucosal damage. In 1997 Immunex reported $188 million in total revenues, with a net loss of $16 million.

Executives in the Biotechnology Industry

One of the remarkable consequences of the rapid growth of the biotechnology industry in the 1980s and 1990s is the appearance of a new breed of managers: biotechnology executives. Many of these executives had careers as venture bankers or scientists; many of the scientists were faculty members at prestigious universities or research institutions. In their younger years they dreamed of academic careers and Nobel Prizes for major contributions in basic sciences. The decrease in the availability of research funds may have influenced their decisions to become entrepreneurs, but other factors were also important. A career in a new industry associated with potentially substantial financial rewards was challenging, glamorous, and capable of satisfying the egos of numerous scientists who left their academic positions to become chief executive officers, presidents, or vice presidents of small biotechnology companies. Some scientists from large pharmaceutical companies also joined the biotechnology industry in expectation that they would have more control over their research programs in a small entrepreneurial company and a better chance to succeed than in a larger organization. Still others were forced to enter the biotechnology industry because of downsizing and mergers among the large pharmaceutical companies.

Only a few of the highly respected scientists who became entrepreneurs in the biotechnology industry managed to retain their academic positions and productive academic laboratories. Among them is Solomon Snyder, professor of neuroscience at Johns Hopkins University, who started Nova Pharmaceutical Company and more recently Guilford Pharmaceuticals in Baltimore. Snyder's academic accomplishments in the field of neuropharmacology led to many new approaches to drug development, which he used in his biotechnology companies. Similarly, Chiron's founder and chairman, William J. Rutter, is still active at UCSF.

All biotechnology companies have many well-known academic scientists who serve on their board of directors or scientific advisory boards, but the degree of involvement of a particular scientist is not always apparent to the public. In some cases it is only "for appearance"—a scientist may visit a company once or twice a year to justify his or her compensation as a member of the advisory board; in other cases academic scientists play an active role in planning research. The scientists occasionally are stockholders in the company. In addition to or instead of monetary compensation they are paid in shares or stock options if their university's conflict-of-interest policy allows such payments.

Common Obstacles for Biotechnology Companies

All biotechnology companies share certain obstacles, from obtaining adequate financing to dealing with the FDA and other regulatory bodies. The most important of these obstacles are outlined below.

FINANCING

Biotechnology companies obtain their financial resources not only from venture bankers, but also from private investors, investment bankers' venture funds, and public offerings. These funds may offset the costs of discovery but not of developing and manufacturing potential products. During the last few years numerous failures of drug candidates in clinical trials disappointed investors and reduced the availability of capital. According to a 1996 survey of the industry, the net cash provided by the financial activities of all publicly held biotechnology companies in 1994 was half of what it was in 1993. Between July 1994 and June 1995 average follow-up offerings were smaller than in the previous year, and only two companies were able to obtain secondary offerings in excess of $25 million. But 1995 was a very active year overall for the biotechnology industry: More than 170 alliances were formed, up from 66 in 1994; and $3.5 billion from public offerings, almost double the $1.8 billion obtained in 1994, further fueled the industry.

Despite this positive trend a quarter of all biotechnology companies will last less than one year with their current burn rate (cash usage), while half of the companies have less than two years' cash supply. Most biotechnology companies do not have sufficient funds to develop and market discoveries and are looking for corporate partners with substantial cash to invest as the only way to ensure survival and to permit product development.

EXPERIENCE IN DRUG DEVELOPMENT

Many leaders of the biotechnology industry initially came from academic institutions or the financial community and had no previous experience in the pharmaceutical industry. Some of the failures of drugs in clinical trials or in the FDA approval process can be attributed to the poor design of trials and managerial inexperience. Biotechnology managers today are much more experienced in selecting drug candidates for development, designing clinical trials, and dealing with the FDA than they were when the industry was just emerging, so errors of inexperience are less likely to be repeated.

Many aspects of drug development in biotechnology and large pharmaceutical companies are now identical. Whether they work for small biotechnology companies or large pharmaceutical houses, medicinal chemists obtain leads for new drugs largely from high throughput screening (HTS) and are required to optimize these leads by synthesizing related compounds. Rational drug design is still in an experimental stage. To select potentially useful chemicals, HTS uses screens designed with the help of molecular biology, such as bacterial or yeast cells transfected with human receptors or ion channels. These screens represent a major contribution of biotechnology to drug development. The secondary evaluation of the potential candidates still involves conventional animal pharmacology, although molecular biologists have developed transgenic animals with disease-causing genes. These animals are not yet used routinely in drug development, but most large pharmaceutical companies are adapting the techniques of molecular biological research to drug discovery and development. As Robert R. Ruffolo, vice president of SmithKline Beecham, recently stated, his company hires scientists who know how to apply the techniques used in molecular biology to pharmacology rather than hiring conventional pharmacologists or molecular biologists.

GOVERNMENT REGULATIONS AND PRICE CONTROLS

The biotechnology industry is subject to the same laws and regulations as large pharmaceutical companies. Potential biotechnology products are tested for efficacy and safety just as are synthetic drugs developed by large pharmaceutical companies. This means that ten years or more might elapse before a product wins approval; the exceptions are drugs that fall into an accelerated approval category, such as drugs for treating AIDS. These rules are particularly hard for small biotechnology companies because of their limited resources. The industry therefore advocates eliminating excessive regulatory burdens in the product approval process.

Because many start-up biotechnology companies are small business concerns, they can apply for small business innovation research grants from the U.S. Department of Health and Human Services. The awards are limited to $100,000 for the first phase and $750,000 for the second phase of a research program leading to commercialization of new technology. This program was established in accordance with the Small Business Research and Development Enhancement Act of 1992. A small business concern is defined as an organization with no more than five hundred employees. At least 51 percent of the voting stock of the concern must be owned by U.S. citizens or permanent resident aliens. These grants represent the only financial assistance the federal government provides directly to biotechnology entrepreneurs. Some states also provide grants or loans to entrepreneurs, although they are usually rather small and difficult to obtain. The Small Business Administration guarantees private bank loans to entrepreneurs who want to start a new business or expand an existing one.

Until it was amended in 1994, the Orphan Drug Act of 1983 was of considerable importance for the biotechnology industry. Designed to encourage the pharmaceutical industry to develop drugs for rare diseases, the act gave some tax incentives and seven-year exclusive marketing rights to drugs developed to treat diseases affecting fewer than two hundred thousand people. The act worked well; between 1983 and 1993 ten times more so-called "orphan" drugs for rare diseases were developed than in the previous decade. The act facilitated the development of new biotechnology-based drugs; companies developed orphan drugs to enter a market niche, hoping to expand into larger markets later. The strategic goal of biotechnology companies was to find new indications for an orphan drug that had not originally been anticipated. It was also easy to sell drugs for diseases for which there was no other treatment.

In the early 1990s a proposal was introduced in Congress to end exclusive marketing rights for an orphan drug if its sales reached a certain level. The industry opposed this proposal, which was modified and passed in 1994. Under the amendment, orphan status of a drug ends when the patient population receiving it exceeds two hundred thousand. The amendment is not retroactive, so drugs on the market or in clinical trial as of 1 March 1994 are not affected. The market exclusivity was reduced to four years, but an additional three years can be granted for drugs with "limited commercial appeal." The Orphan Drug Act is now less attractive, but still remains of some value to the industry.

Pricing of biotechnology products is of interest to all biotechnology companies, even to those who do not yet have any products on the market. Price controls limit future returns and drive investors into other industries with higher returns. The high prices for life-saving drugs are justified if these drugs shorten the duration of hospitalization or subsequent patient recovery more than other available treatments, but they tend to erode hospitals' bottom lines. With the rapid growth of managed-care organizations, the proper pricing of biotechnology products is becoming even more critical. The manufacturer must now provide data on the cost-effectiveness of a new biotechnology product—supported by outcome research and by comparative data on the total cost of therapy using a new drug compared with the alternative therapeutic approaches—to the FDA for consideration during the approval process.

In 1993 some health reform advocates recommended government price controls for the pharmaceutical and biotechnology industries as well as the establishment of an advisory council on breakthrough drugs, which would examine "the reasonableness of the prices of new drugs." Other proposals included denying Medicare coverage for new drugs, a new tax called the "medical rebate tax," and severe limitations on the rights of companies to offer discounts. This interference with the competitive market forces would have severely harmed the biotechnology industry by stifling innovation and drastically reducing the development of new drugs.

The prospect of such legislation alone decreased the stock prices of pharmaceutical and biotechnology companies in the early 1990s and frightened potential investors in the biotechnology industry. This threat of tight government control of prices and its consequences are referred to by many in the industry as the "Hillary factor," a reference to First Lady Hillary Clinton, who actively promoted health reform, which never passed. The biotechnology industry did not begin to recover from this threat until 1995. De-facto price controls were instituted by the companies themselves under pressure from managed-care organizations and insurance companies. A prominent role in setting the standards for drug reimbursements is currently played by the Health Care Financing Administration as well as by third-party payors, such as Blue Cross and health maintenance organizations.

TAX INCENTIVES

The biotechnology industry favors any tax legislation that may serve as an inducement to potential investors, and the commitment of the Republican party to reduce the capital gains tax represents a substantial inducement. Two temporary tax credits are also of interest to the biotechnology industry: the research and the orphan drug tax credits. The industry hopes they will be extended or made permanent. The research tax credit allows a small biotechnology company that loses money during its initial years of operation to subtract these losses from potential profits earned in subsequent years. The industry also favors the restructuring of the orphan drug tax credit so that it can be carried forward by companies free of current tax liabilities.

PROTECTION OF INTELLECTUAL PROPERTY

The most important asset of any biotechnology company is its intellectual property—the technological advances or know-how of its owners or employees that permit the company to obtain an advantage over potential competition in product discovery or development. Without such an advantage the chances that a young company will succeed are greatly diminished. The intellectual property must be protected, and it is best protected by patents. Patents are highly important in the biotechnology industry not only for marketing or licensing potential products but also for attracting badly needed capital from investors. A company that owns unique technology is more likely to obtain public or private funds than its competitor whose technology is unprotected.

The survival of a new biotechnology company is often determined by its ability to protect intellectual property. The companies are advised by their patent attorneys to develop a patent strategy, to decide whether they should patent any technology they can obtain or be highly selective in deciding what technology to patent. Companies must also decide whether the potential products should be protected by composition of matter patents or use patents. How should potential patent interferences be handled? An out-of-court settlement saves often badly needed financial resources, and the loss of a protracted patent interference suit can easily ruin a young biotechnology company.

The biotechnology industry generally favors strong protection of intellectual property. It lobbies for an adequate patent life and strong enforcement of patents in the United States and abroad. The industry was favorably affected by the General Agreement on Tariffs and Trade that was signed by President Bill Clinton in 1994. It provides that patents filed on or after 8 June 1995, are valid for twenty, instead of seventeen, years. In an attempt to streamline the patenting process, the U.S. Patents and Trademarks Office announced in 1995 that it would no longer base decisions on the patentability of a discovery on the availability of clinical data. The so-called "scientifically plausible use" can now be established by in-vitro or in-vivo animal studies. The patenting of genes was facilitated by a recent appeals court ruling that "the structure of the sequence, rather than the method of cloning, makes a gene patentable."

The current American laws covering protection of intellectual property protect inventors, enhance the value of technology, and allow manufacturers to earn sufficient profits to finance further research and development of new technology. Worldwide introduction and enforcement of similar laws would further benefit the biotechnology industry.

Impact of the Biotechnology Industry on the Economy and on Health Care

According to the Ernst and Young Tenth Annual Report on the Biotechnology Industry (Biotech 96), the total annual sales of publicly held biotechnology companies in the United States represents 10 percent of the total sales of the American pharmaceutical industry. Many products developed by biotechnology companies were licensed out to the pharmaceutical industry, and some of

Table 5. Major Products of the Biotechnology Industry Approved by the FDA, from 1985–1995

Company	Product	Indication(s)	Year of Approval
Amgen	Epoetin alfa (Epogen)	Anemia	1989
	Filgrastim (Neupogen)	Neutropenia	1991
Centocor	Abciximab (RheoPro)	Vascular occlusion after angioplasty	1994
Chiron	Interferon B (Betaseron)	Multiple sclerosis	1993
	Aldesleukin (IL-2)	Renal carcinoma	1992
Enzon	Adenosine deaminase (Adagen)	Immunodeficiency	1990
	Pegaspargase (Oncaspar)	Acute lymphoblastic leukemia	1994
Genentech	Alteplase (Activase)	Myocardial infarction	1987
	Gamma interferon (Actimmune)	Chronic granuloma	1990
	Somatrem (Protropin)	Growth hormone deficiency	1985
	DNAse (Pulmozyme)	Cystic fibrosis	1993
Genzyme	Aglucerase (Ceredase)	Gaucher's disease	1991
Immunex	Yeast-derived granulocyte macrophage colony-stimulating factor (Leukine)	Bone-marrow transplants	1991
Interferon Sciences	Interferon alfa-N3 (Alferon N)	Genital warts	1989
Univax Biologicals	Win Rho SD (Adagen)	Thrombocytopenic purpura	1995

the earlier products, such as Epogen, exceeded $1 billion in annual sales in 1996 (Table 5). Most of the major products of the biotechnology industry are either already highly profitable or on the verge of becoming so.

From 1984 to 1994 the biotechnology industry has created more than a hundred thousand specialized jobs, on the average of seventy-five jobs per company. The years 1992 and 1993 were critical for the industry, primarily because of the pending health reform bills with the threats of price control and the failure of several potentially major products to win FDA approval. Obtaining venture capital became more difficult, as did placing stock offerings. Since 1993, however, the industry has managed to get FDA approval for a substantial number of novel drugs. The industry is not yet profitable overall, although some companies, such as Amgen and Genentech, showed substantial profits.

Table 4 compares the revenues, net incomes, and product lines of the eight major biotechnology companies. Despite introduction of novel therapeutics and promising drugs in the pipeline, four of the companies (Genentech, Chiron, Genetics Institute, and Immunex) were either partially or completely taken over by large pharmaceutical concerns (Roche, Novartis, and American Home Products). The takeovers occurred irrespective of profitability. Faced with the uncertain outcome of clinical trials and FDA approval, along with the high cost of development, the stockholders could not resist the generous offers of these large pharmaceutical concerns. The growth of total revenues from 1994 to 1997 was remarkable for most companies, even though it was not always accompanied by commensurate increases in net income. The high cost of development of potential products was the major factor affecting the profitability of these young companies.

The impact of the biotechnology industry should not be judged solely by the number of jobs it has created or by its sales and profits. The industry has contributed substantially to health care by rapidly applying scientific discoveries to the development of pharmaceuticals, by developing drugs for the therapy of rare diseases, and by drawing the attention of major pharmaceutical companies to new approaches in drug development. The biotechnology industry forced the pharmaceutical companies to restructure their research organizations and to revise their research approaches in order to remain innovative and competitive.

The biotechnology industry has also had a substantial impact on the scientific community. Many academic scientists are directly or indirectly involved with the biotechnology industry through consulting or research arrangements. Universities rarely contract to provide services for fees, but they do enter into sponsored research contracts with financial obligations for rights to commercialize university inventions.

Future Perspectives

Predictions are often misleading, but a few points about the likely future development of biotechnology can be made with a fair degree of safety.

The term *biotechnology industry* is probably here to stay, but the industry is not likely to remain the same. Some of the most successful biotechnology companies have already been taken over by large pharmaceutical companies. Even if they retain their own identity and infrastructure, these biotechnology companies are likely to coordinate their research and marketing very closely with their partners. A few companies, such as Amgen and Genzyme, are already fully integrated pharmaceutical companies, operating like major pharmaceutical concerns. Such companies as Alza, Nova, and Liposome Technology, which have valuable technology but no major products, are likely to become pure service companies, offering their testing, research, or manufacturing facilities to the major pharmaceutical or biotechnology companies, while so-called "virtual" companies will continue to contract many of their functions to other organizations. There will be a greater tendency toward the formation of consortia, consisting of a dozen or more biotechnology companies sharing some of their resources. Some of the U.S. biotechnology companies will establish foreign branches and will market their products in Europe and Japan before entering the American market. Within the next two years about a fifth of the currently operating biotechnology companies will not have enough capital to stay in business, and most of them will be taken over by other companies. New companies will replace them, so that the total number of biotechnology companies will probably remain the same or even increase.

The number of biotechnology-based drugs will increase substantially, and it is not unreasonable to expect that by the year 2000 two or three of them will have annual sales in excess of $1 billion each. Investment in biotechnology industry will remain risky, but a few lucky investors in the successful biotechnology companies will enjoy exceptionally high returns.

Sources

G. Allen. *Life Science in the Twentieth Century.* London: Cambridge University Press, 1978.

American Health Consultants. *Biotechnology State of the Industry Report.* Atlanta: BioWorld Publishing Group, 1995.

American Home Products Corporation. 1996 Annual Report. Madison, N.J., 1997.

Amgen. 1996 Annual Report. Thousand Oaks, Calif.: Amgen Inc., 1997.

Anonymous. "Biogen and Anergen: Two Poles of Biotech." *Genesis Report/Rx* 2 (Oct. 1994), 2–9.

Anonymous. "Biotechnology's Top 50: A Looming Cash Crisis." *Genesis Report/Rx* 2 (October 1994), 2–14.

Anonymous. "Biotechnology Medicines and Vaccines Approved and under Development." *Genetic Engineering News* (August 1995), 12–15.

AUTM Licensing Survey and Selected Data FY 1991–FY 1994. Norwalk, Conn.: Association of University Technology Managers (AUTM), 1995.

Biogen. 1997 Annual Report. Cambridge, Mass., 1998.

Robert Bud. *The Uses of Life: A History of Biotechnology.* Cambridge: Cambridge University Press, 1993.

Centocor. 1994 Annual Report. Malvern, Pa., 1995.

Centocor. 1996 Annual Report. Malvern, Pa., 1997.

Centocor. 1997 Annual Report. Malvern, Pa., 1998.

Chiron. 1997 Annual Report. Emeryville, Calif., 1998.

Genetics Institute. 1995 Annual Report. Cambridge, Mass., 1996.

Genzyme. 1994 Annual Report. Cambridge, Mass., 1995.

Genzyme. 1997 Annual Report. Cambridge, Mass., 1998.

V. Glaser. "152 U.S. Biopharmaceutical Firms Lose $1.3 Billion." *Bio/Technology* 13 (1995), 422–425.

W. K. Hallman. "Public Perception of Biotechnology: Another Look." *Bio/Technology* 14 (1996), 35–38.

D. L. Higgins; W. F. Bennett. "Tissue Plasminogen Activator: The Biochemistry and Pharmacology of Variants Produced by Mutagenesis." *Annual Review of Pharmacology and Toxicology* 30 (1990), 91–121.

Immunex. 1994 Annual Report. Seattle, Wash., 1995.

Immunex. 1996 Annual Report. Seattle, Wash., 1997.

R. A. Kaba; D. A. Grossman; J. Tabin. "Intellectual Property in Drug Discovery and Biotechnology." In *Burger's Medicinal Chemistry and Drug Discovery.* Vol. 1, 5th ed. Edited by M. E. Wolff. New York: John Wiley & Sons, 1995.

A. Kornberg. *The Golden Helix: Inside Biotech Ventures.* Sausalito, Calif.: University Science Books, 1995.

K. B. Lee; G. S. Burrill. *Biotech 96: Pursuing Sustainability. The Ernst & Young Tenth Annual Report on the Biotechnology Industry.* Palo Alto, Calif.: Ernst & Young, 1995.

R. Longman. "The Lessons of Centocor." *INVIVO* May (1992), 23–27.

D. Myshko. "High Speed Drug Discovery." *R & D Directions* 1 (June 1995), 24–26.

H. Price; J. Hermann; P. Knight, eds. *The Biotechnology Report 1994/95.* Hong Kong: Campden Publishing, 1995.

H. K. Shah; R. J. Rodgers. "Biopharmaceutical Sales and Forecasts to 1997." *Spectrum* (Decisions Resources, Inc., Burlington, Mass.) 35 (1992), 1–12.

W. A. Strycharz; G. Price. "The History of the Genetics Institute." *BioWorld Today* 4 (1992), 1–3.

Chapter Four

Pharmaceutical Taxonomy of Most Important Drugs

RALPH LANDAU, ARTHUR DESIMONE, AND LEWIS E. GASOREK

Any industry analysis invariably leads to a discussion of the critical success factors and what differentiates winning from losing. This century has seen the pharmaceutical industry grow dramatically by developing products that offered tremendous advances in the treatment and prevention of many diseases and by effectively marketing products that were only modestly superior to existing compounds. A natural objective in studying this industry would be to identify the most "significant" products developed during this period, with the understanding that businessmen, doctors, pharmacologists, public policy makers, and others all have very different ways of evaluating what makes a product significant.

The objective in developing the following drug taxonomy was to identify and characterize the most significant pharmaceutical products developed by this innovative industry during its long history. The explosion in scientific knowledge during the past half century naturally biases the study to this period—the period of greatest pharmaceutical innovation. The taxonomy also evolved into a vehicle for visualizing the many therapeutic approaches and innovations in the treatment of numerous disease categories over time as well as the shifting nature of the innovation process.

To be considered as one of the most significant drugs, the following iterative screening process was used: We began by looking at drugs that were marketed in the United States in 1989 and first approved by the FDA between 1940 and 1988. After completing additional screening steps (discussed below), this process was updated quantitatively to 1995, with selected additions after that time.

To be considered for inclusion, a drug had to be a new chemical entity (NCE). This criterion served as a preliminary and objective screen for innovative new drugs rather than a new indication for an older drug, a mixture of two known drugs, or an improved delivery structure with enhanced therapeutic efficacy. Over-the-counter drugs and nutritional supplements were excluded from the screen.

To be selected as a very important drug, the NCE needed to have a relevant market share of at least 10 percent for its specific therapeutic category (as determined by IMS) by the end of the third year after introduction. The screening categories were later expanded to include such other factors as specific indication of a drug's importance either as a major therapeutic advance for a smaller patient audience or as a significant scientific innovation that led the way for future developments.

After the preliminary screens were done, the drugs were reviewed by panels of pharmacologists, by executives in a number of major and mid-sized pharmaceutical companies, and then by academic and industrial researchers. The screens, coupled with further review, resulted in the selection of the drugs listed in the following taxonomy tables.

The drugs are grouped by therapeutic application area and by class of chemical compound. They are further segregated into chronological periods that roughly follow the changes in methods for pharmacological research.

Some drugs are listed twice because of their importance in differing applications.

Of particular note is the fact that a number of commercially significant drugs are not included because they were either deemed incremental NCE innovations or are the second drugs of a class.

Interpreting the sales data to draw significant conclusions was difficult because of the mix of old and very new drugs. Relative sales volume comparisons across the time periods or within a particular class of drugs are affected by such factors as maturation and substitution of products, breadth of products available on the market, and the increasing importance of chronic (long-term) versus acute (short-term) drug consumption.

The following tables are composed of the top drugs identified and reconfirmed by this entire process. While individuals may have their own preferences and opinions regarding specific compounds, these tables represent an informed consensus as to the technically and commercially most significant drugs developed by this industry.

Group and Subdivision	19th Century	1900–1950
Vaccines	Anthrax Rabies Diphtheria Tetanus	Pertussis Varicella Diphtheria, tetanus, pertussis (DTP, Tri-Immunol)
Antibacterial drugs		
Sulfonamides (see also Diuretics and Antidiabetics)		Sulfamidochrysoidine (Prontosil) Sulfisoxazole (Gantrisin) Sulfasalazine (Azulfidine)
Antibiotics		
Penicillins		Penicillin
Aminoglycosides		Streptomycin
Tetracyclines		Oxytetracycline (Terramycin) Tetracycline (Tetracyn)
Cephalosporins		
Macrolides		
Antitubercular drugs		Isoniazid (INH)
Nitroimidazoles		
Chloramphenicol		Chloramphenicol (Chloromycetin)
Quinolones		Nalidixic acid (NegGram)
Monobactams		
Combinations		
Antiviral drugs		
Antifungal drugs		Mystatin (Mycostatin)
Cardiovascular agents		
Antihypertensive drugs		
Centrally acting		

1951–1965	1966–1980	1981–1995
Polio (Ipol, Orimune) Measles (Attenuvax)	Mumps (Mumpsvax) Rubella (Meruvax, Biavax) *Haemophilus influenzae* (Pedvaxhib, Hibtiter) Measles, mumps, rubella (MMR II)	Hepatitis B (Recombivax HB, Engerix-B) Hepatitis C Varicella (Varivax) Pneumococcus (Pneumovax, Pnu-Immune)
	Sulfamethoxazole + trimethoprim (Bactrim)	
Ampicillin (Polycillin)	Amoxicillin (Amoxil) Methicillin (Staphcillin) Naphcillin (Unipen) Gentamicin (Garamycin) Tobramycin (Tobrex, Nebcin)	
Doxycycline (Vibramycin)		
Cephazolin (Kefzol, Ancef)	Cephalothin (Keflin) Cephalexin (Keflex) Cefaclor (Ceclor)	Cefotaxime (Claforan) Ceftriaxone (Rocephin) Cefprozil (Cefzil)
Erythromycin (Erythrocin) Pyrizinamide (Zinamide) Ethionamide (Nisotin) Ethambutol (Myambutol) Metronidazole (Flagyl)		Azithromycin (Zithromax)
		Ciprofloxacin (Cipro) Norfloxacin (Noroxin) Aztreonam (Azactam) Amoxicillin + clavulanate potassium (Augmentin) Imipenem + cilastatin (Primaxin)
	Idoxuridine (Herplex) Acyclovir (Zovirax)	Zidovudine (Retrovir) Foscarnet (Foscavir) Didanosine (Videx DDL)
Griseofulvin (Fulvicin) Amphoterecin B (Fungizone)	Clotrimazole (Mycelex) Miconazole (Monistat)	Ketoconazole (Nizoral) Fluconazole (Diflucan)
Reserpine (Serpasil) Methyldopa (Aldomet)	Clonidine (Catapres)	

Group and Subdivision	19th Century	1900–1950
Antihypertensive drugs *(cont.)*		
α-Adrenoceptor antagonists		
β-Adrenoceptor antagonists		
ACE inhibitors		
Angiotensin II antagonists		
Ca^{2+} channel antagonists		
Other vasodilators		Hydralazine (Apresoline)
Diuretics		
Cardiac glycosides	Digitalis extract	Digoxin (Lanoxin)
Antianginal drugs		
Nitrites and nitrates	Nitroglycerin, amyl nitrite	
β-Adrenoceptor antagonists (see above)		
Ca^{2+} channel antagonists (see above)		
Anticoagulants and Antithrombotics	Leeches (Hirudin)	Heparin Dicumarol (Dicoumarol) Warfarin (Coumadin)
Hypolipemics		
Central nervous system drugs		
Anesthetics		
General	Chloral hydrate	Diethylbarbituric acid (Veronal)
	Ether	Phenobarbital (Luminal)
	Chloroform	Cyclopropane
Local	Cocaine	Procaine (Novocaine) Lidocaine (Xylocaine)
Analgesics (see also Nonsteroidal anti-inflammatory agents)	Morphine Codeine	Levorphanol (Dromorane)
Analgesic anatagonists		
Antianxiety drugs		Meprobamate (Miltown)

1951–1965	1966–1980	1981–1995
	Prazosin (Minipress)	Terazosin (Hytrin)
Pronethalol (Alderlin)	Propranolol (Inderal)	
	Timolol (Blocadren, Timoptic)	
	Atenolol (Tenormin)	
	Metoprolol (Lopressor)	
	Nadolol (Corgard)	
	Pindolol (Visken)	
	Captopril (Capoten)	Enalapril (Vasotec)
		Losartan (Cozaar)
Verapamil (Isoptin, Calan)		Nifedipine (Adalat, Procardia)
		Diltiazem (Cardizem)
	Minoxidil (Lonitex, Rogaine)	
Acetazolamide (Diamox)	Triamterene (Dyrenium)	
Chlorothiazide (Diuril)	Amiloride (Midamor)	
Hydrochlorothiazide (Hydrodiuril, Esidrex)		
Furosemide (Lasix)		
Chlorthalidone (Hygroton)		
Spironolactone (Aldactone)		
		Transdermal nitroglycerin (Deponit, Nitrodur)
Urokinase (Abbokinase)	Streptokinase (Streptase)	Alteplase (t-PA) (Activase)
Niacin (Niacor)	Clofibrate (Atromid-S)	Gemfibrozil (Lopid)
Cholestyramine (Questran)	Colestipol (Colestid)	Lovastatin (Mevacor)
Secobarbital (Seconal)		
Thiopental (Pentothal)		
Halothane (Fluothane)		
Propoxyphene (Darvon)		
Pentazocine + naloxone (Talwin)		
Naloxone (Narcan)		
Chlordiazepoxide (Librium)	Diazepam (Valium)	Buspirone (BuSpar)
	Triazolam (Halcion)	
	Alprazolam (Xanax)	
	Oxazepam (Serax)	

Group and Subdivision	19th Century	1900–1950
Central nervous system drugs *(cont.)*		
Antidepressants		
Antipsychotics		Chlorpromazine (Thorazine)
Antiparkinsonian drugs		
Drugs affecting the autonomic nervous system		
Adrenoceptor agonists	Epinephrine (Primatene/ Adrenalin)	Norepinephrine (Levophed) Isoproterenol (Isuprel)
Cholinergics	Acetylcholine Pilocarpine HCL Physostigmine	Betanechol (Urecholine) Neostigmine (Prostigmine)
Anticholinergics	Atropine	Scopolamine Dicyclomine (Bentyl)
Antihistamines		
Antiallergics		Diphenhydramine (Benadryl) Tripelennamine (Pyribenzamine) Promethazine (Phenergan)
Antiemetics		Dimenhydrinate (Dramamine)
Anti-ulcer drugs (H_2-receptor antagonists)		
Serotonin antagonists		
Antiallergics		Cyproheptadine (Periactin)
Anti-migraine drugs		Ergotamine (Ergostat)
Hormones		
Sex hormones		
Androgens		Testosterone (Delatestryl) 17-methyltestosterone (Metandren)
Inhibitors of testosterone formation		
Estrogens		Estradiol diethylstilbestrol
Ovulation inducers		
Estrogen antagonists		
Progestins		Progesterone
Insulin		Insulin

1951–1965	1966–1980	1981–1995
Iproniazid (Marsilid)	Desipramine (Norpramine)	Fluoxetine (Prozac)
Isocarboxazid (Narplan)	Trazodone (Desyrel)	Sertraline (Zoloft)
Imipramine (Tofranil)		
Amitriptyline (Elavil)		
Thioridazine (Mellaril)		Clozapine (Clozaril)
Haloperidol (Haldol)		
	Levodopa (Larodopa)	Selegiline, L-deprenyl (Eldepryl)
	Levodopa + carbidopa (Sinemet)	
	Albuterol (Proventil, Ventolin)	
	Terbutaline (Bricanyl, Brethine)	
Edrophonium (Tensilon)		Tacrine (Cognex)
Propantheline (Pro-Banthine)		
Diphenoxylate + atropine (Lomotil)		
Chlorpheniramine (Chlor-Trimeton)	Terfenadine (Seldane)	
Hydroxyzine (Atarax)		
Meclizine (Bonine)		
	Cimetidine (Tagamet)	Ranitidine (Zantac)
Methysergide (Sansert)		Sumatriptan (Imitrex)
Oxandrolone (Anavar)		
		Finasteride (Proscar)
Clomiphene (Clomid)		
	Tamoxifen (Nolvadex)	
Norethynodrel (present in Enovid)		
Norethindrone (Norlutin)		
		Recombinant human insulin (Humulin, Novolin)

Group and Subdivision	19th Century	1900–1950
Hormones *(cont.)*		
Corticosteroids		Cortisone (Cortone)
Nonsteroidal analgesic and anti-inflammatory agents	Aspirin Aminopyrine Acetanilide	Acetaminophen (Tylenol)
Immunosuppressants		
Drugs for metabolic diseases		
Antidiabetics		
Sulfonamides		
Anti-gout drugs	Colchicine	Probenecid (Benemid)
Cancer chemotherapeutics		
Folic acid antagonists		Aminopterin
Alkylating agents		Mechlorethamine (Mustargen)
Antibiotics as anticancer drugs		Actinomycin C (Sanamycin)
Antimitotic drugs		
Sex hormones as anticancer drugs		Diethylstilbesterol (Synestrin)
Miscellaneous		
Miscellaneous biotechnology products		

1951–1965	1966–1980	1981–1995
Hydrocortisone (Cortril)		
Prednisone (Deltasone)		
Dexamethasone (Decadron)		
Methylprednisolone (Medrol)		
Phenylbutazone (Butazolidin)	Sulindac (Clinoril)	
Indomethacin (Indocin)	Piroxicam (Feldene)	
	Ibuprofen (Motrin)	
	Naproxen (Naprosyn)	
	Ketoprofen (Orudis)	
Mercaptopurine (Puri-Nethol)	Cyclosporin A (Sandimmune)	Tacrolimus (Prograf)
Azathioprine (Imuran)		
Tolbutamide (Orinase)	Glipizide (Glucotrol)	
Chlorpropamide (Diabenese)	Glyburide (Micronase)	
Allopurinol (Zyloprim)		
Mercaptopurine (Purinethol)		
Fluorouracil (Efudex)		
Chlorambucil (Leukeran)		
Cyclophosphamide (Cytoxan)		
	Doxorubicin (Adriamycin PFS or RDF)	
	Vinblastine (Velban)	
	Vincristine (Oncovin)	
	Etoposide (Vepesid)	
	Tamoxifen (Nolvadex)	
Cisplatin (Platinol)		Paclitaxel (Taxol)
		Docetaxel (Taxotere)
		Aldesleukin (Proleukin)
		Recombinant growth hormone (Protropin)
		Interferon-gamma 1b (Actimmune)
		Alfa epoetin (Epogen)
		Filgrastim (Neupogen)
		Recombinant antihemophilic factor (Kogenate)

Chapter Five

The Economics of Drug Discovery

IAIN COCKBURN AND REBECCA HENDERSON

Pharmaceutical research presents extraordinary economic challenges. Drug discovery and development is a highly risky and expensive investment, whose returns—if any—are realized only after lengthy periods of research, development, testing, and government approval and through commercial success in an intensely competitive and closely regulated market. The revolutionary changes in the technologies of drug research have intensified these pressures. Drug discovery has become more and more costly, but the payoff to successful innovation—in the profitability of leading firms and the benefits to the consumers of their products—has also risen. In this environment the research performance of drug companies—the efficiency with which they use resources expended on drug discovery and development—is a matter of critical concern not only for their stockholders but also for public policy. This chapter surveys our research on the managerial and economic forces determining pharmaceutical firms' innovative performance and capabilities, how they have changed over time, and their implications for understanding the evolution of the industry.[1]

Over the last few decades real spending on research and development by pharmaceutical companies has increased enormously, while the number of new drugs introduced every year has not. In 1971 the members of the Pharmaceutical Manufacturers Association spent about $360 million on R&D, but in 1991 they spent $8.9 *billion*, a real increase of more than 2,300 percent in twenty

[1] For additional and broader discussion of the role of R&D in the economics of the industry, see the U.S. Congress Office of Technology Assessment study (1993), Caglarcan et al. (1978), Comanor (1986), Gambardella (1995), Schwartzman (1976), Spilker (1989), and Temin (1980). See also Jensen (1987), Grabowski and Vernon (1990), DiMasi et al. (1991), Graves and Langowitz (1993), and Wiggins (1979).

years.[2] But this enormous escalation of investment in R&D has not been accompanied by a similar increase in the introduction of new drugs. During the same twenty-year period the number of new drugs introduced per year has remained more or less constant. Furthermore, although it is hard to control for changes in the quality of individual patents, by some measures the number of important patents generated by the industry has actually fallen. Because one might expect the new technologies of drug discovery to have greatly increased the efficiency of research, at first glance the industry's apparent decline in output per research dollar is surprising.

We focus first on changes in the costs of research and in industry productivity over time. We explore the degree to which the apparent decline in output per research dollar can be explained by a variety of competing hypotheses, including an overall decline in scientific opportunity, a shift from "easy" toward "more difficult" fields, greatly accelerated rivalry within the industry, and the problems inherent in measuring "output" with any precision. We conclude that, while the changing nature of drug research has considerably increased the costs of discovering new drugs and introducing them to the marketplace, this increase has probably been matched by comparable increases in the average economic value or therapeutic significance of each drug, so that the real productivity of the industry may have actually increased.

We then turn to a detailed examination of the internal economics of research within the firm. Our work on the productivity of research at the level of individual research programs inside a sample of firms[3] generated several intriguing results. The most important determinant of research productivity is historical success, a result we attribute to the importance of idiosyncratic knowledge and to the role played by particularly highly skilled individuals in particular fields. We also find, however, that firm productivity is significantly shaped by research strategy—particularly by the size and scope of the research portfolio—and by the ways in which research is managed.

Large pharmaceutical companies have several important advantages in the conduct of research. Both size and experience appear to be important sources of advantage in the industry. This superior performance, however, appears to flow as much from economies of scope (e.g., from the ability to share common inputs among different programs and from the ability to take advantage of internal spillovers across programs) as it does from economies of scale (e.g., from specialization and the sharing of fixed costs).

---

[2] Here and throughout the chapter all amounts are denominated in constant 1991 U.S. dollars, unless otherwise noted.

[3] These ten firms include both European and American companies. Unfortunately, for reasons of confidentiality, we cannot list their names. But we can say that they include both large and small and successful and unsuccessful firms, and between them they account for approximately 28 percent of U.S. R&D and sales and a somewhat smaller proportion of worldwide sales and research.

Our results also suggest that the new technologies of drug research have put an increasing premium on the ability to access and use knowledge from outside the firm. In our data the benefits of pure scale have fallen since the advent of rational drug design, while those attributable to spillovers and scope economies have risen.

A second theme of this chapter is the influence of organizational choices on research productivity. Idiosyncratic firm effects or unique competencies are important drivers of research productivity. Firms in which publication in the open literature is an important criterion for promotion of individual researchers appear to be more productive than their rivals, as are firms that use committees to allocate resources rather than relying on a single "dictator." These results support the view that firms that actively manage the exchange of knowledge both across and within the boundaries of the firm are more successful than those firms that manage knowledge in a more fragmented manner.

Our results throw light on another puzzling aspect of industry economics: the fact that dramatic changes in the technology of pharmaceutical research do not appear to have had equally dramatic effects on the composition of the industry. Although some firms have found it quite difficult to adopt the new techniques, few new companies have attempted to enter the industry, with the notable exception of the biotechnology sector. Despite recent mergers and consolidation brought about by changes in the technology of performing R&D and the impact of managed care upon demand, the industry remains dominated by firms whose position was first established before World War II. In contrast to the impact of microelectronics on computing and related industries, for example, the revolutionary changes in biomedical science appear to have reinforced rather than undermined incumbent firms. The chapter closes with a discussion of the degree to which our analysis of the roots of research productivity sheds light on this continued dominance of the industry by historically well-established firms.

Trends in Industry Productivity over Time

As previously stated, pharmaceutical research is expensive. In a study using data from 1963 to 1975, Hansen (1979) found that the average cash outlay per successful new drug introduced was $65.5 million (in 1990 dollars), while DiMasi et al. (1991), using similar methods and data from 1970 to 1982, found that total cash outlays per successful drug introduction had increased to $127.2 million (in 1990 dollars). And research has apparently become even more costly since then. Figures 1 and 2 show the evolution of R&D expenditures and two measures of output: regulatory approval of investigational new drug applications (INDs), which mark the introduction of new drugs into clinical trials, and new drug approvals (NDAs), which mark the introduction of new drugs into the marketplace. Figure 1 shows total research spending by the members

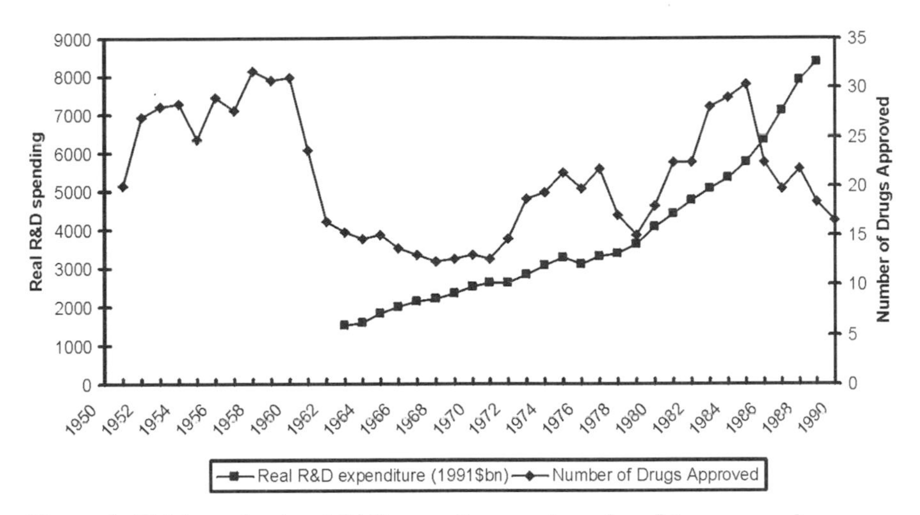

Figure 1. PMA members' real R&D expenditures and number of drug approvals.

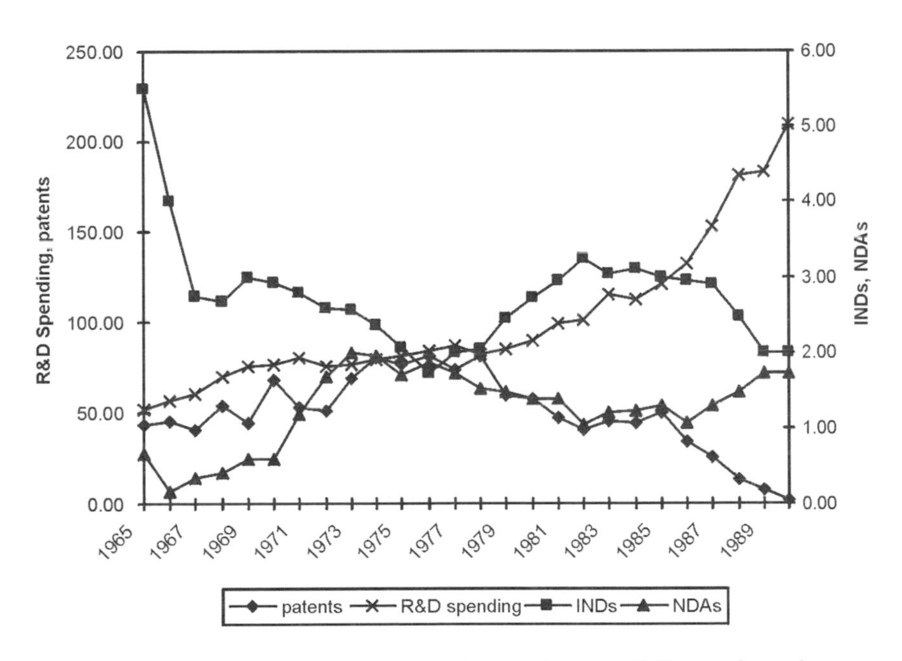

Figure 2. R&D patents, investigational new drug applications (INDs), and new drug approvals (NDAs) for sample firms.

of the Pharmaceutical Manufacturers Association compared with total NDAs approved from 1965 to 1990. Spending increased by an average of 14 percent a year during the period, while output, at least as measured by the number of new drugs introduced, increased slightly between 1970 and 1985, but fell quite dramatically afterward.

We take a closer look at these developments using data on research inputs and outputs gathered as part of a detailed analysis of the research strategies and performance of the previously mentioned sample of major research-oriented pharmaceutical companies. Although the sample contains only ten firms, we believe it is representative of the industry as a whole. These ten firms span the industry's full range of size, research performance, and financial performance, and includes European as well as North American firms. Figure 2 shows mean research expenditures, mean number of NDAs and INDs filed, and mean number of "important" patents granted for the ten firms in the data set. Because, on average, more than ten years elapse from initial discovery of a drug in the laboratory to the time it reaches the market, these data must be interpreted with care. But they do suggest that total cash costs for drugs introduced in the 1980s may have increased to as much as $175 million (in 1991 dollars).[4]

Disaggregating these numbers throws some light on the nature of the underlying dynamics that may be driving these trends. Figure 3 shows increases in mean rates of spending per firm on drug discovery versus clinical development. This split immediately suggests that the lion's share of the increase in pharmaceutical research expenditures is a function of the increasing amounts firms spend on development: While mean spending on discovery nearly tripled between 1965 and 1990, mean annual spending on development increased from roughly $30 million per firm to approximately $140 million per firm during the same period.

What are the underlying dynamics responsible for these trends? How has the shift in research technology from so-called random methods of drug discovery to more rational techniques changed the economics of pharmaceutical research? What is responsible for the apparent decline in research productivity over time?

INCREASING COSTS: DECREASING SCIENTIFIC OPPORTUNITY

One possible explanation for declining productivity is that scientific opportunity in the industry is decreasing—the easy drugs have been found. The qualitative evidence is clearly mixed: The techniques of modern research are incontrovertibly more expensive, but they may also be more efficient. For example, the use of more rationally designed screens may increase the odds of

[4] This is an approximate number, calculated by assuming that cash costs for a drug have risen in line with mean research expenditures over the period.

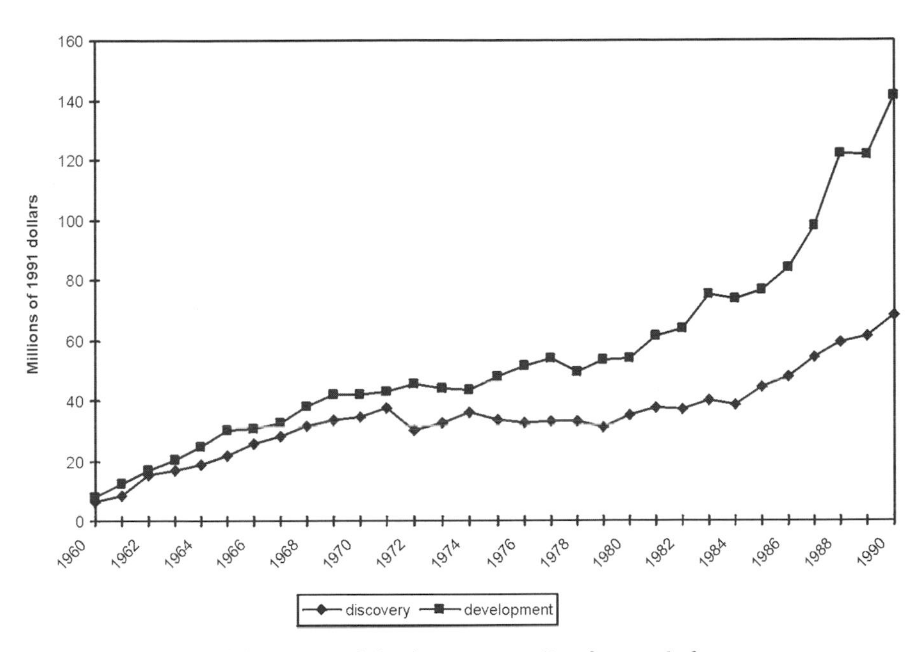

Figure 3. Mean real discovery and development spending for sample firms.

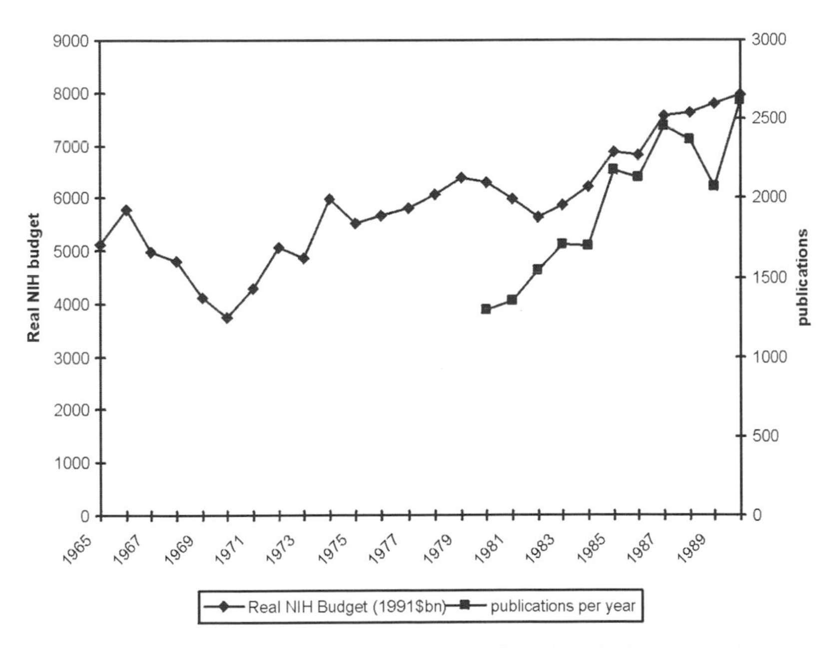

Figure 4. Real spending by the National Institutes of Health (NIH) versus publication activity of sample firms.

finding therapeutically useful compounds, or the use of combinatorial chemistry may dramatically increase the productivity of synthetic chemists.[5]

The pool of publicly available knowledge, at least as measured by public spending or by publication rates, has clearly increased during the last twenty years (Figure 4). At face value these trends make it seem unlikely that scientific opportunity has decreased. Indeed, there is some evidence that firms with tighter links to the public research community are more productive than their rivals (Cockburn and Henderson, 12 Nov. 1995), which suggests that this knowledge can be usefully translated into new drugs. Some observers, however, have argued that much of this knowledge remains too remote to serve as the basis for new drug discovery (Wurtman and Bettiker, 1994).

CHANGES IN THE COMPOSITION OF RESEARCH

Another possibility is that research productivity may have decreased because research has shifted from areas in which it has historically been quite successful, such as cardiovascular drugs and antibiotics, toward more difficult areas, such as oncology and gerontology. Wiggins (1979) first demonstrated the importance of distinguishing between therapeutic classes in modeling the determinants of productivity in the industry. Table 1 begins the process of pulling apart the aggregate numbers to reveal the heterogeneity of pharmaceutical research. It shows the ratio of cumulative outputs to cumulative inputs by therapeutic class for 1975 to 1990. These numbers must be approached with caution because they are subject to both left and right censoring.[6] They nonetheless illustrate the huge variation across different drug classes that is hidden by aggregating the data. The number of important patents obtained per million dollars invested in research, for example, varies from a high of 2.6 for dermatologic drugs to a low of 0.2 for anti-infectives. The ratio of INDs obtained for each $1 billion spent varies from 25 for anti-infectives to 81 for dermatologic drugs, and the ratio of NDAs to cumulative R&D spending varies from a low of 6 for each $1 billion in musculoskeletal research to a high of 34 in dermatology. Figure 5 further explores this phenomenon by showing trends in productivity, as measured by important patents per research dollar, for five key therapeutic areas.

These variations translate into significant differences in the average cost per drug in each class. Making the very crude assumption, for example, that investment in each program is constant across the sixteen-year period and that the time value of money is 9 percent, these differences translate into approximately

[5] The term *combinatorial chemistry* refers to the use of robots to synthesize a very large number of small molecules using various "combinations" of small molecule "ingredients."

[6] It is certainly the case, for example, that some of the discovery and development spent during this period has yet to yield fruit and that some of the output results from investment made before 1975.

Table 1. Ratio of Cumulative Outputs to Inputs, by Therapeutic Class, for Ten Pharmaceutical Firms, 1975–1990*

Class	Important Patents per Million Dollars Spent on Research	INDs per Billion Dollars Spent on Research	NDAs per Billion Dollars Spent on R&D
Gastrointestinal	1.7	34	13
Hematology	0.8	32	8
Cardiovascular	1.0	41	11
Dermatology	2.6	81	34
Anti-infective	0.2	25	13
Hormones	1.9	96	16
Oncology	0.6	61	25
Musculoskeletal	1.1	36	6
Central nervous system	1.6	49	14
Respiratory	1.7	43	9
Sensory organs	1.1	38	32

Taken from raw data from Derwent Publications, Inc.; Tufts Center for Drug Discovery; and the Food and Drug Administration.
* IND = investigational new drug; NDA = new drug approval.

$370 million for a musculoskeletal NDA, around $200 million for a cardiovascular NDA, and around $66 million for a dermatologic NDA.[7]

Figures 6a and 6b provide further evidence on the extent to which a shift between classes is responsible for changes in research productivity: Figure 6a shows the mean share of the discovery portfolio by therapeutic class over time, while Figure 6b shows the mean share of the development portfolio. Both graphs suggest that shifts in portfolio composition are unlikely to drive increases in research costs. Investment in research has generally been characterized by a switch away from anti-infectives toward cardiovascular drugs. Given the data in Table 1 and Figure 5, this shift should have *increased* research productivity. In development, the firms in our sample have been shifting out of research on drugs for the central nervous system toward work in cardiovascular drugs, while the share of resources devoted to anti-infective work has remained almost constant. These trends suggest that, if anything, shifting resources across therapeutic classes should have left research productivity more or less unchanged.

Thus, although research productivity differs systematically across therapeutic classes, little evidence in our data suggests that shifts from "cheap" to "expensive" therapeutic classes are at the root of a long-term decline in the productivity of pharmaceutical research and development. On the contrary, the firms in our sample shifted resources into areas where technological opportunity, in the sense of developments in the underlying science base, was increasing.

[7] These numbers are reassuringly in line with those calculated by DiMasi et al. (1991). However, they are *very* approximate and indicate no more than an order of magnitude. The DiMasi study uses much more sophisticated methods and more detailed project data to calculate the cost per drug.

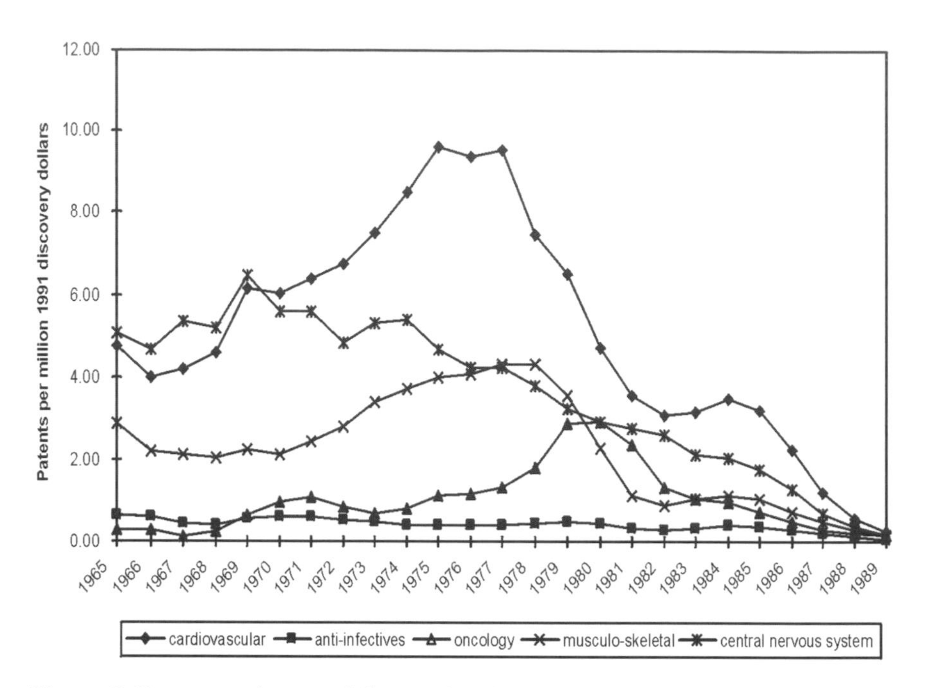

Figure 5. Patents per discovery dollar in selected therapeutic classes.

OVERINVESTMENT IN RESEARCH? THE DYNAMICS OF INVESTMENT BEHAVIOR

A third possible explanation for escalating real research costs is that they reflect increasing competition and overinvestment in research. If we liken the search for new drugs to the search for a few fish in a deep pond, as the number of boats fishing on the pond increases, the average catch per boat is likely to fall.

Indeed, although the extensive theoretical literature exploring the relationship between competitive dynamics and investment strategy is ultimately inconclusive, the majority of these models show that free entry into R&D competition will result in more investment in research than is beneficial to either the investor or society. In deciding to invest in research, firms apparently consider only their own marginal returns and do not take into account the externality that they impose on other firms by reducing their chances of success. In the extreme, these models suggest that entry into R&D will occur until all expected profits are dissipated (Dasgupta and Stiglitz, 1980; Loury, 1979; and Reinganum 1982, 1989)

Cockburn and Henderson (1994) explore the degree to which this factor is a plausible explanation for reduced industry productivity. Unfortunately these ideas are difficult to test: Models that attempt to incorporate all of the relevant variables quickly become dauntingly complex, and there are still no general results about the relationship between market structure and scientific or technological regime or the relationship between realized and optimal levels of research investment (Baldwin and Scott, 1987; Harris and Vickers, 1987; Reinganum, 1989). Thus excess investment cannot be entirely ruled out as an explanation of apparently declining industry productivity, but several factors indicate that it may not be very important.

First, there appear to be significant complementarities in research output across firms. The research productivity of any single firm is positively correlated with the research output of its rivals. To use the fishing analogy again, when competitors are successful, the knowledge that they acquire about the habits of fish increases the fishing proficiency of everyone else on the pond, and each boat catches more fish.

Second, there is little evidence of the kind of short-term "racing" behavior that some economic models suggest reduces returns to R&D through excess entry. The primary determinant of research investments appears to be historical levels of investment. In our econometric models, after controlling for technological opportunities, investment in research shows some correlation with sales, but almost none at all with competitors' investments (Cockburn and Henderson, 1994).

These results are not of themselves sufficient to reject the hypothesis that increasing competition in the industry may be pushing up real research costs. Nonetheless, they suggest that the more extreme forms of the dissipation of economic returns to investment in R&D identified in the literature probably poorly characterize the reality of competition in the industry. Moreover, some compelling aggregate evidence—the dramatic increase in private rates of return to research discussed below—suggest that it is unlikely to be an important explanation.

DIFFICULTIES IN MEASURING OUTPUT

A final possible explanation for these trends is that important patents, INDs, and NDAs are simply not good measures of the industry's output. It may be the case that advances in the science underlying drug research are such that the average value of each drug has dramatically increased. While random drug discovery led to larger numbers of drugs of widely varying qualities being introduced, more targeted research may have led to the introduction of fewer, but much more significant "breakthrough" drugs. At the same time sharply increased costs of marketing and regulatory approval and a variety of competitive pressures have increased the incentives for firms to be much more selective about the leads pursued.

Table 2 lends some support to this idea. In the ideal case one would measure the "social return" or "social value" of each drug introduced. Because that has proved extraordinarily difficult (Greenberg, Finkelstein, and Berndt, 1995), we

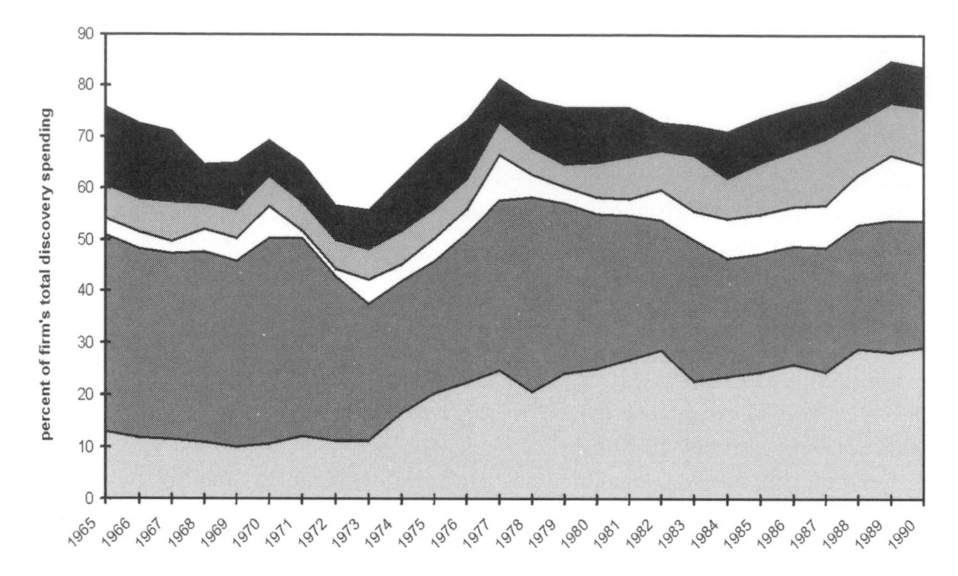

Figure 6a. *Mean share of discovery spending in selected therapeutic classes. The black area at the top shows central nervous system drugs; the gray area beneath the black shows musculoskeletal drugs; the white area shows oncologic drugs; the large, darker gray area shows anti-infectives; and the area at the bottom shows cardiovascular agents.*

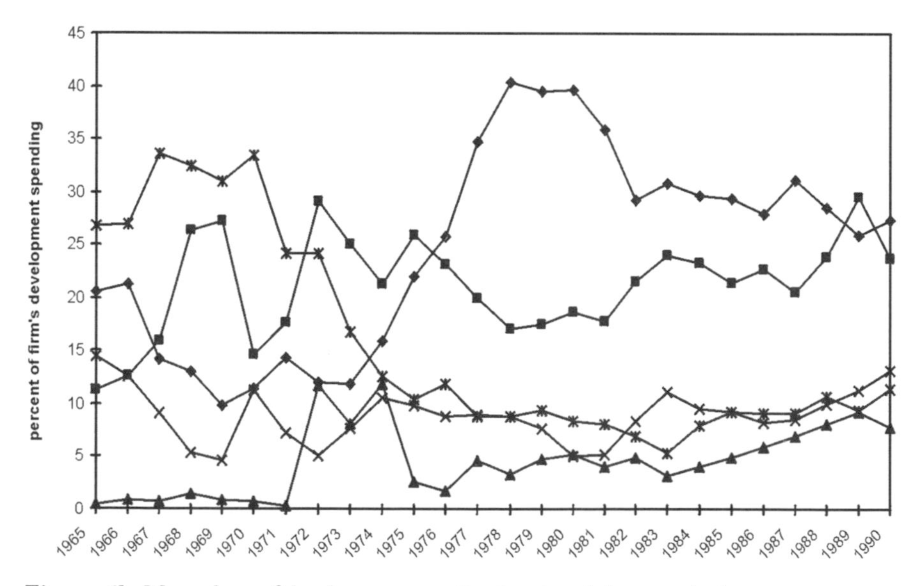

Figure 6b. *Mean share of development spending in selected therapeutic classes. Diamonds show cardiovascular agents; squares, anti-infectives; triangles, oncologic drugs; Xs, musculoskeletal drugs; and asterisks, central nervous system drugs.*

use two approximate measures of the value of new drugs: sales revenues and the "importance" measure, calculated by Ward and Dranove (1995).[8] Sales for the industry as a whole have increased dramatically over the period, at least as demonstrated by the ten firms in our sample, where sales per NDA granted have skyrocketed (Table 2). Ward and Dranove's importance measure for NDAs increases steadily between 1960 and 1990 (although their data are somewhat scanty after 1985). Moreover, R&D spending as a percentage both of sales and of the market value of pharmaceutical firms appears to have increased even faster than sales revenue (Figure 7), suggesting that both the firms themselves and the stock market, at least, expected research to become increasingly profitable.

CONCLUSIONS: FALLING PRODUCTIVITY?

The most plausible explanation for the apparently flat or declining productivity of the industry is that simple count measures of output are misleading and that the average value of each patent, IND, and NDA has probably increased significantly over the last twenty years. Evidence of any dramatic decline in technological opportunity is scant, as is evidence of any shift away from easy toward more difficult fields or of competition that is so intense that it has driven down marginal returns to investment in drug discovery. The explosion in activity that has accompanied the revolution in modern medical science appears to have resulted in real economic benefit for the industry, at least for those firms that have prospered under the new regime.[9]

Determinants of Research Productivity

In the pharmaceutical industry three important factors appear to drive research productivity: the historical success of a firm in any given area; the firm's research strategy—particularly the size and scope of its research portfolio; and the ways in which research is managed within the firm. These effects account for the presence of very large research-oriented firms, despite sharply decreasing marginal returns to increasing spending on individual research programs. Measuring research productivity is a difficult exercise. Some of the issues are discussed here briefly; interested readers should refer to the individual papers cited for a richer discussion of the methodology. We focus on the determinants

[8] We would like to express our appreciation to Michael Ward and David Dranove for making these data available to us.

[9] We run the risk of significant sample selection bias in interpreting our results on this point, because by definition the ten firms in our data set were not forced to exit the industry between 1960 and 1990. To the degree that firms excluded from our data set either had much lower rates of return or were forced to exit the industry, overall rates of return to pharmaceutical research may have been much lower. The ten firms in our data set, however, also displayed a seeming "decline" in productivity as measured by simple counts of patents, INDs, or NDAs per research dollar.

Table 2. Trends in Output Measures

Period	Average Sales in Constant 1991 U.S. Dollars (in $100 Millions) for Sample Firms	Average Ward and Dranove "Importance" Measure for NDAs*
1960–65	196	–0.49
1965–70	458	0.38
1970–75	634	0.54
1975–80	740	1.81
1980–85	800	1.01
1985–90	895	6.93

Data from IMS America and Ward and Dranove (1995).

* Ward and Dranove used a variety of metrics to construct their measure of "importance," including the FDA ranking of the drug, its sales, and the number of pages describing its properties in some standard medical references. See Ward and Dranove (1995) for a full description.

of productivity in discovery. Our work in development is at a much more preliminary stage, but we summarize it briefly at the end of this section.

MEASURING RESEARCH PRODUCTIVITY

We define *research productivity* as the efficiency with which research inputs (expenditures) are translated into research outputs. In our econometric work we estimate a production function relationship between inputs to the research process (R&D expenditure) and its output—in this case, important patents, that is, those filed in at least two out of three major jurisdictions worldwide. We test for the impact of various factors, such as the diversity of the overall research effort in mediating this relationship. By using detailed data at the level of individual research programs within the firm, we are able to control for several important effects, such as differences across therapeutic classes, and to perform a much richer analysis than is possible using data aggregated to the level of the firm.

SIZE, SCOPE, EXPERIENCE, AND RESEARCH PRODUCTIVITY

Pharmaceutical firms typically invest in ten to fifteen distinct research programs. Each program is targeted toward the discovery of therapies for a particular disease area—hypertension or depression, for example—and may have anywhere from one to several hundred full-time researchers associated with it. Research programs may be grouped in turn into such therapeutic areas as cardiovascular disease or conditions of the central nervous system.

Conventional wisdom in the industry suggests that under most circumstances little can be gained from increasing the size of an individual research program beyond a minimum threshold. The fact that the firms in our data set sustain a large number of relatively small programs is consistent with this belief. Because pharmaceutical research often requires investment in substantial fixed costs, however, and because the complexity of the underlying science offers considerable scope for specialization, we hypothesized significant economies of scale at the level of the entire research effort. For example, to the degree that such inputs as libraries, computer resources, or large pieces of equipment are fixed costs, we hypothesized that a larger research effort could realize economies of scale by spreading such fixed costs over a larger base of research activity. Similarly a larger firm may also be able to obtain economies of scale through its ability to support highly specialized scientific personnel.

We identified two kinds of returns to the scope of the firm's research effort. In the standard analysis of production, economies of scope are present when the costs of conducting two or more activities jointly are lower than they would be if they were conducted separately. This occurs most obviously when activities can share such inputs as knowledge capital at no additional cost, which is an important aspect of pharmaceutical research. A clear distinction can be drawn between these economies of scope and a second source of returns to the diversity of the research effort: internal spillovers of knowledge among programs that enhance each other's productivity. Economies of scope relate to research expenditures, whereas internal knowledge spillovers affect output irrespective of expenditures. Our qualitative work suggested that both types of returns to scope are critically important in shaping research productivity. Consider, for

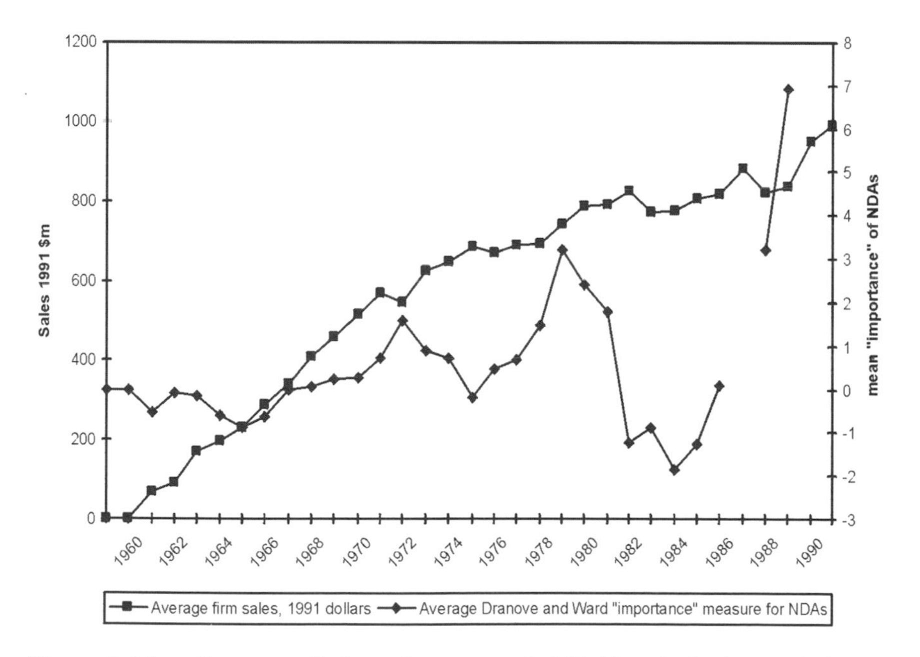

Figure 7. Mean "importance" of new drug approvals (NDAs) and sales for sample firms.

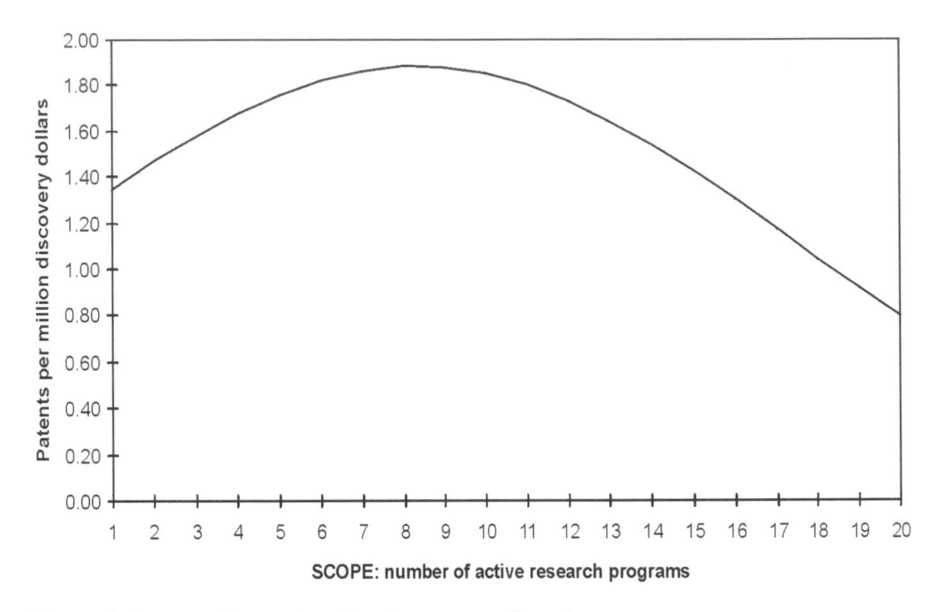

Figure 8. *Impact of increasing diversity on research performance.*

example, the benefits of investing in a centralized laboratory devoted to peptide chemistry. Economies of *scale* exist if the costs of the laboratory are partially fixed and if the lab can serve a larger and larger research effort for a less-than-proportionate increase in cost. They will also exist if the laboratory can become more efficient as it has more work to do, possibly through the specialization of its members. Economies of *scope* exist if the expertise of a group of peptide chemists is potentially relevant to a wide range of applications and can be used in any one of them without diminishing its usefulness to the others. Benefits of diversity may also arise if discoveries made in one program stimulate the output of another through cross-fertilization of ideas or other forms of knowledge spillovers. Several important treatments for central nervous system disorders, for example, were discovered during searches for drugs active in the cardiovascular system. Hence we expected there to be significant returns to scope in discovery.

The econometric analysis tests of these hypotheses resulted in the following conclusions. (A more detailed discussion and supporting regression results can be found in Henderson and Cockburn [1994 and 1996] and Cockburn and Henderson [1994].) First, although increasing the resources committed to any individual research program produces only limited returns, larger firms are significantly more productive in drug discovery. They are able to realize economies of scale in R&D by spreading fixed costs and by specialization and to realize economies of scope by diversifying into related research programs. Second, these economies of scope are only present up to a point; too much diversification reduces research productivity (Figure 8). Third, substantial spillovers of knowledge occur both within the firm and across firms. Fourth, the productivity of any given research program is enhanced by the success of related programs within the same firm and by the success of competitors' research programs (Figure 9). The most statistically important determinant of a research program's success, however, is its past productivity. The keys to this determinant are the "knowledge capital" accumulated by the program as an organizational unit as well as the skills and experience of individual scientists.

The impact of the shift in the technology of pharmaceutical R&D in the late 1970s appears most strikingly in changes in the relative importance of these effects over time. The sample is divided in 1978, the year after the publication of a seminal paper announcing the synthesis of an orally active angiotensin-converting enzyme inhibitor (Cushman et al., 1977), often cited as one of the first examples of successful rational drug design. Experimenting with alternative cut-off dates had little effect on the results. After 1978, returns to scale per se appear to have fallen, while returns to scope and the importance of spillovers appear to have increased. As pharmaceutical research has moved from a regime of random research to one of rational drug design, the evolution of biomedical science has placed an increasing premium on the ability to exchange information within the firm and to evaluate and use information from external sources (Gambardella, 1995). The primary advantage of size has become the ability to exploit internal returns to scope—particularly the ability to exploit internal spillovers of knowledge—rather than any economy of scale per se. As Cohen and Levinthal (1989) have pointed out, the benefits of spillovers can be realized only by incurring the costs of maintaining "absorptive capacity"—the ability to

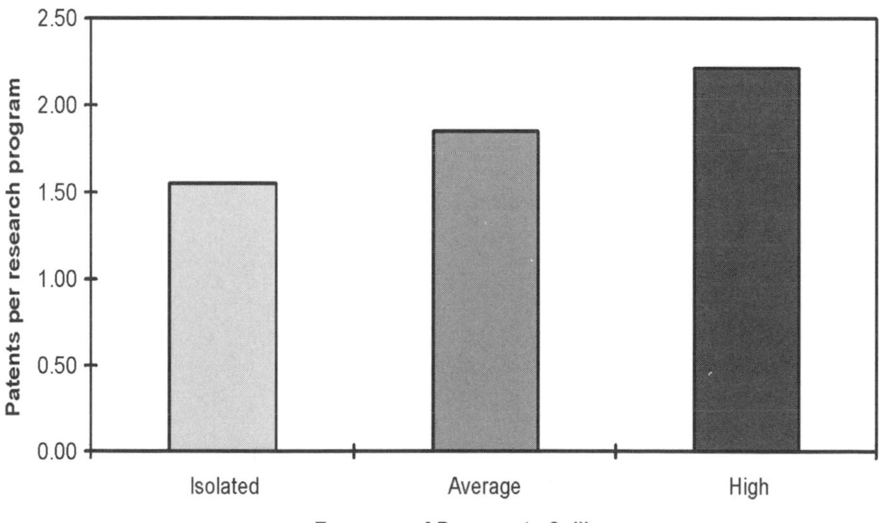

Figure 9. *Impact of spillovers on research productivity.*

capture and use spillovers—which takes the form here of large numbers of small and apparently unproductive programs that large firms are able to maintain.

THE DETERMINANTS OF PRODUCTIVITY IN CLINICAL DEVELOPMENT

Thus far our discussion has focused on the determinants of research productivity in drug discovery. In some preliminary work (Cockburn and Henderson, September 1995) we have extended our analysis to the very different case of clinical development. There are a number of reasons for hypothesizing that development might be characterized by economies of both scale and scope. Economies of scale arise when the costs of doing any single activity can be spread out over a larger activity base or when undertaking an activity on a larger scale permits the adoption of more effective techniques. Because development is a complex, multifaceted activity requiring the coordination of a wide range of specialist expertise, both sources of scale economies may well be present. On the one hand a firm may be able to amortize the costs of "generic" specialized expertise that can be applied to a wide variety of diseases over several candidates for development. Expertise in biostatistics, for example, or in dealing effectively with regulatory authorities across the world may have this property. On the other hand, as the size of its development effort grows, a firm may be able to afford to employ significantly more specialized expertise. For example, a small firm may be able to employ only one clinician to supervise all of its clinical trials, while a larger firm can hire specialized cardiologists, oncologists, and so on to supervise trials of potential treatments for heart conditions, cancer, and other diseases and disorders. Similarly, smaller efforts may be forced to use simple, general-purpose software to monitor the progress of their trials, while clinical programs conducted on a larger scale may be able to invest in specialized software that is particularly tailored to the needs of the firm. Thus we expect that there were significant returns to scale in development at the firm level.

Economies of scope arise when a firm can make use of knowledge generated within one area of the firm at a low or zero additional marginal cost in another. Here again are two central reasons for believing that development activities might demonstrate economies of scope. First, larger firms may be able to transfer general knowledge about the efficient running of clinical trials across projects within the firm. Large clinical trials are costly and complex, requiring the tracking and integration of complex data across thousands of patients. Expertise gleaned from running large trials on antidepressant drugs, for example, may reduce the costs of running large trials on drugs for hypertension or musculoskeletal disorders. Second, because few systems in the human body operate in complete isolation, particular medical knowledge gained through work in one area may be useful in another. For example, knowledge gained through work with antianxiety drugs may be of direct relevance to the conduct of trials of antidepressant drugs, while work in hypertension may be of immediate relevance to work in arrhythmia or cardiotonics. Thus we also expected significant returns to scope in development.

Several other reasons, however, suggest that pharmaceutical development may not be subject to economies of scope or scale. Economies of scale arising from the sharing of a fixed cost, for example, arise only if the cost is truly fixed. Whether some specialist skills such as biostatistics can be applied to multiple disease conditions, compounds and diseases are diverse enough that clinical trials may be highly idiosyncratic, and the design of each new trial may require a roughly equivalent input of biostatistical skill. Similarly, economies of scale arising from the ability to adopt superior techniques occur only when these techniques cannot be traded in the open market. Some significant fraction of clinical trials are outsourced, however, and in principle any firm can make use of the skills of the extensive infrastructure of specialist firms.

Economies of scope in the management of clinical trials are likely to arise only if the knowledge gained during their conduct is not "sticky," or not instantaneously transmitted across the boundaries of the firm. Although such transmission is unlikely in the case of pharmaceutical discovery, it is much less so in the case of pharmaceutical development, because responsibility for the conduct of the actual trials themselves remains largely in the hands of physicians operating outside the firm. Trials are designed and paid for by the individual pharmaceutical firms, but most are carried out at hospitals by physicians under contract. Knowledge about the conduct of clinical trials thus tends to diffuse fairly rapidly across the medical community. This is particularly likely with those diseases that are less well understood, because preliminary trials of compounds designed to address these conditions are often supervised by groups of leading physicians in the field who are well known to each other and who publish freely.

In contrast to drug discovery our preliminary results support the hypothesis that drug development is unlikely to benefit from economies of scale or scope. There is little apparent advantage from size in drug development. Firms do show marked performance differences in their ability to pick winners in the clinic, but these do not appear related to easily measurable aspects of their development efforts.

ORGANIZATIONAL EFFECTS

Our exploration of the effect of firms' size and scope on their research productivity highlighted the role that idiosyncratic firm effects have on such productivity. Despite controlling for size, scope, therapeutic class, and other quantifiable aspects of the research portfolio, we still found surprisingly large and persistent differences in research performance across firms. One possible explanation is that these effects reflect differences in the ways in which research is organized within the firm. In Henderson and Cockburn (1994) we focused

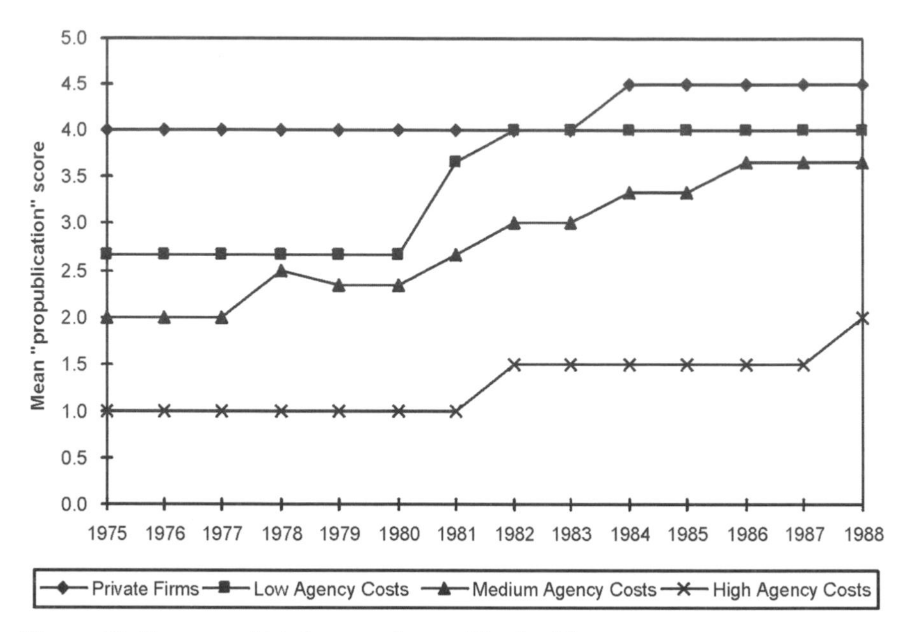

Figure 10. *Degree to which firms are "pro-publication," by agency costs.*

particularly on the degree to which differences in research productivity across firms are driven by what we term "architectural competencies"—by the ability to integrate knowledge in new and flexible ways both within and across the boundaries of the organization (Henderson, 1994). We hypothesized that firms with the ability to encourage and maintain an extensive flow of information across the boundaries of the firm and among scientific disciplines, therapeutic specialties, and organizational units would have significantly more productive drug discovery efforts.

Our econometric tests broadly confirmed these hypotheses. Firms characterized as "pro-publication," that is, those that encouraged their researchers to participate in the wider scientific community and used publication in the open peer-reviewed literature as a criterion for compensation and promotion, were better able to sustain a rich information flow across organizational boundaries and were significantly more productive. By contrast, using a "dictator" to allocate R&D resources as opposed to a peer-review committee hampers information flow and results in significantly lower research productivity.

The organizational capabilities required to take advantage of rational drug discovery emerged in some firms in the industry quite early (Cockburn and Henderson, May 1995). Figure 10 shows changes in the pro-publication measure for the ten firms in the data set. Having pro-publication internal incentive mechanisms and organizational structures in place—or adopting them quickly—conferred substantial advantages. Curiously, these firms show large differences in the speed and extent to which they adapted their organizations to the new regime of rational drug discovery. One explanation for this may lie in problems of corporate governance. In Figure 10 firms are grouped by a measure of agency costs—the extent to which management is disciplined by capital markets—and firms with high agency costs, that is, those in which managers had weaker incentives to pursue shareholders' interests, responded to the dramatic change in their environment more slowly and less effectively.

The Continued Dominance of Established Firms

This exploration of the economics of pharmaceutical research gives some insight into one of the remarkable features of the industry, documented elsewhere in this volume: Despite enormous shifts in its core technologies, it continues to be dominated by firms founded before World War II.

Incumbent firms in the industry enjoy several important advantages over potential entrants, such as experience in dealing with regulatory authorities and ownership of highly specialized marketing and distribution assets. Focusing exclusively on R&D, however, our econometric results reveal some substantial barriers to entry into the mainstream arena of ethical drug discovery. One factor is the importance of the historical experience of leading firms, as evidenced by the highly significant effect of accumulated "knowledge capital" on the productivity of individual research programs. Our findings also confirm that economies of scope and scale generate significant advantages for large firms; firms whose research efforts are too small or insufficiently diversified pay a heavy price in terms of R&D performance. These barriers alone may help to explain why there has been so little new entry into the industry.[10]

Another important factor, however, may be the role played by the organizational capabilities of leading science-oriented firms in maintaining the resilience of the industry. As established above, propublication firms have been significantly more productive in drug discovery, and when the technology of research changed in the late 1970s, those firms that already had the kinds of internal incentives and organizational structures that promoted closer connections with the wider scientific community—or moved more quickly to adopt them—outperformed those that did not. These close connections to the wider scientific community, and their accompanying internal incentive mechanisms, appear to have allowed the industry's leading firms to adapt very successfully to the changing research environment. Unlike other industries where incumbent firms responded slowly and ineffectually to revolutionary changes in their underlying technologies, thus generating opportunities for new entrants, here

[10] The sector in which there has been great entry—biotechnology, or the production of high-molecular-weight drugs—is arguably driven by quite different dynamics that are likely to make the role of scale, scope, and historical experience much less important. (See Cockburn, Henderson, Orsenigo, and Pisano, forthcoming.)

incumbent firms have been able to turn the impact of technological change to their advantage. By rapidly and effectively embracing the new technology, and by having in place a set of internal incentives that promoted organizational change and the development of assets complementary to the new technology, successful incumbent firms presented further challenges to entrants. To compete effectively in this industry, entrants face a lengthy, costly, difficult process of investing heavily in basic research capabilities and organizing themselves to be able to use public-sector research. With the exception of a fringe of mostly very small and so far largely unsuccessful biotechnology firms with close connections to university science (and apparently little expectation on the part of investors of near-term profitability), few potential entrants have been willing or able to undertake these investments.

Conclusions

The economics of drug discovery confer substantial advantages upon large, established, historically successful firms. In general, research programs conducted inside these firms generate significantly more patents per R&D dollar. In the 1960s and 1970s, when the dominant mode of drug discovery was random screening, these productivity differences stemmed from larger firms' ability to employ highly specialized resources and personnel and to spread fixed costs over a larger base of activity. As science-based rational drug design has become more important, the basis of large firms' advantage has changed: Their superior performance reflects the benefits of diversity as much as size per se. In the research environment of the 1980s and 1990s the ability to exploit spillovers of knowledge from inside and outside the organization and to take advantage of economies of scope flowing from the repeated application of science-based expertise and knowledge capital are increasingly important determinants of research productivity.

In this light the recent wave of consolidation and mergers in the industry can be seen (in part at least) as a search for economies of scope and scale in research. To compete effectively, firms must make substantial and sustained investments in basic science and in accumulating knowledge capital and must be able to employ these assets in a sufficiently wide enough range of applications to justify their cost.

The experience of the last three decades suggests that hand in hand with fundamental changes in firms' research "hardware" the new technologies for pharmaceutical research also require changes in the managerial "software" used to deploy research resources. Flexible, forward-looking management and visionary leadership, able to sustain a long-run commitment to the institutions of "open science" in the face of intensifying product market competition, are likely to play a significant role in determining research performance as the impact of the revolution in biomedical sciences continues to shape competitive forces in this industry.

References

M. N. Baily; A. K. Chakrabarti. *Innovation and the Productivity Crisis.* Washington, D.C.: Brookings Institution, 1988.

W. L. Baldwin; J. T. Scott. *Market Structure and Technological Change.* Chur, Switzerland: Harwood Academic Publishers, 1987.

E. Caglarcan et al., editors. *The Pharmaceutical Industry: Economics, Performance and Government Regulation.* New York: Wiley, 1978.

I. Cockburn; R. Henderson. "The Determinants of Research Productivity in Ethical Drug Discovery." In *Competitive Strategies in the Pharmaceutical Industry.* Edited by Robert B. Helms. Washington, D.C.: American Enterprise Institute, 1996.

———. "Do Agency Costs Explain Variation in Innovative Performance?" Mimeo, MIT, May 1995, presented to the National Bureau of Economic Research 1995 Summer Institute Industrial Organization Conference.

———. "Public-Private Interaction in Pharmaceutical Research." *Proceedings of the National Academy of Sciences* 93:23 (12 Nov. 1996), 12725–12730.

———. "Racing to Invest? The Dynamics of Competition in Ethical Drug Discovery." *Journal of Economics and Management Strategy* 3:3 (1994), 481–519.

———. "Scale and Scope in Drug Development: Unpacking the Advantages of Size in Pharmaceutical Research." Mimeo, MIT, September 1995, presented to the Conference on Industrial Organization of Health Care, Boston University, September 1995.

I. Cockburn; R. Henderson; L. Orsenigo; G. Pisano. "The Pharmaceutical Industry and the Revolution in Molecular Biology: Exploring the Interactions between Scientific, Institutional and Organizational Change." In *The Sources of Industrial Leadership.* Edited by David Mowery and Richard Nelson. Cambridge: Cambridge University Press, forthcoming.

W. M. Cohen; D. A. Levinthal. "Innovation and Learning: The Two Faces of R&D." *Economic Journal* 99 (1989), 569–596.

W. Comanor. "The Political Economy of the Pharmaceutical Industry." *Journal of Economic Literature* 24:3 (1986), 1178–1217.

D. W. Cushman; H. S. Cheung; E. F. Sabo; M. A. Ondetti. "Design of Potent Competitive Inhibitors of Angiotensin-Converting Enzyme." *Biochemistry* 16 (1977), 5484–5491.

P. Dasgupta; J. Stiglitz. "Industrial Structure and the Nature of Innovative Activity." *Economic Journal* 90 (1980), 266–293.

J. DiMasi et al. "The Cost of Innovation in the Pharmaceutical Industry." *Journal of Health Economics* 10 (1991), 107–142.

A. Gambardella. *Science and Innovation: The U.S. Pharmaceutical Industry.* Cambridge: Cambridge University Press, 1995.

H. G. Grabowski; J. M. Vernon. "A New Look at the Returns and Risks to Pharmaceutical R&D." *Management Science* 36:7 (1990), 804–821.

S. B.Graves; N. Langowitz. "Innovative Productivity and Returns to Scale in the Pharmaceutical Industry." *Strategic Management Journal* 14 (1993), 593–605.

P. Greenberg; S. Finkelstein; E. Berndt. "Economic Consequences of Illness in the Workplace." *Sloan Management Review* 36:4 (1995), 26–38.

Z. Griliches. "Patent Statistics as Economic Indicators: A Survey." *Journal of Economic Literature* 28:4 (1990), 1661–1707.

R. W. Hansen. "The Pharmaceutical Development Process: Estimates of Development Costs and Times and the Effect of Proposed Regulatory Changes." In *Issues in Pharmaceutical Economics.* Edited by R. A. Chien. Lexington, Mass.: D.C. Heath and Company, 1979.

C. Harris; J. Vickers. "Racing with Uncertainty." *Review of Economic Studies* 54:177 (1987), 1–22.

R. Henderson. "The Evolution of Integrative Competence: Innovation in Cardiovascular Drug Discovery." *Industrial and Corporate Change* 3:3 (1994), 607–630.

R. Henderson; I. Cockburn. "Measuring Competence? Exploring Firm Effects in Pharmaceutical Research." *Strategic Management Journal* 15:Winter Special Issue (1994), 63–84.

———. "The Routinization of Radical Innovation: Pharmaceutical Firms and the Biomedical

Revolution." Mimeo, MIT, prepared for the American Management Association meetings, Vancouver, August 1995.

———. "Scale, Scope and Spillovers: The Determinants of Research Productivity in Drug Discovery." *RAND Journal of Economics* 27:1 (1996), 32–59.

E. J. Jensen. "Research Expenditures and the Discovery of New Drugs." *Journal of Industrial Economics* 36 (1987), 83–95.

G. Loury. "Market Structure and Innovation."*Quarterly Journal of Economics* 93 (1979), 395–410.

J. Mairesse; M. Sassenou. "R&D and Productivity: A Survey of Econometric Studies at the Firm Level." *STI Review* (Paris: OECD), 8 (1991), 9–43.

Office of Technology Assessment, United States Congress. *Pharmaceutical R&D: Costs, Risks and Rewards.* Washington, D.C.: U.S. Government Printing Office, 1993.

J. F. Reinganum. "A Dynamic Game of R and D: Patent Protection and Competitive Behavior." *Econometrica* 50 (1982), 671–688.

———. "The Timing of Innovation: Research, Development and Diffusion." In *Handbook of Industrial Organization.* Vol. 1. Edited by R. Schmalensee and R. Willig. Amsterdam: North Holland, 1989.

D. Schwartzman. *Innovation in the Pharmaceutical Industry*. Baltimore: Johns Hopkins University Press, 1976.

B. Spilker. *Multinational Drug Companies: Issues in Drug Discovery and Development.* New York: Raven Press, 1989.

P. Temin. *Taking Your Medicine: Drug Regulation in the United States.* Cambridge, Mass.: Harvard University Press, 1980.

M. Ward; D. Dranove. "The Vertical Chain of R&D in the Pharmaceutical Industry." *Economic Inquiry* 33 (1995), 1–18.

S. Wiggins. "Regulation and Innovation in the Pharmaceutical Industry." PhD dissertation, MIT, 1979.

R. J. Wurtman; R. L. Bettiker. "How to Find a Treatment for Alzheimer's Disease." *Neurobiology of Aging* 15 (1994), S1–S3.

Chapter Six

Clinical Champions as Critical Determinants of Drug Development

CHRISTOPHER R. FLOWERS AND
KENNETH L. MELMON

Drug development is a complex, multifaceted, interdisciplinary process that is central to the success of the pharmaceutical industry and the practice of medicine. Developing new drugs is a time-consuming, expensive, and risky endeavor, with the costs and time required continuing to increase. Because the average development times have lengthened since the mid-1960s, the effective patent life has been reduced. A number of investigators have estimated that about twelve years or more and approximately $231 million (in 1987 dollars) are required to develop a new drug, from the time of synthesis to approval by the Food and Drug Administration (FDA). It seems that the regulatory process is no longer the only key element determining the speed of development of a new chemical entity. But despite barriers to development, pharmaceutical companies have successfully created a broad range of innovative and therapeutically important drugs.

If the time necessary for pharmaceutical innovation was to be reduced, examining the entire process of drug discovery and development in depth might yield clues as to how it could be done. New drugs are discovered by a variety of methods, including serendipitous findings, broad screening, modification of existing compounds, rational drug design, and breakthrough discoveries. Candidate compounds are examined in animal and in-vitro studies for activity and toxicity before testing in humans. Drug development involves demonstration of the safety and efficacy of drugs in clinical settings. Priorities for developing promising chemical entities are commonly set based on their availability and production cost, their patent status, their potential spectrum of activity, their ability to satisfy regulatory requirements, and their positioning in the corporate portfolio and in the marketplace. Pharmaceutical research and development (R&D) programs typically involve researchers from a wide range of disciplines, including organic and medicinal chemists, chemical engineers, biochemists, molecular biologists, microbiologists, pharmacologists, toxicologists, pharmacists, and clinicians. Modern pharmaceutical R&D is increasingly becoming more interdisciplinary, involving X-ray crystallographers, physicists, computer scientists, and other researchers.

Information drives pharmaceutical companies and other research-based organizations. Consequently, successful pharmaceutical innovation, especially in light of the increasing diversity of R&D project teams, depends on the effective management of numerous sources of information and the individuals who produce it. Some authors have advocated rational drug design (setting development priorities for chemical entities based on structure-activity relationships) as a means to reduce drug development time. However, it is not well understood how to effectively translate these pharmacologic data into drugs, or whether understanding of structure-activity relationships is the most significant influence on development time. The factors that affect the translation of experimental data from basic scientific studies into clinical applications may be the most important determinants of the time required to develop drugs. This study focuses on three important factors that influence the time required for drug development: the availability of relevant scientific information, the information management infrastructure of a pharmaceutical organization, and the roles of researchers who provide and use the information.

Medical Technology Transfer

The increasing dependence of modern medicine on new or improved technologies is widely acknowledged. Advances in chemotherapeutics, biologicals, and procedures continually redefine what is clinically feasible. Transfer of scientific data into a medical technology is a process through which prototype technologies are generated from scientific findings and are applied to clinical situations. But the factors that influence the successful transfer of a fundamental observation to the development of medical technology are not often analyzed. The discovery of a medical technology involves the generation of ideas, the creation and testing of innovations, the examination of results, and the dissemination and adoption of the novel effect. Each innovation passes through at least eight phases: basic research, applied research, technology development, evaluation, demonstration of efficacy, adoption and approval, widespread clinical application, and obsolescence. Reviews directed by the Institute of Medicine Committee on Medical Innovation have examined factors related to the adoption and obsolescence of medical technologies and their use in clinical practice. This study focuses on the first five phases of the development of a medical technology, which constitute the five stages of medical innovation. In general, medical innovation refers both to the act of designing a novel physical item (an invention) and to the successful application of that invention into a defined medical situation. Although a broad range of definitions exists in the literature for the term *innovation*, S. Globe, G. W. Levy, and C. M. Schwartz best conceptualized what is meant in this study. Innovation, they wrote,

> is a complex series of activities, beginning at "first conception," and proceeding through interwoven steps of research, development, engineering, design, market analysis, management decision making, and ending at "first realization" when an industrially successful product, which may actually be a thing, a technique, or a process, is accepted in the marketplace. The term *innovation* also describes the process itself, and, when so used, it is synonymous with the phrase *innovative process*. (Globe, Levy, and Schwartz, 1973, p. 8)

Medical innovation includes the discovery of new basic scientific knowledge, the application of this knowledge to one or more clinical settings, and the examination and demonstration of the safety and efficacy of the medical technology for one or more specific clinical indications. As presently defined, the process of innovation extends over a bounded interval from the point when the technology is originally conceived to the point when the product for a given clinical indication is realized. This innovative period offers an encapsulated time frame for analysis of factors that influence the speed and success or failure of drug development. This study examines scientific, organizational, and individual factors during the innovative period in five case studies of chemically related compounds that ultimately were applied to diverse clinical indications.

Selection of Case Studies

This investigation focuses on innovations that developed from basic science research related to purine and pyrimidine analogs. Between 1942 and 1983 George Hitchings and Gertrude Elion performed research at Burroughs Wellcome Research Laboratories that enabled the development of drugs to prevent rejection of transplanted organs and to treat gout, leukemia, and viral diseases. In 1988 Hitchings and Elion shared the Nobel Prize for this research. We studied five drugs developed from Hitchings and Elion's basic science research findings. The case studies are of four drugs developed at Burroughs Wellcome—6-mercaptopurine (Purinethol), azathioprine (Imuran), allopurinol (Zyloprim), and acyclovir (Zovirax)—and of a chemically related drug developed at Syntex, ganciclovir (Cytovene). The four cases of drug development at Burroughs Wellcome were jointly selected by the authors and an Institute of Medicine working group on innovation in medicine. The Hitchings group at Burroughs Wellcome also discovered and developed drugs by investigating the properties of pyrimidine derivatives; these compounds are not discussed in this study. The case study on the development of ganciclovir was added to provide opportunity for comparison across companies. This R&D project was contemporary to the development of acyclovir and involved similar scientific barriers in its conceptualization and subsequent development. Figure 1 shows the chemical structures of the five purine analog drugs that we studied. This figure

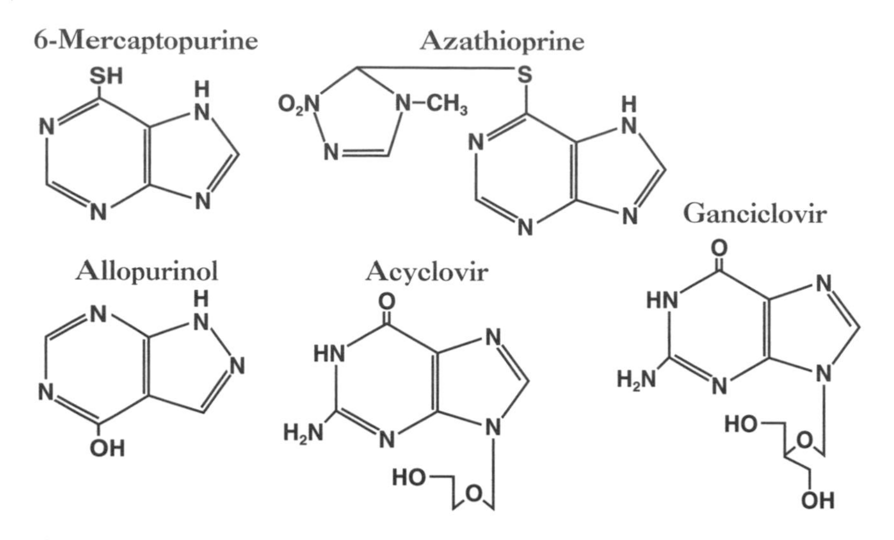

Figure 1. *The chemical structures of five purine analog drugs.*

illustrates the structural similarities between compounds that were developed for a variety of indications.

Syntex and Burroughs Wellcome followed different paths to the discovery of these compounds. The drugs developed at Burroughs Wellcome emerged from investigation of compounds based on their predicted ability to interact with cellular targets of interest. Ganciclovir, on the other hand, was discovered through modification of purine analogs to produce an effective antiviral, given the known properties of acyclovir. The cases of innovation at Burroughs Wellcome provide an opportunity to examine factors that influenced differences in the time required to develop drugs when rational drug design was used as the method of discovery. As was previously mentioned, selection of compounds to investigate based on structure-activity relationships has been proposed as a factor in accelerating drug discovery. These case studies provide the opportunity to examine how this and other methods of managing pharmacologic information influence the rate of innovation. In these case studies we specifically examine how motivating scientific discoveries, the roles of individuals, and organizational structure influence the success and the speed of drug discovery and development.

The development of purine analogs was selected for this in-depth study of pharmaceutical innovation for several reasons. Since most modern pharmaceutical R&D involves at least some interdisciplinary research, examining the factors that influence the time required for drug development can be quite difficult: Determining the relationships among relevant data across several scientific fields can overwhelm the study. The selected case studies involve basic scientific findings that were closely linked to pathologic endpoints. The development of chemotherapeutics from purine analogs involved researchers in a circumscribed area of basic science (biochemistry and pharmacology) and a few clinical domains. Although the innovations at Burroughs Wellcome are described as four case studies, all occurred within a single organization and were initiated by a single core group of researchers.

When compared to other medical innovations, the development of purine analog drugs involved basic science research that was closely linked to clinical endpoints. The basic scientists who enabled drug development began their research on compounds that were known or hypothesized to be active in the clinical indications for which the drugs were eventually developed. Moreover, the research for these innovations occurred primarily at Burroughs Wellcome, and the same basic science researchers were involved in the discovery and development of each drug. These factors have allowed us to examine closely the factors that influenced the transfer of information from the enabling research group to several medical domains. This examination would be more difficult had the research been as interdisciplinary as most projects employing modern products of biotechnology. Through a comprehensive review of the recorded literature and interviews with the primary researchers involved in the drug development process, we documented the chronology of each innovation. These historical records were compiled and analyzed using the methods described below.

Methods

Our first task was to review the literature on critical factors (including people, organizational structure, and information management methods) shown to have influenced discovery and development in the pharmaceutical industry. We reviewed the Nobel addresses of George Hitchings and Gertrude Elion and other articles covering their research on purine analogs, and we interviewed Elion to understand the primary researchers' perspectives on the discovery and development of the four drugs at Burroughs Wellcome. A review article on the development of ganciclovir by Julien Verheyden and interviews with him served as similar resources for the case study of innovation at Syntex. Interviews were conducted with both researchers to verify or refute findings drawn from the literature and to provide additional information that could not be gleaned from the literature. All sources were used to detail the basic scientific and clinical knowledge possessed by the researchers at specific times in the drug development process.

The historical record was subjected to three analyses: identification of all significant events and their timing for each of the case studies; search through the historical record for the timing of the enabling observations (see below for definition; these were verified in interviews with Elion and Verheyden); and examination of the functional roles of individuals and organizational structures that influenced the time required for drug development (again, these were verified in the interviews). The data collected from reviews of the literature and interviews were used to construct chronologies of the innovative process at

Burroughs Wellcome and Syntex between 1940 and 1988. Apparent inconsistencies within the case histories were examined and reconciled through interviews.

The data were used to list and codify the significant and decisive events in the innovation process in each case study. Here the terms *event, significant event,* and *decisive event* follow the technical definitions of Globe, Levy, and Schwartz. A significant event is defined as one that encapsulates a critical activity in the history of an innovation; a decisive event is "an especially important significant event that provides a major and essential impetus to the innovation." These authors add, "In judging an event to be decisive, one should be convinced that, without it, the innovation would not have occurred or would have been seriously delayed" (Globe, Levy, and Schwartz, 1973, p. 9). Although the innovative process includes numerous events that may happen sequentially or concurrently, certain events can be identified that encapsulate the progress of the innovation. Their selection, though somewhat arbitrary, reflects the best judgment of the investigators of the historical record compiled from the literature and interviews.

Because research findings often are published up to several years after the studies have been completed, the dates that are reported in review articles and symposia proceedings about when research findings occurred were used to mark the start and completion of events. When these data were unavailable, the dates by which findings were published were used and noted. The approximate dates at which decisive events were determined to occur were plotted, and these findings were compared for each case study.

The second part of this analysis involved determining the existence of one particular type of decisive event, the *enabling observation.* An enabling observation is a decisive event that creates the opportunity for existing basic scientific knowledge to be applied to a specific clinical indication. This opportunity can arise even if an innovation is not immediately envisioned. An enabling observation is a form of recognition of scientific opportunity that motivates the acquisition of new fundamental knowledge to promote innovation. When the enabling observation exists, it defines the earliest point of origin of the drug development process. Three criteria determined what was an enabling observation: It had to be a basic scientific discovery or finding; it had to precede the clinical application discussed; and it had to create the potential for applying existing basic scientific knowledge to clinical disease states.

In the third part of the analysis we examined the case histories to discover which individual roles and organizational structures, if any, were associated with reducing or increasing the time required for drug development. The case histories that follow detail the key basic scientific and clinical observations (significant events) that led to the discovery and development of compounds for particular indications.

Case Studies

THE ENABLING OBSERVATION

In 1950 George Hitchings succinctly stated the research goals of his lab group at Burroughs Wellcome.

> A study was begun in these laboratories in 1942, of the relationships between chemical structure and the ability of certain pyrimidine derivatives to serve as precursors for or to modify nucleic acid synthesis . . . It was felt that such studies might lead to fundamental knowledge of the role of pyrimidine and purine bases in growth and of the part played by folic acid in the synthesis of these bases. It was felt that new chemotherapeutic agents might be discovered by this means since, it was argued, parasitic tissues in general depend for survival on a more rapid growth, hence a more rapid synthesis of nucleic acid, than host tissues. This argument applies equally well to bacterial, viral, rickettsial, and neoplastic diseases, so that, in a sense, one might say we have been searching for a philosopher's stone, the universal panacea of the ancients. (Hitchings, Elion, Falco, et al., 1950, p. 1318)

Hitchings envisioned the potential for development of several innovations at approximately the same time; he foresaw two of the eventual clinical applications for purine analogs—as antineoplastic and antiviral drugs. His statement also shows his understanding that agents might be discovered that modified other cell types, such as stimulated lymphocytes and bacteria, that proliferate more rapidly than most host tissues. Hitchings's statement meets each of the criteria for an enabling observation. More important, his statement was an enabling observation for not one but at least four drugs used for several unrelated clinical indications! His decisive fundamental finding is the point at which a connection was made between basic science research on purines and the potential for the development of drugs to treat bacterial, viral, rickettsial, and neoplastic diseases. In fact, because lymphocytes were known to turn over reasonably rapidly and also to mediate immunity, he could have targeted immunomodulation as yet another potential indication.

The scientific knowledge base commonly limits technological innovation and inhibits foresight into the potential applications of basic scientific research findings. Hitchings's enabling observations distinguish these case studies as important to investigate because they provide an opportunity to research the factors that influenced the rate of drug development when the basic science researchers could foresee several attractive and medically important applications of their work simultaneously. It seems reasonable to state that the enabling observation let all the horses out of the gate for the management of several seemingly unrelated diseases. Any major variation in the rate of development of purine analog drugs for the above indications must therefore be associated with factors other than the insight of the enabling scientists. Factors that might explain variations in the rate of drug development include differences in the clinical knowledge bases, organizational competencies relative to particular indications, perceptions of the importance of the indications, time and attention available to pursue particular therapies, and the roles that individuals, both inside and outside the pharmaceutical company, played in the innovative process.

By the time that Hitchings proposed that purine analogs could be used to treat a variety of diseases, researchers had already uncovered the cellular or molecular pathophysiology of each of these disease states. Therefore, differences in the clinical knowledge available at the time of the enabling observation cannot explain the differences in development time for purine analog drugs. By the date that the enabling observation was made, sufficient clinical knowledge existed for Burroughs Wellcome to justify pursuing any of Hitchings's hypothesized indications. In 1845 Rudolf Virchow described and defined the pathophysiology of leukemia. As early as 1848 it was understood that gout arose from hyperuricemia; in 1876 Robert Koch established the role played by bacterial agents in infectious disease. The U.S. Army Commission established that a virus was the pathogen involved in yellow fever in 1900. In 1942 Karl Landsteiner and Chase proposed that cells mediated the immune response, and in 1945 P. B. Medawar proved the immunologic basis of allograft rejection. Thus, by the date that George Hitchings made his enabling observation, Burroughs Wellcome might have begun testing purine analogs for their ability to treat gout, cancer, transplant rejection, and bacterial, parasitic, and viral diseases.

CASE 1: THE DEVELOPMENT OF 6-MERCAPTOPURINE

Chronology

Initial investigation of the properties of purine analogs began in the early 1940s. Research concerning the biochemical properties of purine analogs emerged from a core group of researchers within the biochemistry department at Burroughs Wellcome. In 1942 Hitchings began work as the sole member of this department. Around 1943 Elvira Falco, who had been an assistant in the bacteriology department, joined him, and in 1944 Hitchings hired Gertrude Elion. When Hitchings and Elion began in the early 1940s, few scientists had studied nucleic acid chemistry, and few tools existed for their investigation. Other researchers were not actively pursuing chemotherapeutics related to purine analogs because of the limitations in technique and a lack of fundamental knowledge about purines and pyrimidines. Although purines and pyrimidines were known to exist in nucleic acids, their structures, functions, and metabolism were relatively unknown. Furthermore, "convenient methods for the determination of RNA and DNA did not become available until 1945," and "although a number of enzymes for the degradation of nucleic acids were known at that time, essentially nothing was known about their biosynthesis" (Elion, 1976, p. 4). Such research tools as ion and paper chromatography that later were used commonly were not yet available to biological scientists in the 1940s.

Although few tools and techniques existed for the study of purines, the group at Burroughs Wellcome began their research based on the finding that sulfonamides could serve as antagonists of para-aminobenzoic acid, which was an essential nutrient for many bacteria. In their Nobel addresses Elion and Hitchings cited the antimetabolite theory proposed by D. D. Woods and Paul Fildes as a chief motivation for their investigation of purine antagonists. This theory, developed to explain the action of sulfonamides, stated that chemotherapeutics could be developed by selectively developing substrates and inhibitors of enzymes. Elion believes that "this theory made possible a biochemical approach to chemotherapy that could, hopefully, result in agents with known mechanisms of action" (Elion, 1976, p. 5). Hitchings, Elion, and others at Burroughs Wellcome knew that sulfonamides were antimetabolites of para-aminobenzoic acid and believed that analogs of purines and pyrimidines could be synthesized that might be active antagonists of nucleic acid synthesis. In the beginning, however, according to Elion, they

> didn't really have a concept as to how the chemicals would work—how they would get into the nucleic acid—and if they got into the nucleic acids how they would interfere with function or prevent the normal metabolite from getting into the nucleic acid. These were big open questions. But the exciting thing was that as we began to unravel what the compounds did, we could deduce what enzymes were there even before we knew that the enzymes existed. (Elion, 1994)

Although the antimetabolite theory sparked the interest of the researchers, support from Burroughs Wellcome R&D management was necessary to permit investigation of purine analogs. Sir Henry Wellcome believed that he should support work in the basic sciences, and in 1904 hired Sir Henry Dale to work on anything he wished. That philosophy, to allow certain researchers to pursue scientific interests without precise institutional direction, remained even after Wellcome's death in 1936. According to Elion, one of the secrets of the success of the biochemistry department was the fact that nobody pushed their work; in fact no one even paid serious attention to their work until they discovered a compound that they considered particularly important. Although the permissive organizational structure of Burroughs Wellcome apparently served to promote the initial investigation of purine analogs, as becomes evident from these case studies, the absence of in-house medical expertise in transferring enabling observations into clinical applications created an institutional barrier to drug development in several areas.

Hitchings and Elion first applied the antimetabolite theory to purine metabolism in 1944 by examining purines as substrates or inhibitors of the enzymes guanase and xanthine oxidase. At that time xanthine oxidase was one of the few catabolic enzymes whose purification had been accomplished. It therefore served as one of the fundamental tools that the laboratory used to study the effects of structural modifications on purines. In their early work the group studied the relative effects of bromouracil, nitrouracil, and aminouracil on folic acid, thymine, and purine metabolism. In 1943 George Hitchings and Elvira Falco grew *Lactobacillus casei* with a growth factor from liver known as the *L. casei* factor and with a mixture of thymine and a purine. They then used these bacteria as a screen to determine if a compound could substitute for thymine. This screen was important for several reasons. It was used to test all future purine derivatives for activity, to study structure-activity relationships, and to provide insights into loci of action. Moreover, it indicated the effectiveness of a purine as an enzyme substrate or inhibitor by its ability to inhibit the growth of bacteria.

By 1947 at least six researchers were studying purine metabolites at Burroughs Wellcome. In his Nobel address Hitchings recounted that by this date his research group observed, "Now we have the chemotherapeutic agents; we need only to find the diseases in which they will be active" (Hitchings, 1988, p. 303).

In retrospect Hitchings's statement raises key questions concerning the process of innovation at Burroughs Wellcome. If this research group held this belief, how did they propose to carry out the experiments that would determine the indications for which these compounds might be effective? What motivated the order in which indications were pursued? Why was the development of antiviral agents relatively slow when such agents were one of the first likely indications for purine analogs? Why were purine analogs never developed as antibacterial agents at a time when few effective bactericidals were available for life-threatening diseases?

The researchers began to apply their knowledge of modifiers of purine metabolism to clinical indications in 1947, when Sloan Kettering Institute arranged to collaborate with Hitchings to test whether purine and pyrimidine analogs had antineoplastic activity. Indeed the first application of Hitchings's enabling observation was promoted by C. P. Rhoads and other clinicians at Sloan Kettering. Despite Hitchings's hypotheses it took just such a clinical interest to initiate the studies of clinical efficacy for the new chemical entities discovered at Burroughs Wellcome. In fact Sloan Kettering paid for the investigation, not Burroughs Wellcome. This suggests that knowledge of clinical medicine was necessary to promote or give high enough priority to the translation of biochemical findings to initiate investigation of their clinical applications. In this case clinicians directed the expansion of in-vitro data to support antileukemic development, driving effective transfer of information from the laboratory to clinical settings.

During this period Hitchings made similar arrangements with the laboratories at Wellcome in the United Kingdom to test the purine analogs for antibacterial and antiviral uses, as will be discussed below. Researchers at Burroughs Wellcome began collaborating with outsiders because they had little laboratory space and equipment, few in-house scientists, very few screening systems, and no clinical expertise. They were screening compounds in *L. casei* and some pathogenic bacteria, but did not have any animal screening systems, and had only three or four clinicians in the medical department whom they shared with other research groups. At this time Sloan Kettering had tumor cells growing in human tissue cultures and animal tumor and leukemia models. So Hitchings began to send Sloan Kettering compounds that his group thought would be active based on its studies in *L. casei*.

Hitchings claims that the association with Sloan Kettering was a major impetus for the growth of Burroughs Wellcome. Hitchings believed that the arrangement with the institute established productive and satisfying research collaborations especially because of the efforts of C. P. Rhoads, Joseph Burchenal, C. C. Stock, F. Philips, and D. Hutchison. Sloan Kettering even provided financial support that enabled Hitchings's group to hire additional chemists to search for antitumor agents, at a time when Burroughs Wellcome could not or would not fully support the group. According to Hitchings, this arrangement resulted from Rhoads's realization that the compounds Hitchings was investigating were of special interest because of their pharmacological properties as antitumor agents and their potential for use in cell biology research. Sloan Kettering's support allowed the Burroughs Wellcome biochemistry department to expand to about fifteen researchers—support that continued for several years until Burroughs Wellcome furnished Hitchings's group with additional funding. This unusual reverse financial relationship between academia and industry can be commented on in two ways. First, it appears that the facilitator of the progress of drug development was not the pharmaceutical company but academia and the academic clinicians in particular. Second, had these clinicians not been interested in the application of purine analogs, the insights of Hitchings and Elion into leukemia might have been no more effective than their insights into the potential value of purine analogs as antiviral agents, an idea that waited more than thirty years to be developed. This point becomes clear in the case study of acyclovir.

Among the first few compounds that Sloan Kettering Institute examined was 2,6-diaminopurine, which proved active against both viruses and cancer and later produced remissions in acute leukemia. Around 1948 Hitchings's group found that 2,6-diaminopurine inhibited *L. casei* growth and was reversed by adenine but not by the other natural purines. After careful investigation of the pharmacologic properties in animal models by F. S. Philips and J. B. Thiersch, Joseph Burchenal found that diaminopurine inhibited mouse tumors and mouse leukemia; two years later he reported that 2,6-diaminopurine produced clinical remission of chronic granulocytic leukemia in two adults, but caused severe nausea and vomiting. These findings were sufficient to establish development of cancer chemotherapeutics as a primary goal of the biochemistry department.

The second compound that the Sloan Kettering Institute found active was the pyrimidine derivative 2-4-diamino-5 (3',4'-dichlorphenyl)-6-methyl pyrimidine (DDMP). This substance showed activity against mouse leukemia and produced remissions in childhood acute leukemia much like methotrexate; unfortunately it proved highly toxic. With this finding, work on DDMP was discontinued, although later studies demonstrated that, unlike methotrexate, DDMP crosses the blood-brain barrier and had potential for use in treating patients with leukemia who had developed central nervous system–meningeal involvement. From 1949 to 1951 Elion and others at Burroughs Wellcome synthesized and tested over a hundred purines in the *L. casei* screen. The group discovered that substitution of oxygen and sulfur at the 6-position of guanine and hypoxanthine produced inhibitors of purine utilization. Two substituted purines, 6-mercaptopurine (6-MP; Purinethol) and 6-thioguanine (TG), were tested at Sloan Kettering and found to be active against a wide spectrum of rodent tumors and leukemias. For example, D. A. Clarke and colleagues found that 6-MP inhibited growth and occasionally caused regression of sarcoma 180. Despite their success several basic questions remained as to the fundamental mechanisms of the drug's effect and the reasons for its differential effect on neoplastic cells. How could the differential effect be improved? By this time, however, the clinical effects of these compounds were apparent.

Subsequent studies with a 6-MP–resistant strain of *L. casei* identified 6-MP as an antagonist of hypoxanthine and suggested a biochemical pathway for the metabolism of purines. In 1952 Hitchings and Elion filed to patent 6-MP. After preclinical studies of its pharmacology had been completed, the Sloan Kettering group followed up Clarke's findings with studies in humans. In 1953 Burchenal and colleagues reported that 6-MP produced complete remissions in one-third of children with acute leukemia resistant to methotrexate. At a daily dose of 2.5 mg/kg of body weight, mercaptopurine caused leukemic white cells in the peripheral blood and the marrow to disappear but permitted the return of normal white cells, red cell precursors, and platelets. The drug therefore was considered rather specific for leukemic cells at that dosage. In January 1953 Memorial Sloan-Kettering held a meeting to share these results with other clinical investigators to inform them about the use of this compound. By September 1953, when the findings appeared in the literature, much confirmatory evidence already existed. In the same year the FDA approved 6-MP for clinical use. This approval came only a little more than two years after the synthesis of the drug and investigation of its fundamental pharmacologic effect. When 6-MP treatment became available, the median survival for children with acute leukemia increased from about eight to twelve months.

Synopsis

The discovery and development of 6-MP exemplifies rapid pharmaceutical innovation. The progression from synthesis and in-vitro screening of the drug to animal and clinical trials was far faster than the average of eleven to fifteen years that is now quoted as "the time necessary to develop a new chemical entity." Examination of the flow of information between Burroughs Wellcome and Sloan Kettering and the roles that key individuals in each of these institutions played reveals factors that influenced the time required for drug development. The early events of this innovation involved support for purely scientific research by the biochemistry department at Burroughs Wellcome. Hitchings's group was permitted to pursue undirected research because R&D management was sympathetic to fundamental curiosity and confident that the work would lead to useful outcomes for the company. Burroughs Wellcome had confidence in the science, but no one in the medical group predicted or uncovered clinical applications for the work on purine analogs. Burroughs Wellcome did not drive clinical transfer of Hitchings's enabling observation, even though they were dependent on drug development to be successful. The point that clinicians treating the disease and not the company producing the chemical entity may be the primary or sole driver of technology transfer is reinforced by each of the case studies.

Fortunately, the organizational structure of Burroughs Wellcome permitted the screening of compounds by an external group as a part of the search for activities that might lead to clinical indications. Burchenal at Sloan Kettering provided the clinical data that became integral to the discovery and development of 6-MP. This occurred before the purine analogs had been evaluated internally as chemotherapeutics. According to Elion, neither the medical department nor the marketing department at Burroughs Wellcome even knew of their discovery until after the compound had demonstrated efficacy in the Sloan Kettering clinical trials! Burroughs Wellcome's permissive information management philosophy lowered the barriers between internal and external researchers and facilitated sharing data between Hitchings's group and outside clinicians at Sloan Kettering who were interested in the development of antileukemic drugs. Howard Skipper, one of the collaborators, said that the sharing of basic science and clinical information among researchers greatly facilitated the development of 6-MP and other antineoplastic drugs that followed. He described the research relationships that they established.

> We all worked independently, but our views and goals were quite similar. We met often and pooled what we had learned or thought we had learned. There may have been a little competitive spirit, but it was very friendly competition with rapid and open exchange of all data and working concepts.
>
> One of the good things about those days was that none of us, most especially George [Hitchings], had yet discovered that there was anything "wrong" with admixing the most fundamental organic, biochemical and biological research we could dream of with the most practical and applied research we could conceive (e.g., in animals and man). (Skipper, 1976, p. 12)

It appears from this statement that integration of basic science and clinical information was an essential component of the rapid development of 6-MP.

Skipper also implies that while seeming natural these research collaborations were indeed unusual and far from spontaneously developed. In fact it was Sloan Kettering that sought and paid for the collaboration, not Burroughs Wellcome. Few if any such relationships exist today. Each party in the collaboration needed to understand and respect the other's expertise and function in the interface between the disciplines to promote effective information transfer. Key individuals at Burroughs Wellcome (Hitchings) and Sloan Kettering Institute (Burchenal and Rhoads) were critical in facilitating the transfer of information. But the primary driver of technology transfer was an academic clinical researcher, not the company.

On the basic science side of the interface Hitchings actively shared results and compounds with inside and outside research groups and assimilated the knowledge gained from outside resources into the ongoing research in the biochemistry department. At that time Hitchings also was making arrangements to collaborate with other researchers to screen purine and pyrimidine analogs for activity as antibacterial and antiviral agents. He was clearly investigating several means to apply his enabling observation. Hitchings appears to have performed his role by establishing and maintaining strong connections with both internal colleagues and external researchers. Skipper proposes that Hitchings's "overall philosophy and his technique for admixing good basic science and very practical biomedical research" promoted the successful application of his observations (Skipper, 1976, p. 12). This case study supports Skipper's conclusions. The case study also suggests, however, that the rapid development of 6-MP depended on the drive by outside clinical investigators interested in creating effective treatments for leukemia. In particular, two clinicians at Sloan Kettering, Rhoads and Burchenal, championed the development of the drug, collecting and creating the data to support the efficacy of purine analogs as antileukemic agents. These clinical champions were two key individuals who were critical determinants of the rapid development of 6-MP.

The permissive information management infrastructure at Burroughs Wellcome and Hitchings's enabling observation were necessary but not sufficient conditions for the rapid development of 6-MP. As was previously pointed out, the medical department within the company was not structured to promote the development of chemotherapeutics by the biochemistry department. Burroughs Wellcome required a pull by clinical investigators at Sloan Kettering to develop purine analogs as antileukemics. According to Elion, Rhoads had probably seen some of their first publications, became interested in their research, and wanted to examine their compounds for antineoplastic activities. The initial discovery of activity against leukemic cells was followed up by focused research aimed at determining the means to refine the compound for this target. Sloan Kettering provided Burroughs Wellcome with basic researchers, funding, and strong clinical advocates for developing chemical entities into useful chemotherapeutics. Sloan Kettering's support indicated a critical interest and mind-set on the part of clinicians to facilitate transfer of fundamental observations about the purines and pyrimidines to a clinically useful antileukemic drug. Rhoads and Burchenal (not Burroughs Wellcome) had the insight to develop the data to transfer Hitchings's enabling observation into a useful drug. In a series of studies they expounded upon in-vitro pharmacological data on purine analogs to produce the data that proved the value of 6-MP for the treatment of leukemia. The critical role of the clinician-champion in drug development repeats throughout these case studies and illustrates the surprisingly important role of "outside" clinicians in fundamental drug development.

CASE 2: THE DEVELOPMENT OF AZATHIOPRINE

Chronology

The second application pursued by the biochemistry department was to use purine analogs as immunosuppressive drugs. According to Robert Schwartz, a clinical scientist at the New England Medical Center, immunosuppressive drugs were developed when relatively few of the fundamentals of immunology were well understood. The fact that lymphocytes mediated the immune response was a developing concept. On the basis of the discovery of acquired tolerance in mice by R. E. Billingham, L. Brent, and P. B. Medawar (1956), the transplantation of foreign cells or tissues was considered clinically feasible by the late 1950s. Schwartz characterized the immunologic research arena of the late 1950s as "a curious mixture of profound ignorance and enormous enthusiasm" in which "clinicians provided a substantial stimulus to fundamental immunologic research by insisting that their patients could be treated by immunologic means" (Schwartz, 1976, p. 39). For example, the New England Medical Center Hospital attempted to transplant bone marrow in patients with aplastic anemia, but disastrous results with total body irradiation forced the investigators to seek alternative means to suppress the recipient's immunologic response to donor tissue. Schwartz was drawn into this research as a hematology fellow under the direction of William Dameshek, who was then chief of hematology at the New England Medical Center. Schwartz was assigned the task of finding a substitute for total body irradiation with the guiding principle that "immunologically competent cells were stimulated to proliferate on contact with antigen" (p. 39). At that time this concept was a relatively novel idea without much data to support it. Nevertheless, Schwartz and Dameshek hypothesized that drugs known to block cell proliferation should be able to serve as effective immunosuppressive agents. Here again clinical investigators were responsible for extrapolating from the basic scientific findings to the clinically useful therapies and championing the use of a fundamental observation to manage disease (Schwartz, 1976).

In 1958 Schwartz and Dameshek proposed to investigate the effect of methotrexate and 6-MP on the immune response, reasoning that the proliferative activity of immunoblastic lymphocytes resembles that of leukemic lymphocytes. Both agents were being used to treat leukemia, but neither drug was commercially available when Schwartz requested the compounds from Lederle Laboratories and Burroughs Wellcome. Lederle never responded to his request, but Elion's group sent Schwartz a generous supply of 6-MP with instructions concerning its administration. That same year Schwartz and colleagues reported that rabbits given 6-MP were unable to mount an immune response against a foreign antigen, and that 6-MP was most effective when administered at the same time as the antigen. By 1959 Schwartz and Dameshek demonstrated that animals could be made tolerant to a particular antigen and remain immunoreactive to others after using 6-MP.

The observation that 6-MP had immunosuppressive activity prompted Schwartz to press for close collaboration with Hitchings and Elion. Hitchings soon sent Schwartz a series of compounds to test in the rabbit model. Included in this sample was Burroughs Wellcome #57-322 (azathioprine), which was an ineffective immunosuppressive in rabbits because they metabolized it in a peculiar manner. According to Schwartz, Hitchings insisted that azathioprine had immunosuppressive properties in rodent models and encouraged its continued investigation.

A few years later, Roy Calne, a British surgeon, was prompted by Schwartz's findings to examine the effect of 6-MP on rejection of transplanted kidneys in dogs. In 1960 he reported that 6-MP could modify renal transplant rejection in dogs, possibly producing less immunological damage than total body irradiation. Again a physician outside of Burroughs Wellcome had promoted the investigation of new applications for purine analog drugs. In the same year Hitchings and Elion filed to patent azathioprine.

Calne left England to study at the Peter Bent Brigham Hospital in Boston and stopped at Burroughs Wellcome Laboratories en route. Elion and Hitchings provided Calne with purine and pyrimidine analogs to investigate for immunosuppressive properties. Calne later found that azathioprine was superior to 6-MP in preventing rejection of canine kidney homografts. He applied the drug to the treatment of human patients with kidney grafts at the Peter Bent Brigham Hospital. Three years later, in 1963, J. E. Murray and colleagues found immunosuppression with azathioprine and prednisone to allow successful transplantation of kidneys in unrelated human donors. In 1968 the FDA approved azathioprine (Imuran) for clinical use.

Synopsis

As was the case in developing indications for 6-MP, the chronology of azathioprine illustrates that rapid drug development is stimulated by open communication between industrial and academic settings and initiated by clinicians interested in discovery of new treatments for perplexing clinical problems that they want to overcome. In this case it appears that three physicians—Schwartz,

Dameshek, and Calne—were critical to defining clinical applications for the technologies that Burroughs Wellcome made available to them. Indeed the initial hypothesis for the use of azathioprine appears to have come largely from clinician-investigators external to Burroughs Wellcome. Although Hitchings's guiding statement generally included any clinical state in which target cells proliferate more rapidly than host tissues, immunosuppression was not an indication originally considered by Hitchings or other members of the biochemistry department. The biochemistry department was prompted to focus on immunosuppression because of Schwartz's interest, and later by Calne's desire to prevent transplant graft rejection. Again, in this instance, academic clinicians promoted the application of Hitchings's enabling observation to clinical indications resulting in rapid drug development. The critical clinical insight for the development of azathioprine was Schwartz and Dameshek's hypothesis that immunologically competent cells were stimulated to proliferate on contact with antigen and that drugs known to prevent rapid cell proliferation might serve as effective immunosuppressive agents. Promulgation of this hypothesis was the decisive event that promoted the investigation of purine analogs for immunosuppression. According to Elion, if Schwartz and Dameshek had not pushed her group to develop an animal model for testing the immunosuppressive activity of purine analog drugs, the biochemistry department at Burroughs Wellcome might not have pursued this indication (Elion, 1994).

In March 1960 Hitchings and Elion filed to patent azathioprine as one of a series of heterocyclic derivatives of 6-mercaptopurine that had antibacterial properties against some pathogenic organisms in combination with sulfonamides. Use of purine analogs for this indication was never aggressively pursued, in spite of the fact that the market for antibiotics was wide open and the clinical needs were profound. If clinical champions had emerged (as Rhoads and Burchenal did in the case of applications to leukemia, and Dameshek, Schwartz, and Calne did in the case of immunosuppression), would Burroughs Wellcome have pursued the development of antibacterials? Our study leads us to believe that had there been a champion the development of purine analog antibacterials might have been investigated more aggressively. Without the clinical champion even a patent was not sufficient to push the investigation.

The case study of azathioprine confirms that clinical champions are critical determinants of the speed of drug development. Schwartz, Dameshek, and Calne, the outside experts in clinical disease, were crucial to the successful development of azathioprine. They, not Burroughs Wellcome, drove the process of innovation. Through their relationships with Burroughs Wellcome, Schwartz and Dameshek defined the potential for purine analogs to serve as immunosuppressive drugs. They presented the clinical data that made it appear plausible for purine analogs to be refined for immunosuppression. Calne followed up on their findings by applying azathioprine to an indication in which he was intensely interested—renal transplantation. Because of their exposure to the purine analogs within the context of their clinical practices, these physicians were uniquely positioned to recognize the importance of the available information on these drugs to a host of seemingly unrelated clinical problems. As a side note Calne also performed the first clinical studies on cyclosporin, and he, not Smith Kline, may have been the key determinant of the development of yet another immunosuppressive drug for organ transplantation.

Thus development of 6-MP and azathioprine was drawn out by outside clinicians, not by researchers or the medical department at Burroughs Wellcome. The company was not structured to transfer its own basic scientific observations into marketable drugs. Neither did Burroughs Wellcome identify these clinical champions; in each case the outsiders approached Burroughs Wellcome for compounds to test. Hitchings's enabling observation presented the researchers in the biochemistry department with the opportunity to develop purine analogs for several distinct indications, but they were unable to effect lines of research to pursue without outside help. They needed clinical investigators with intense interests in particular diseases in order to transfer their enabling observations to useful drugs. They needed clinical champions who understood the limitations of the available technology and the opportunities that fundamental science presented for new treatments.

Burroughs Wellcome was sufficiently flexible to allow outsiders to experiment with their compounds, but they could not perform these studies unaided. The biochemistry department did not conceive the idea of immunosuppression for transplantation; clinical investigators did. The company did not limit advances in the therapeutic field, but they plainly depended on outside investigators to develop clinical applications of their primary observations. Without the permissive information-management structure that governed the scientific pursuits of the group, the biochemistry department at Burroughs Wellcome would not have been able to collaborate effectively with outside researchers. However, open communication channels were far from sufficient for rapid drug development. Burroughs Wellcome depended on the expertise, insights, and persistence of academic clinicians. The case studies of 6-mercaptopurine and azathioprine suggest that the critical role of clinical champions is not a unique phenomenon and may be more the rule in rapid drug development than the exception.

CASE 3: THE DEVELOPMENT OF ALLOPURINOL

Chronology

In 1944 Elion and Hitchings began to examine purines as potential substrates for and inhibitors of xanthine oxidase, an enzyme that metabolizes purines to uric acid. The first and clinically most attractive inhibitor they found was allopurinol. They used this drug to extend the life of 6-MP in vivo, which was necessary because the quantities of 6-MP were limited by its complicated synthesis. Although early use of allopurinol was likely to have shown decreases in uric acid concentration (or equivalent animal metabolites) in plasma or urine, no strong conclusions were drawn regarding the use of purine analogs to treat gout or other forms of hyperuricemia. R. Wayne Rundles, a physician who performed much of the early clinical evaluation of allopurinol, stated that in retrospect he was certain that neither Hitchings "nor his collaborators thought much about gout or hyperuricemia in those days" (Rundles, 1976, p. 27). Nevertheless, Doris Lorz and colleagues in the biochemistry department at Burroughs Wellcome performed a prolonged study of xanthine oxidase that laid the groundwork for the development of allopurinol. By 1950 Lorz had identified many substrates and inhibitors of xanthine oxidase. By 1956, of the several hundred purine and pyrimidine derivatives investigated, twenty to thirty were identified as strong inhibitors of xanthine oxidase, including 4-hydroxypyrazolo [3,4-d] pyrimidine, known as allopurinol (Zyloprim), which was both an inhibitor and a substrate for the enzyme.

When the Burroughs Wellcome researchers found that 6-thiouric acid was one of the principal products of 6-MP catabolism, it seemed possible that inhibiting xanthine oxidase might interfere with this oxidation and thus increase the effect of 6-MP. To test xanthine oxidase inhibition in vivo, Elion's group chose a compound that had no inhibitory effects on bacteria or tumors, was nontoxic, and was a potent inhibitor of xanthine oxidase. This compound was the hypoxanthine analog, allopurinol. When administered to mice treated with 6-MP, allopurinol inhibited 6-MP oxidation, increased its immunosuppressive effects three- to four-fold, and doubled its toxicity. Not surprisingly, allopurinol intensified the effect of 6-MP for any indication for which 6-MP was used. Instead of studying the utility of allopurinol per se, the company turned its attention to exploring the clinical utility of the thiopurines to improve upon the pharmacology of 6-mercaptopurine and to better understand their mechanisms of action. After some biological and pharmacological investigation Burroughs Wellcome identified twenty or more thiopurines to investigate for clinical utility.

In the late 1950s more researchers began to collaborate with Burroughs Wellcome's biochemistry department. During this period at Duke University, Wayne Rundles was researching the hematologic effects of purines, pyrimidines, and folic acid metabolites in patients with megaloblastic anemia and primary refractory anemia. After meeting Elion and Hitchings at a conference in Tuckahoe, New York, Rundles began evaluating the antileukemic effects of purine analogs in patients with chronic granulocytic leukemia. In these studies 6-mercaptopurine served as a standard by which experimental compounds were compared to determine a ranking of clinical effectiveness. As Rundles states, "When it became evident that compounds showing considerable antitumor activity in experimental animals, especially rodents, were often ineffective in man, an experimental, clinical and metabolic comparison of selected compounds was carried out" (Rundles, 1976, p. 29).

The Burroughs Wellcome researchers used this expanded knowledge of thiopurine metabolism to investigate methods to modify the pathways of 6-MP inactivation. Since the oxidative degradation of 6-MP by xanthine oxidase appeared to be the most significant route of drug loss from the body, Hitchings and Elion proposed to investigate inhibitors of this enzyme. In a study published in 1966, Elion demonstrated that allopurinol served as a substrate and a competitive inhibitor of xanthine oxidase. The first studies of this nature involved simultaneous administration of 6-MP and allopurinol in mice with adenocarcinoma. In 1962 Elion and colleagues reported that the antitumor activity and the toxicity of 6-MP increased with simultaneous administration of allopurinol. While the amount of free 6-MP excreted in the urine increased, the amount of thiouric acid excreted was greatly reduced. After toxicity studies in several animal species demonstrated favorable therapeutic indices and good tolerance, Elion's group collaborated with Rundles to begin studies in humans. These studies ultimately revealed that allopurinol inhibited oxidation of 6-MP in a dose-related manner in patients with chronic granulocytic leukemia. At the same time allopurinol increased the antileukemic activity of 6-MP proportionally (but also enhanced toxicity). Allopurinol produced no serious adverse effects of its own. In the first test subject, a woman with chronic granulocytic leukemia, oral allopurinol reduced the percentage of free thiouric acid excreted in the urine. "An additional finding was documented at this time, namely, that with the continued administration of 4-HPP [allopurinol], the concentration of uric acid in the serum fell, as did the amount of urate excreted in the urine, concomitantly with a rise in urine xanthine and hypoxanthine" (Rundles, 1976, p. 30).

This finding made it clear that allopurinol might be useful as an agent to "reduce uric acid in diseases manifested or complicated by hyperuricemia, hyperuricosuria or urinary urate stone formation" (Rundles, 1976, p. 31). Initially it was known that inhibition of xanthine oxidase could promote the accumulation of uric acid precursors (hypoxanthine and xanthine), and it was believed that those intermediates might form stones in the urinary tract or tophi in tissues equivalent to those produced by uric acid. Several researchers subsequently reported that allopurinol could prevent uric acid crystal formation in the joints and kidneys and that hypoxanthine and xanthine did not precipitate in the tissues. These investigators ultimately found that in patients with primary gout allopurinol led to the disappearance of tophi with no evidence of xanthine deposition or iron deficiency, as had been previously postulated. All of these studies were initiated by independent clinical investigators. In 1966 the FDA approved allopurinol for clinical use. In 1967, after the clinical data demonstrating the value of this compound was obvious, Hitchings and Falco filed to patent Zyloprim (allopurinol) for the treatment of hyperuricemia. This sequence suggests that allopurinol was not even envisioned to be clinically useful until after it had been shown to be efficacious in humans in tests that were not initiated by the company. Once again the determination of the efficacy of purine analogs for a new clinical indication was permitted but not promoted by Burroughs Wellcome's open communication with outside clinicians.

Synopsis

Although research on xanthine oxidase inhibitors began earlier than the work on azathioprine, it took Burroughs Wellcome twenty years to apply their findings to a treatment for gout, whereas azathioprine received FDA approval ten years after Schwartz and Dameshek began their investigation. Decreases in natural purine metabolites (allantoic acid and uric acid) were probably observed when allopurinol was given to animals. Allantoin, a degradation product of uric acid in animals, most likely was measured before Rundles noted decreases in uric acid concentration in humans; however, this discovery went unremarked. In their earliest research on allopurinol, Elion's group examined only changes in thiouric acid levels in animals since allopurinol was evaluated specifically for its ability to modify 6-MP metabolism. But when allopurinol was administered to human subjects, they reviewed the data from ongoing toxicology studies in animals and found changes in allantoin levels in the urine. These findings were not published until after Hitchings and Falco filed to patent allopurinol, so outside researchers did not have easy access to the data. When the group finally realized that allopurinol induced decreased uric acid concentration in humans, a use patent was filed for the compound that specified gout as one of its indications.

Why was the clinical pull to explore the value of allopurinol so slow and weak? Surely it could not be that there was no medical value or market for preventing acute gout attacks. In the 1950s physicians were intensely interested in the metabolic basis of disease. A hotbed of activity surrounded Hitchings's enabling observation concerning the purines and pyrimidines. It seems logical that clinical scientists who understood which precursors and enzymes were involved in hyperuricemia would have grabbed the ball as soon as it was available. What inhibited the timely investigation of the use of purine analogs for the treatment of gout?

Clearly, the biochemistry department at Burroughs Wellcome may have been actively pursuing the development of other compounds at this time, and the utility of xanthine oxidase inhibitors may not have been clear to the basic scientists. In light of the previous case studies discussed, however, one major reason that development of allopurinol lagged appears to be that clinicians did not prompt the basic scientists to consider or promote development of a drug to treat gout. Elion believes that clinicians may not have recognized the potential utility of purine analogs in the treatment of hyperuricemia because relatively few researchers were involved in the study of gout. However, research on the biochemical pathogenesis of the disease had been under way since 1848. The findings of Elion's group might not have been well known to researchers in this area, but the group's earlier studies of xanthine oxidase inhibition were clear and published. During that time fear of iatrogenic effects from xanthine oxidase inhibition may have stifled clinical interest. Researchers suggested that inhibition might induce xanthine oxidase and promote the accumulation of precursors that could also produce crystals, both of which could exacerbate gout. But Elion's group did perform experiments to address these fears, though these studies do not appear to have significantly hastened development. The delay is intriguing because it shows that, even though Burroughs Wellcome unearthed the potential for allipurinol's clinical application, that insight alone was not enough to produce a drug to treat gout. A disease-oriented clinician needed to emerge before the hypothesis was tested for this indication.

No such champion emerged to force the investigation of allopurinol use for gout at Burroughs Wellcome. Development lagged until a clinician was struck by the potential use of the drug for an unpredicted indication. Because no clinician emerged as an external director of research on gout, the group at Burroughs Wellcome did not push to investigate the use of purine analogs for this indication until the obvious implications that allopurinol reduced uric acid concentrations in humans finally became too great to ignore. Although Rundles appeared as if he might serve as a clinical champion, his involvement with the development of allopurinol progressed from the study of the metabolism of mercaptopurine and mercaptopurine analogs to the effects of allopurinol in some of his leukemic patients who were receiving 6-MP. He incidentally demonstrated that allopurinol might be used to treat hyperuricemia, but he did not anticipate this finding, analyze the supporting data, or promote the development of allopurinol for gout until the data suggested that the drug must have an effect on uric acid production. In the absence of an external clinical champion the development of allopurinol did not proceed as rapidly as the previous two drugs discussed.

CASE 4: THE DEVELOPMENT OF ACYCLOVIR

Chronology

The development of purine analogs as antiviral agents took much longer than the development of the three drugs previously discussed. In 1947 Hitchings arranged to collaborate with outside labs to test whether purine analogs possessed the antiviral activity that he had hypothesized. At the same time that Hitchings's group sent compounds to Sloan Kettering, they provided the same purine and pyrimidine analogs to Randall Thompson, a researcher in the Department of Microbiology at Indiana University Medical Center. In 1949 Thompson and colleagues reported that analogs that were active in the *L. casei* screen were also potent inhibitors of vaccinia virus replication. In particular, 2,6 diaminopurine had strong antiviral activity. Active purine compounds also produced anorexia, anemia, and changes in the bone marrow in animal models;

but viral inhibition occurred at relatively low doses, and 2,6 diaminopurine toxicity could be reversed by purine administration. Nevertheless these early adverse effects and the pull by clinicians for antileukemic drugs led Burroughs Wellcome to abandon development of antiviral agents for almost twenty years.

According to Burchenal, once Hitchings's group had demonstrated that pyrimidine and purine derivatives could selectively inhibit growth of neoplastic cells in humans and suggested that they could be used for antiviral purposes as well, other investigators followed. This work produced several drugs: idoxuridine (many trade names) and trifluridine (Viroptic) for herpes keratitis, 5-fluorouracil (Adrycil) for various types of cancer, and cytarabine (many trade names) for acute leukemia in adults. Other researchers, including F. Sorm and colleagues, C. K. Chu, K. A. Watanabe, and J. J. Fox, and A. Giner-Sorolla, presumably stimulated by the work of Hitchings and Elion, also synthesized pyrimidine analogs that had antineoplastic and antiviral activity. What delayed Burroughs Wellcome in developing purines as antivirals when outside researchers readily adapted their findings to this indication? We believe that this might be an instance in which the absence of clinical champions slowed drug development. This factor along with the presumed (but unproven) toxicity of the antivirals prevented Burroughs Wellcome from acting on the competitive advantage that they held in the 1950s. Since no clinical champions pulled Burroughs Wellcome into the antiviral market, the company waited until the data from outside groups pushed them into drug development.

Burroughs Wellcome's research on the antiviral activities of purine analogs was inactive for nineteen years, until Schabel discovered that arabinosyladenine (Ara-A) inhibited the growth of DNA and RNA viruses. Elion believed that although this drug "was toxic, its therapeutic index was not bad, and it was better than anything that had come before" (Elion, 1994). Therefore she hypothesized that purine analogs could successfully be developed as antiviral agents. Elion had spent three years studying diaminopurine and knew that it was an adenine antagonist. She postulated that "if an adenine arabinoside would serve as an antiviral, a diaminopurine arabinoside should do so as well" (Elion, 1994). Elion's group first investigated whether the arabinoside of 2,6-diaminopurine would be as active as Ara-A. In the late 1960s Janet Rideout of Burroughs Wellcome synthesized the diaminopurine arabinoside, Ara-DAP. Since the Elion group did not have a virus lab at the time, the compound was sent to John Bauer at Wellcome Research Laboratories in the United Kingdom for antiviral screening. In a few weeks Bauer reported that Ara-DAP was highly active against both herpes simplex virus (HSV) and vaccinia virus and less cytotoxic to mammalian cells than Ara-A. Perhaps if there had been compelling arguments from clinicians about the utility of antivirals, as there had been for antileukemics and immunosuppressives, this particular research would have been taken up two decades earlier.

In 1970 the biochemistry department (renamed "experimental therapy" after 1967), now headed by Elion, moved to North Carolina and increased from twenty to fifty-five researchers from a variety of disciplines. By this time the group that composed experimental therapy was as interdisciplinary as most modern pharmaceutical R&D groups. That same year Howard Schaeffer joined Burroughs Wellcome to head the organic chemistry department. Schaeffer focused primarily on the field of nucleic acids, studying adenine compounds, and had no interest or experience with antiviral agents. Before joining Burroughs Wellcome, he investigated the minimal components of the ribose ring of adenosine that would inhibit adenosine deaminase. When he arrived at Burroughs Wellcome, the compounds he had synthesized were screened broadly, including for their antiviral properties. In the early 1970s Elion's group studied purine arabinosides by exploring structure-activity relationships, synthetic methods, and metabolic studies in mice. They found that Ara-DAP could be deaminated to arabinosylguanine (Ara-G), which also had antiviral activity, but the investigators were not certain whether these findings warranted the full-scale development of Ara-DAP or Ara-G.

In the mid-1970s Schaeffer and Lila M. Beauchamp synthesized acyclic purine nucleoside analogs, including acyclovir, in spite of the prejudice that predicted that acyclic compounds would have no pharmacologic activity. Bauer and Peter Collins tested the antiviral potential of these new chemical entities, and Elion's group studied their mechanisms of action, enzymology, and in-vivo metabolism. When Elion decided that Burroughs Wellcome needed to study viruses on site, Hitchings provided a virus lab with Phillip Furman as its head. In one of their initial studies they examined the selectivity of acyclovir (ACV) on herpesvirus. Shortly thereafter Schaeffer filed a patent on ACV. In 1978 Schaeffer reported that ACV was not cytotoxic to mammalian cells, even at concentrations more than a hundred times those required for antiviral activity, and that ACV was active against HSV-1, HSV-2, varicella-zoster virus, and other herpes-like viruses. In 1982 the FDA approved acyclovir (Zovirax) for intravenous use and in 1985 for oral use.

Synopsis

The chronology of the development of acyclovir shows it to be a long and interrupted process. Barriers to innovation clearly included limitations in the interpretation of available scientific information. For instance, the biochemistry department had discovered purine analogs with antiviral activity as early as 1947; the early compounds, however, were considered too toxic to be generally useful, although this was not proved. Testing stopped between 1949 and 1968 despite the early positive leads. In 1970 Schaeffer synthesized the acyclic compounds using entirely different methods from the work that Elion's group had been performing. These scientists focused on distinctions between the acyclic compounds and other purine analogs, but had Schaeffer or other researchers at Burroughs Wellcome been subjected to the clinical pull of the need for antiviral agents, there was no scientific barrier that would have precluded much earlier synthesis and testing.

Limited clinical involvement early in the process appears to have slowed this innovation. The development of acyclovir seems to have been driven almost completely by basic scientific curiosity rather than drawn out by clinical interest in the enabling field. No clinician, either inside or outside of Burroughs Wellcome, approached the biochemistry department with an interest in antiviral agents. This may have been a factor that slowed progress toward the first clinical trials of purine analogs for this indication. This case study serves as another example of delayed drug development in which no clinician emerged to apply Burroughs Wellcome's compounds. In this case the basic scientists hypothesized that purine analogs could be used as antiviral agents twenty years before they were seriously investigated and over forty years before acyclovir was developed. This case accentuates the critical role of clinical champions in rapid drug development. In the absence of such champions it appears that scientists in the enabling field do not pursue development aggressively or effectively until outside researchers provide the clinical information to support full-scale development. This fact is clearly illustrated in the next example of similar research done at Syntex.

CASE 5: THE DEVELOPMENT OF GANCICLOVIR

Chronology

Ganciclovir was developed at Syntex through a lengthy process, primarily because of the push of chemists in the Syntex Institute for Molecular Biology (SIMB). In 1961 Julien Verheyden, then a postdoctoral fellow, joined the newly formed SIMB in Palo Alto, California. According to Verheyden, SIMB was established by Joshua Lederberg and Carl Djerassi "to explore the emerging field of molecular biology" (Verheyden, 1993, p. 2). This institute was composed of three departments—biology, enzymology, and chemistry—and "was unusual in the industry" in that fundamental research was its main goal (p. 2). It was established during a period when Syntex made funding available for research that was not to be directed toward specific product development. At that time Syntex Institute was composed of eight chemists, four biochemists, and four enzymologists. From 1961 until 1970 John Moffatt headed this group, which was considered to be one of the top laboratories in the nucleoside field. The research group at SIMB was devoted to synthesis of 4'-substituted nucleosides since the 4'-position of the ribose ring had not yet been extensively explored. Although SIMB was not to have a preconceived clinical target to shoot at, neither did it have clinical oversight or insight to determine whether its results had clinical implications (a clear contrast to the philosophy of Henry Wellcome).

In his first work, published in 1964, Verheyden extended the synthetic methods of N. K. Kotchetkov and A. I. Usov to produce iodinated pyrimidine and purine nucleosides. This method permitted the synthesis of a 4',5'-unsaturated nucleoside, which in turn presented the opportunity to synthesize nucleocidin, a naturally occurring 4'-fluorinated nucleoside. Nucleodin was a broad-spectrum antibacterial and antitrypanosomal agent. However, according to Verheyden, nucleocidin was of great interest to SIMB "because it was the first naturally occurring nucleoside having a fluorine atom in its glycosidic moiety; moreover, the fluorine was located at the 4'-position, a position that had not been manipulated by nucleoside chemists" (Verheyden, 1993, p. 6). This statement demonstrates that researchers at SIMB were obviously more interested in challenging syntheses than in investigating the chemotherapeutic effects of purine analogs. These investigators were not oriented toward translating Hitchings's enabling observation into useful antibiotic or antiviral agents, although the compounds they were manipulating were suspected of having these properties.

In the mid-1960s other chemists inside and outside of Syntex synthesized 4',5'-unsaturated nucleosides and related compounds in an attempt to discover an agent with antimicrobial activity. In 1964 Verheyden began screening a collection of 5'-deoxy-5'-iodonucleosides, but did not discover any pharmacologically exciting antimicrobials. The institute subsequently continued investigating a variety of 4'-substituted nucleosides, which at that time no other researchers had synthesized in abundance. In 1969 Ian Jenkins joined the Syntex Institute as a postdoctoral fellow and began the synthesis of nucleodin. Early efforts were only marginally successful, until 1971, when J. Bornstein and L. Skarlos described a reagent that allowed SIMB to improve its yield of nucleodin. By 1971 Geoffrey R. Owen, another postdoctoral fellow at Syntex, had synthesized a nucleoside analog of nucleodin. When Verheyden became aware that the 4' position of nucleosides could be modified successfully, the group began to screen these compounds to examine the effect that various 4' substituents had on the antimicrobial properties of nucleoside analogs.

Verheyden set out to synthesize 4' methoxyuridine as a model compound in the search for antimicrobial agents. Verheyden embarked on what was to be a lengthy synthesis that was later shortened by another chemist at SIMB, Nghiep Le-Hong. By the late 1970s, 4'-substituted nucleosides had been explored by a number of researchers inside and outside of Syntex Institute. At this point Verheyden explored 4'-hydroxymethylated nucleosides since they were likely to be stable in acid and base, more resistant to enzymatic degradation, and easier to isolate. Following their previous work, Verheyden's group prepared a series of 4'-substituted purine and pyrimidine nucleosides. Each nucleoside synthesized underwent biological screening in Japan for antiviral, antibacterial, antitumor, and antifungal properties. None of these compounds demonstrated interesting pharmacologic properties.

In over fourteen years of research at SIMB only one compound, cyclo-cytosine, showed activity, but it had unacceptable central nervous system toxicity. During this period the Syntex Institute was not concerned with how the compounds they synthesized were assayed, nor were they driven to discover drugs for any particular indication. The group was primarily focused on the chemical synthesis of new entities and did not concern themselves with the mechanisms by which these compounds functioned. Scientists at Syntex were encouraged to publish and patent, but the organization was not structured to emphasize the transfer of enabling observations to clinically useful drug entities.

Although the Syntex Institute generated several chemically interesting compounds, within the framework of the company, these years of research were considered nonproductive. The existence of this group was eventually threatened, and the researchers were forced to refocus. According to Verheyden, "In late 1974, Syntex management, somewhat discouraged by the lack of potential drugs coming out of our synthetic effort, asked us to redirect our attention to fields that were more likely to generate new products" (Verheyden, 1993, p. 16).

John Freid, head of Syntex R&D at that time, was dismayed by the lack of lead compounds coming from this group and became much more interested in the bottom line than previous managers apparently had been. Freid divided the researchers into two groups: one to study peptides, and the other, aminoglycosides. Freid wanted Verheyden's group to synthesize a compound similar to neomycin, which needed to be ten times as effective as gentamicin to be marketable. However, the clinical resources needed to promote rapid development of aminoglycoside antibiotics was not any more accessible to SIMB than similar resources were when they were studying purine and pyrimidine derivatives. Without these data and a strong pull by clinicians for new treatments, SIMB had neither the infrastructure nor the clinical drive necessary to promote aminoglycoside development.

Verheyden worked on aminoglycosides for five years to no avail. In the following years antiherpetic agents were discovered by other labs. Those compounds bore some structural similarity to the compounds that the Syntex Institute had synthesized. These included fluoroiodoaracytosine, brivudine, and acyclovir. In view of these developments Syntex management thought it was appropriate for the group to resume their research on nucleosides. But support for further research on purines only came after others had shown the way. At this point Syntex again changed the focus but not the structure of the institute. Now the researchers were directed to find a nucleoside that would serve as an effective antiviral agent. However, SIMB was not restructured to attract the clinicians to provide the data that could prove the efficacy of these compounds. They were still without a medical champion to help them overcome the barriers to demonstrating the clinical value of these compounds. Syntex's potential competitive advantage in discovering an antiviral agent was lost by the lack of an infrastructure that could promote recognition of the clinical potential of fundamental science. There was no clinician-investigator focused on viral diseases to investigate the value of Syntex's compounds, until clinicians began clamoring for a treatment for human immunodeficiency virus and its sequelae.

Because the field of antivirals was vast and the likelihood of finding a broad-spectrum antiviral was remote, Verheyden wanted to focus the group's search. Following the advice of Syntex's marketing experts, they began to work on the development of inhibitors of herpesviruses. To forward this work, Verheyden read hundreds of papers on the biology, biochemistry, and treatment of HSV and engaged himself in investigating a number of chemical structures that were proposed to have some activity against the virus, including Tromantadine, benzimidazoles, thiosemicarbazones, benzylidene hydantoins, amidines, polyphenols, 2-deoxyglucose, long-chain fatty alcohols, and nonsteroidal anti-inflammatory drugs. Before this Verheyden had not been actively involved in the literature outside of synthetic chemistry and biochemistry. The interdisciplinary approach he finally took may have been critical to the successful development of ganciclovir, but it still lacked the strong pull of a clinician-champion.

Verheyden initially shied from synthesizing a compound structurally related to acyclovir because he felt that such a compound was likely to be too toxic to be useful. In 1979 acyclovir was shown to be the first nucleoside that was effectively nontoxic at therapeutic doses. In November 1979 Verheyden's group submitted its first agent to be tested for antiherpetic activity, a benzylidene thiohydantoin derivative, which was the first of a long series of inactive compounds. In 1980 Verheyden conceived of the structure 4'-hydroxymethyl-acyclovir (ganciclovir). In developing a pathway for the synthesis of this compound, he and John Martin decided that "this compound should be considered a secoguanosine rather than a 4'-substituted nucleoside" (Verheyden, 1993, p. 19). At this point Martin embarked on the synthesis of the compound. A few months later RS-21592, now known as ganciclovir, was synthesized. In July 1980 Verheyden's group submitted this compound for biological evaluation, which showed that the drug was quite active in vitro against HSV-1 and HSV-2. Syntex soon learned, however, that Burroughs Wellcome's patent on acyclovir held a generic claim covering variations on acyclovir, potentially including RS-21592. Verheyden and colleagues were extremely concerned that they might not receive credit for their discovery. This situation could have been prevented had SIMB been prompted to modify and investigate the antiviral properties of modified purines almost twenty years earlier when they were first synthesized. Syntex relinquished its competitive advantage in part because investigators did not emerge to establish connections between the chemical structure of these early compounds and their potential as antiviral agents.

By early 1981 ganciclovir had been tested in vivo in a variety of biological systems. In a mouse model with systemic HSV-2 infection, ganciclovir showed greater antiviral activity than acyclovir and fluoroiodoaracytosine. It eventually became evident that ganciclovir was broadly active against HSV-l, HSV-2, varicella-zoster virus, Epstein-Barr virus, and cytomegalovirus (CMV). The antiviral properties of ganciclovir were important because at this time the acquired immunodeficiency syndrome (AIDS) was emerging as a significant clinical problem. Since CMV infection is nearly ubiquitous in patients with human immunodeficiency virus and can cause retinitis, pneumonia, disseminated disease, and death, development of a treatment for CMV carried significant clinical value for this patient population. On the basis of these findings Syntex filed a patent in May 1981 to cover gangiclovir. In March 1982 the U.S. Patent Office allowed the ganciclovir patent.

Howard Rhingold, then director of the institute of biological science at Syntex, appointed Verheyden to head the project team developing ganciclovir. Rhingold, who trained as a chemist but worked most of his career as a biologist, was an experienced research coordinator who was able to use his influence in the company to achieve goals quickly when the need arose. He supported the development of ganciclovir despite the potential infringement of the Burroughs Wellcome patent. Rhingold served as an effective manager of information transfer in the late phases of development, but in an unusual role. He worked between the disciplines of chemistry and biology in the scale-up process. Rhingold joined the project after the compound had already been shown to be an effective antiviral agent, and he facilitated development by managing the sharing of chemical, biological, and medical data.

In August 1983 Syntex became aware of the gonadal toxicity of ganciclovir and decided to develop it for a more limited indication—life-threatening CMV. In May 1984 a champion finally appeared for Syntex. Verheyden "received a call from Clyde Crumpacker at the Beth Israel Hospital requesting ganciclovir for compassionate treatment of a young mother who had undergone bone marrow transplantation and was suffering from CMV pneumonia" (Verheyden, 1993, p. 21). Crumpacker was actively investigating treatments for viral diseases and came to Syntex once information on the potential clinical value of ganciclovir became available. Soon other clinicians requested ganciclovir to prevent blindness in AIDS patients with CMV retinitis. The great number of requests for compassionate use of ganciclovir kept Syntex's Institute of Clinical Medicine busy for many months and raised ethical concerns for performing randomized, controlled clinical trials. Later a careful study was done by D. A. Jabs and colleagues that reported on the effectiveness of ganciclovir in patients with retinitis. Based on these and other findings, in 1989, the FDA approved ganciclovir for the treatment of sight-threatening CMV retinitis.

Other researchers followed on the trail of acyclovir. In July 1982 K. O. Smith and colleagues described the potent antiviral activity of BIOLF62, a compound identical in structure to ganciclovir, which was reported to be active against HSV-1, HSV-2, varicella-zoster virus, and Epstein-Barr virus. However, the authors failed to describe the potent activity of BIOLF62 against CMV, "probably because a fluorescent antibody against the immediate early antigen was used to determine viability of the virus" (Verheyden, 1993, p. 21). In October 1982 Richard L. Tolman and Kevin Ogilvie at Merck described a potent antiherpetic that they called 2'-nordeoxyguanosine (2'-NDG), which was also identical in structure to ganciclovir. Furthermore, in 1972, a few years before acyclovir was patented, U. K. Pandit and coworkers had synthesized a carba derivative of ganciclovir that was later shown to be active against HSV-1 and HSV-2. But this group never tested the compound for antiviral activity. Had they gained access to clinical information demonstrating the efficacy of their compound, development of an antiviral agent might have occurred several years sooner.

This case study shows that several groups of researchers appear to have discovered the same compound independently. The groups differed from each other, however, in the infrastructure surrounding the researchers, which either facilitated understanding of the potential pharmacology of the drug or impeded transfer of this information. All of these companies also lacked a serious champion who understood viral diseases and supported the cause of drug development. The development of ganciclovir was painfully slow, but the barriers to innovation did not arise solely within the company. Development was also slowed by a lack of demand for antiviral agents from outside clinical investigators.

Synopsis

The development of ganciclovir demonstrates some of the barriers to the rapid creation of new drugs. As mentioned above, other groups also conceived of ganciclovir or closely related compounds independently. Thus this case study offers an opportunity to speculate not only on factors that may have led to successful but very slow innovation at Syntex but also on factors that may have sped or slowed their research on antiviral agents. In this case, drug development appears to have been slowed because available information on the potential clinical applications of purine analogs was not used by the company to direct the synthesis of new chemical entities. SIMB's attention clearly was focused on the synthesis of new compounds rather than on the discovery of antiviral, antibacterial, or other drugs. By the time they began their series of syntheses, sufficient data existed suggesting that purine and pyrimidine analogs could be applied to these and other indications. The Syntex Institute, however, was not structured to use its skilled chemists to rapidly translate the available information on the properties of purine analogs into useful drugs. This organization had no enabling observation to guide the examination of novel compounds. Little interdisciplinary exchange seems to have occurred between the chemists and biologists or clinicians, and no broad connections were made early on between the structures of the nucleosides synthesized and a multitude of potential clinical indications for which they could be used. The compounds that SIMB synthesized were broadly screened but without anticipating how they could be applied clinically.

Verheyden's ability to handle interdisciplinary resources may have been one factor that allowed the group at Syntex ultimately to succeed in this highly competitive area. After the approval of acyclovir Verheyden read much of the existing literature across disciplines that had been published on viruses and investigated all compounds reported to have antiviral activity. According to Verheyden, "with the ever-increasing amount of information, the real art is to assimilate this large body of often disparate data and to come up with that unique perspective that constitutes a breakthrough" (Verheyden, 1993, p. 22). But Verheyden was primarily interested in chemistry, not in the application of nucleoside compounds to chemotherapeutics. This case study suggests that he should have added that the assimilation must include knowledge of disease. The one area where reading was insufficient was in the extrapolation to clinical medicine.

In the absence of clinical interest or pull the discovery and development process proceeded much more slowly than in the cases of 6-MP or azathioprine. Instead of being drawn into the clinic by perceived need, ganciclovir was slowly pushed through the stages of development by chemists from SIMB. Only after the drug and its effects had been discovered did clinicians finally demand its use for CMV retinitis. But the question remains why Syntex did not seek Crumpacker earlier as a key member of the research team. His interest in investigating antiviral agents was clear and well established in the literature. Perhaps less time would have been required to develop ganciclovir had Syntex sought outside clinicians interested in antiviral therapy to aid in refining their compounds until they proved efficacious. It seems that in this case establishing relationships between basic scientists within the company and clinical investigators may have provided the opportunity to transfer compounds from the lab to clinical practice more rapidly. Bridges between basic science and medicine might function best if deliberately forged.

Conclusions

The time required to develop purine analog drugs for different medical indications varied widely. Table 1 shows the date of the enabling observation, patent, and approval for each drug. A standard and glib explanation for the differences in the time to patent and to new drug approval (NDA) by the FDA is that purine analogs simply are toxic drugs and priorities were systematically set to develop drugs first for more severe conditions. Although the issue of toxicity is important when discussing the development of purine analog drugs, infectious diseases in the 1940s had severe morbidity and mortality, and there were few therapies to alter their course. For instance, in 1942, penicillin had not yet been developed, and there were no effective antiviral agents. Thus great medical and economic incentives existed for the development of any drug, including purine analogs, that could be used for these treatment modalities.

Table 1. Comparisons of the Time Required to Develop Purine Analog Drugs

Type of Drug	Clinical Investigation Began	Date of Action (by year)		
		Enabling Observation	Patent Application	FDA Approval
Antibacterial agent (Burroughs Wellcome)	1876	1942	—	—
Antiviral agent (Syntex)	1900	1942	1979	1988
Antiviral agent (Burroughs Wellcome)	1900	1942	1977	1985
Immunosuppressant (Burroughs Wellcome)	1956	1958	1960	1970
Antigout (Burroughs Wellcome)	1848	1942	1968	1968
Antileukemic agent (Burroughs Wellcome)	1845	1942	1951	1951

Furthermore, allopurinol was quickly discovered to have limited toxicity, yet its use in gout was remarkably delayed. The toxicity of the drugs should not have been sufficient to prevent their testing for efficacy. Moreover, careful examination of the events involved in the innovative process reveals other factors that influence the rate of drug development more significantly.

Hitchings's enabling observations created a situation in which the opportunity arose to develop drugs simultaneously for various seemingly unrelated clinical problems. Therefore, differences in the rate of drug development must be associated with factors other than limitations in the foresight of basic scientists. Table 1 displays the dates when clinicians began to understand the pathogenesis of each of the diseases for which purine analog drugs were developed. When Hitchings's enabling observations were published in 1950, sufficient clinical information existed for clinicians to predict the potential therapeutic benefits of purine analogs for each hypothesized indication. In every case except azathioprine use for immunomodulation, relevant clinical data were available decades before Hitchings's findings were published. The case studies of pharmaceutical innovation at Burroughs Wellcome and Syntex suggest that the marked differences in the time required for drug development across indications arises because of differences in the transfer of information across the interface between basic science and clinical research. The most important factor influencing the flow of information across this interface is the presence or absence of clinician-champions who pull the application of basic findings toward particular treatments.

In all of the case studies examined, the presence of a clinical champion sped the transfer of enabling observations to the development of purine analog drugs. The efforts of Burchenal and Rhoads in the case of 6-MP and Schwartz, Dameshek, and Calne in the case of azathioprine clearly indicate that the pull of interested clinicians can accelerate drug development. Push by basic scientists or other industry professionals does not appear to be sufficient for rapid drug development. Despite Hitchings's early insights into the potential uses for purine analogs, the development of drugs for each indication appears to have been driven by the interests of clinicians. Surprisingly, in the case of 6-MP and azathioprine, Burroughs Wellcome's priorities for development were set by researchers outside of the company. In the cases of allopurinol and acyclovir, when no clinician emerged as a strong advocate for new therapies, drug development occurred more slowly. Furthermore, although we have not examined antibacterial agents directly in this study, it appears that the absence of a clinical champion completely stifled the development of purine analogs for this indication. (Some might argue that the toxicity of the drug and possibly Burroughs Wellcome's development of a competing drug, trimethoprim, were more significant impediments, but the relative importance of these and other factors remains debatable.)

After completing this study, we uncovered further evidence supporting the importance of clinical champions in drug development. One of the key clinical champions involved in the development of purine analog drugs for use in transplantation was Roy Calne. His persistent interest in this subject as well as his persistent aggressiveness in furthering the field led to his critical role in the development of another compound for the same indication. Calne continues to search for new modalities to facilitate organ transplantation. A recent review of his MEDLINE citations from 1966 to 1995 revealed over three hundred articles published on the subject. Clearly he has dedicated himself to investigation of methods to prevent organ transplant rejection. Pertinent to the present discussion is his involvement in the development of cyclosporin.

After his work with Burroughs Wellcome, Calne continued to investigate the use of immunosuppressive drugs. Calne became intrigued by data suggesting that Sandoz had a compound, cyclosporin A, with immunosuppressive effects. In 1976 an associate of Calne's, David White, wrote to Borel of Sandoz requesting cyclosporin A to test in rat models. Note that the correspondence was initiated by Calne, not Sandoz. Correspondence between White and Borel between May 1976 and December 1977 clearly demonstrates that Calne and White pushed for the development of cyclosporin for prevention of allograft rejection. Calne and White were ultimately and somewhat reluctantly given enough cyclosporin A to test its effects in their in-vitro assays of allograft rejection and in in-vivo studies of rat and porcine heart graft survival. Calne and colleagues then demonstrated that cyclosporin was useful for preventing allograft rejection. This preliminary evidence suggests that Calne served as a clinical champion for the development of cyclosporin years after his seminal work with purine analogs. In spite of his critical role in the development of azathioprine pharmaceutical companies (and specifically Sandoz) did not seek his help in the development of cyclosporin—probably because they did not recognize his importance.

Though relatively few studies have focused on the roles of individuals in drug development, champions have been identified as important influences on the success of innovations in other industries. Several studies have identified individuals who are strong proponents of innovation. In other industries champions have been shown to determine the objectives and goals for a project and play a dominant role in overcoming technical and organizational obstacles, primarily by getting R&D management sufficiently interested in the project. Champions perform this role by identifying relevant scientific data, maintaining strong connections with colleagues inside their own organization as well as external sources of information, and effectively translating between the two systems. The clinical champions described in this study differ from champions in other industries in that they are academic investigators external to the innovating company.

Other findings suggest that effective transfer of information across the interface between basic science and clinical research is an important determinant of success in the pharmaceutical industry. Rebecca Henderson has shown that the most successful pharmaceutical companies of the 1990s achieved competitive advantage by enhancing their ability to innovate in information-intensive environments. These companies were tightly connected to the larger scientific community and effectively allocated resources to stimulate rapid transfer of information. Henderson also demonstrated that successful pharmaceutical firms are characterized by constant attention to the integration of knowledge across disciplinary and organizational boundaries, firms' most critical competence being their ability to continually re-examine the possible linkages between relevant scientific and medical disciplines. The idea that information flow across disciplinary and problem boundaries is essential to scientific problem solving is not new and is widespread in the literature on effective management of research-based organizations. The willingness to share information across disciplinary boundaries and openness to sources of information outside of the immediate organization also is associated with successful Japanese firms in other industries.

Henderson's findings accentuate the importance of methods for sharing preclinical and clinical information and suggest that clinicians who perform this function are critical determinants of innovation. Samuel Thier proposes that without clinical investigators "the relevant scientific questions that feed back into the biomedical system will not be asked" (1994, p. 542). This seems to be precisely the case in the slow development of allopurinol, acyclovir, and ganciclovir. Without clinical pull the data demonstrating the activity of purine analogs for the treatment of gout and viral diseases took much longer to develop. Thier stresses the need to develop clinical investigators versed in the "integrative sciences" to connect fundamental observations to clinical problems. An Institute of Medicine committee found that this role for clinicians is infrequently acknowledged, encouraged, or supported, but appears to be crucial to the development of novel diagnostic and therapeutic modalities.

If the presence of a clinical champion promotes more rapid drug development, what functions does this clinician perform? Clinician-champions possess a deep understanding of the pathogenic mechanisms of diseases and integrate this knowledge of disease pathogenesis with information on the proven or proposed pharmacology of new chemical entities that show potential for addressing the underlying mechanism of disease. These clinicians serve as information filters of basic scientific data, matching relevant pharmacologic findings with fundamental knowledge of pathogenesis. The development of azathioprine is a clear example of this process. Schwartz and Dameshek were familiar with treating patients with leukemias and autoimmune diseases; they recognized the similarities in their underlying pathology and hypothesized that antileukemic drugs might also be used as immunosuppressive agents. Similarly Calne recognized that immunosuppression with the same compounds might also be beneficial in preventing transplantation rejection. Each of these clinicians was intensely interested in a specific disease, and each was willing to promote the development of therapeutics related to this disease focus. A similar approach should promote more rapid drug development in today's pharmaceutical industry. Physicians' expertise in testing drugs for off-label therapies is commonly recognized. That testing is responsible for producing many new indications for existing drugs and combination therapies. But clinicians' role in promoting development of new chemical entities is not commonly appreciated or fostered.

Successful clinician-champions exploit therapeutic opportunities. In the case studies of 6-MP and azathioprine clinicians placed demands on basic scientists to provide compounds and data and were willing to develop (and sometimes pay for) these data to promote transfer. These clinical champions subtly set Burroughs Wellcome's priorities for development and demanded continual modifications of the compounds to address effectively particular indications when the initial drugs tested were not efficacious or were too toxic. The clinician-champions bridged the broad interface between basic research and clinical medicine. In the case studies examined, clinicians performed this role while simultaneously spanning the interface between academia and industry.

The conclusions drawn from these case studies of innovation at Burroughs Wellcome and Syntex are equally applicable to the modern era of rational drug discovery. The collection of case studies of drug development spans nearly five decades, from the early 1940s to the late 1980s. Furthermore, evidence suggests that Calne served as a clinical champion in the recent development of cyclosporin. Although the tools and techniques available for drug discovery and development have improved over time, these case studies illustrate that the presence of a clinician-champion continues to be an important determinant of the time required for drug development.

Pharmaceutical innovation has become a complex, interdisciplinary process, typically involving a variety of activities and researchers from numerous scientific domains. As this process becomes increasingly complex and interdisciplinary, understanding the factors that limit knowledge sharing between basic scientists and clinicians will become more and more important. Burroughs Wellcome was able to achieve a relative competitive advantage because its biochemistry department permitted open exchange of data and external clinician-champions drove the development process. Pharmaceutical companies that do not explicitly acknowledge and examine the roles of clinician-champions in drug development and the information management infrastructure that surrounds this process are susceptible to the same delays experienced in the cases of acyclovir and allopurinol. Although ignoring the importance of clinician-champions does not necessarily prevent successful development, as these cases show, it can be a significant impediment in situations in which time to market is crucial. Similarly, modern pharmaceutical companies must examine the information channels through which data are shared between basic and clinical researchers. Information management systems that facilitate the transfer of enabling basic scientific knowledge to critically important clinicians may greatly speed the development process.

If indeed the presence of a clinical champion promotes more rapid drug development, several questions remain concerning the nature of this role. Can champions be deliberately sought, trained, and placed in positions to promote drug development? In this preliminary study we have identified some of the characteristics of successful clinician-champions. Clinical investigators who are well-suited for performing this role clearly exist. Most of the functions of clinician-champions appear to be ingrained in the traditional training of clinician-investigators. Their roles in promoting and detecting fundamental investigation and as drivers of the technology transfer process, however, distinguish the clinician-champion from other clinical investigators. These roles are rarely encouraged or taught in traditional academic settings. Therefore, although many clinician-investigators are positioned to serve as clinician-champions, most do not. In order for this role to become more commonplace, the functions of clinician-champions must be appreciated and encouraged, and the skills necessary to transfer enabling scientific observation to clinical applications must be taught.

Training in such modern methods of information management as biomedical informatics may help prepare clinicians to integrate fundamental scientific data with understanding of disease pathogenesis. As the amount of relevant data increases and research methods become more interdisciplinary and computer based, this type of training will become essential in teaching clinicians how to recognize opportunities to apply existing information to new drug development and how to act on them. From the case studies examined, it appears that clinician-champions are most likely to arise and function in academic settings. Medicine must make efforts to find, recount, characterize, and emulate individuals who perform these roles and train clinician-investigators to function effectively at the interface between basic and clinical research. For individuals to be deliberately placed in these roles, close collaboration is needed between academia and the industrial setting in which drugs are most likely to be developed. Thus there must be open communication channels that promote two-way sharing of data across this interface. Finally, in order for clinical champions to be widely effective, this role must be legitimized in the eyes of both the industrial and academic communities. Further examination of this role may also suggest means for improving the organizational structure and the information filters involved in the drug development process.

Our study offers insight into the management of information in drug development. The development of drugs of various chemotherapeutic classes from research on purine analogs has demonstrated several principles about the nature of medical technology transfer and raised a series of new questions. First, this study demonstrates the development of a circumscribed enabling observation leading to the development of drugs for use in cases of leukemia, immunosuppression, transplantation, gout, and viral disease. However, it appears that the identification of targets and biochemical pathways is not sufficient to promote rapid drug development. The time required to develop the purine analog drugs studied illustrates that the rate at which technological innovations occurs is not entirely or even substantially dictated by the scientific knowledge base or by traditionally cited methods for setting research priorities. These variations in the rate of innovation are related to one important and vastly underappreciated factor: the presence or absence of a clinician-champion who is charged with managing the disease and therefore is likely outside of the industry.

The focus on clinical management is the key feature that distinguishes clinician-champions from physicians who work within the pharmaceutical industry. In these case studies this focus appears to drive certain clinicians to make connections between basic scientific findings and disease management and pull drugs into development. Physicians inside of pharmaceutical companies are not continually presented with patients' problems, which, for the clinical champions in our study, provide the impetus for promoting rapid drug development. Clinician-champions appear to accelerate pharmaceutical innovation by defining and vigorously exploring novel targets for compounds with known apparent mechanisms of action. But clinical investigators do not routinely function in this capacity. To reduce drug development time, pharmaceutical companies should attract and develop relationships with clinicians who demonstrate vision for improving the management of disease. Future research on the process of integrating basic and clinical information in drug development and on the role of the clinician-champion may elucidate criteria for training researchers to perform this role.

References

T. J. Allen. "Managing the Flow of Technology." *Technology Transfer and the Dissemination of Technological Information within R&D.* Cambridge, Mass.: MIT Press, 1977.

J. M. Armstrong; P. E. Hicks. "An Industrial Approach to Drug Discovery: A Pharmacologist's Eyeview." *Clinical and Experimental Pharmacology and Physiology* 19 (1992), 51–56.

W. T. Ashton; J. D. Karkas; A. K. Field; R. L. Tolman. "Activation by Thymidine Kinase and Potent Antiherpetic Activity of 2'-nor-2'-deoxyguanosine (2'NDG)." *Biochemical and Biophysical Research Communications* 108:4 (1982), 1716–1721.

J. F. Borel. Letters to D. J. White. 19 January, 3 February, and 15 June 1977. David White's personal collection.

J. F. Borel; H. Friedli. Letter to D. J. White. 6 May and 15 November 1976. David White's personal collection.

J. H. Burchenal. "The Friendly Antagonists—Pyrimidine Derivatives from Bench to Bedside." In: *Design and Achievements in Chemotherapy: A Symposium in Honor of George H. Hitchings.* Research Triangle Park, N.C.: Science & Medicine Publishing, 1976: 18–26.

J. H. Burchenal; A. Bendich; G. B. Brown, et al. "Preliminary Studies on the Effect of 2,6-diaminopurine on Transplanted Mouse Leukemia. *Cancer* 2 (1949), 119–120.

J. H. Burchenal; S. K. Goethchius; C. C. Stock; G. H. Hitchings. "Diamino Dichlorophenyl Pyrimidines in Mouse Leukemias." *Cancer Research* 12 (1952), 251.

J. H. Burchenal; D. A. Karnofsky; E. M. Kingsley-Pillers, et al. "The Effects of the Folic Acid Antagonists and 2,6-diaminopurine on Neoplastic Disease: With Special Reference to Acute Leukemia." *Cancer* 4 (1951), 519–569.

J. H. Burchenal; M. L. Murphy; R. R. Ellison, et al. "Clinical Evaluation of a New Antimetabolite, 6-Mercaptopurine, in the Treatment of Leukemia and Allied Diseases." *Blood: The Journal of Hematology* 8:11 (1953), 965–999.

R. Y. Calne. "Cyclosporin." *Nephron.* 1980;26(2)57-63.

———. "The Development of Immunosuppressive Therapy." *Transplantation Proceedings* 13:1 Suppl 1 (1981), 44–49.

———. "Immunosuppression for Organ Grafting—Observations on Cyclosporin A." *Immunological Reviews* 46 (1979), 113–124.

———. "Pharmacological Immunosuppression in Clinical Organ Grafting. Observations on Four Agents: Cyclosporin A, Asta 5122 (Cytimun), Lambda Carrageenan and Promethazine Hydrochloride." *Clinical and Experimental Immunology* 35:1 (1979), 1–9.

———. "The Rejection of Renal Homografts: Inhibition in Dogs by 6-Mercaptopurine." *Lancet* 1 (1960), 417–418.

R. Y. Calne; G. P. J. Alexandre; J. E. Murray. "A Study of the Effects of Drugs in Prolonging Survival of Homologous Renal Transplants in Dogs." *Annals of the New York Academy of Sciences* 99 (1962), 743–761.

R. Y. Calne; K. Rolles; D. J. White, et al. "Cyclosporin A in Organ Transplantation." *Advances in Nephrology from the Necker Hospital* 10 (1981), 335–347.

R. Y. Calne; K. Rolles; D. J. White, et al. "Cyclosporin A Initially as the Only Immunosuppressant in 34 Recipients of Cadaveric Organs: 32 Kidneys, 2 Pancreases, and 2 Livers." *Lancet* 2:8151 (1979), 1033–1036.

R. Y. Calne; D. J. White. "Cyclosporin A: A Powerful Immunosuppressant in Dogs with Renal Allografts." *IRC Medical Science: Surgery and Transplantation.* 5 (1977). 595.

———. "The Use of Cyclosporin A in Clinical Organ Grafting." *Annals of Surgery* 196:3 (1982), 330–337.

R. Y. Calne; D. J. White; D. B. Evans, et al. "Cyclosporin A in Cadaveric Organ Transplantation." *BMJ* 282:6268 (1981), 934–936.

R. Y. Calne; D. J. White; B. D. Pentlow, et al. "Cyclosporin A: Preliminary Observations in Dogs with Pancreatic Duodenal Allografts and Patients with Cadaveric Renal Transplants." *Transplantation Proceedings* 11:1 (1979), 860–864.

R. Y. Calne; D. J. White; K. Rolles; D. P. Smith; B. M. Herbertson. "Prolonged Survival of Pig Orthotopic Heart Grafts Treated with Cyclosporin A." *Lancet* 1:8075 (1978), 1183–1185.
R. Y. Calne; D. J. White; S. Thiru, et al. "Cyclosporin A in Patients Receiving Renal Allografts from Cadaver Donors." *Lancet* 2:8104-5 (1978), 1323–1327.
R. Y. Calne; D. J. White; S. Thiru, et al. "Cyclosporin A in Clinical Kidney Grafting from Cadaver Donors." *Proceedings of the European Dialysis and Transplant Association* 16 (1979), 305–309.
A. K. Chakrabati. "The Role of Champion in Product Innovation." *California Management Review* 17:2 (1974), 58–62.
A. K. Chakrabati; J. Hauschildt. "The Division of Labour in Innovation Management." *R&D Management* 19:2 (1989), 161–171.
K. Clark; T. Fijimoto. *Product Development in the World Automobile Industry.* Boston: Harvard Business School Press, 1991.
D. A. Clarke; F. S. Philips; S. S. Sternberg, et al. "6-Mercaptopurine: Effects in Mouse Sarcoma 180 and in Normal Animals. *Cancer Research* 13 (1953), 593–604.
F. Damanpour; W. M. Evan. "Organizational Innovation and Preference: The Problem of Organizational Lag." *Administrative Science Quarterly* 29:3 (1984), 392–409.
A. deVries; M. Frank; U. A. Liberman; O. Sperling. "Allopurinol in the Prophylaxis of Uric Acid Stones." *Annals of the Rheumatic Diseases* 25 (1966), 691–693.
J. A. DiMasi. "Risks, Regulation, and Rewards in New Drug Development in the United States." *Regulatory Toxicology and Pharmacology* 19:2 (1994), 228–235.
J. A. DiMasi; N. R. Bryant; L. Lasagna. "New Drug Development in the United States from 1963 to 1990." *Clinical Pharmacology and Therapeutics* 50:5 Pt 1 (1991), 471–486.
J. A. DiMasi; R. W. Hansen; H. G. Grabowski; L. Lasagna. "Cost of Innovation in the Pharmaceutical Industry" [see comments]. *Journal of Health Economics* 10:2 (1991), 107–142.
J. A. DiMasi; M. A. Seibring; L. Lasagna. "New Drug Development in the United States from 1963 to 1992." *Clinical Pharmacology and Therapeutics* 55:6 (1994), 609–622.
G. B. Elion. "Historical Perspectives." In *Design and Achievements in Chemotherapy: A Symposium in Honor of George H. Hitchings.* Research Triangle Park, N.C.: Science & Medicine Publishing, 1976, 4–6.
G. B. Elion. Personal communication. 1994.
G. B. Elion; S. W. Callahan; G. H. Hitchings; R. W. Rundles; J. Lazlo. "Experimental, Clinical, and Metabolic Studies of Thiopurines." *Cancer Chemotherapy Reports* 16 (1962), 197–202.
G. B. Elion; S. W. Callahan; H. Nathan, et al. "Potentiation by Inhibition of Drug Degradation: 6-Substituted Purines and Xanthine Oxidase." *Cancer Chemotherapy Reports* 8 (1963), 47–52.
G. B. Elion; S. W. Callahan; R. W. Rundles; G. H. Hitchings. "Relationship between Metabolic Fates and Antitumor Activities of Thiopurines." *Cancer Research* 23 (1963), 1207.
G. B. Elion; P. A. Furman; J. A. Fyfe; et al. "Selectivity of Action of an Antiherpetic Agent, 9-(2-Hydroxyethoxymethyl) Guanine." *Proceedings of the National Academy of Sciences of the United States of America* 74:12 (1977), 5716–5720.
G. B. Elion; G. H. Hitchings; H. VanderWerff. "Antagonists of Nucleic Acid Derivatives: VI. Purines." *Journal of Biological Chemistry* 192 (1951), 505–518.
G. B. Elion; J. L. Rideout; P. de Miranda; P. Collins; D. J. Bauer. "Biological Activities of Some Purine Arabinosides." *Annals of the New York Academy of Sciences* 255 (1975), 468–480.
G. B. Elion; S. Singer; G. H. Hitchings. "The Purine Metabolism of a 6-Mercaptopurine Resistant *Lactobacillus casei.*" *Journal of Biological Chemistry* 204 (1953), 35–41.
R. R. Ellison; J. F. Holland; M. Weil, et al. "Arabinosylcytosine: A Useful Agent in the Treatment of Acute Leukemia in Adults." *Blood* 32 (1968), 507.
P. Fildes; M. B. Camb. "A Rational Approach to Research in Chemotherapy." *Lancet* 1 (1940), 955–957.
W. A. Fischer; B. Rosen. "The Search for the Latent Information Star." *R&D Management* 12:2 (1982), 61–66.

D. Gerwin. "Control and Evaluation in the Innovation Process: The Case of Flexible Manufacturing Systems." *IEEE Transactions in Engineering Management* 28:3 (1981), 62–70.

S. Globe; G. W. Levy; C. M. Schwartz. "Key Factors and Events in the Innovation Process." *Research Management* Jul (1973), 8–15.

C. Heidelberg; N. K. Chanduri; P. Danneberg, et al. "Fluorinated Pyrimidines, a New Class of Tumor-Inhibitory Compounds." *Nature* 179 (1957), 663.

R. Henderson. "Managing Innovation in the Information Age." *Harvard Business Review* 72:1 (1994), 100–105.

R. M. Henderson; I. Cockburn. "Measuring Competence? Exploring Firm Effects in Pharmaceutical Research." *Strategic Management Journal* 15 (Special issue) (1994), 63–84.

G. Hitchings. "Nobel Lecture in Physiology or Medicine—1988. Selective Inhibitors of Dihydrofolate Reductase." *In Vitro Cellular and Developmental Biology* 25:4 (1989), 303–310.

G. H. Hitchings; G. B. Elion; E. A. Falco; P. B. Russell; H. VanderWerff. "Studies on Analogs of Purines and Pyrimidines." *Annals of the New York Academy of Sciences* 52 (1950), 1318–1335.

G. H. Hitchings; E. A. Falco. "The Identification of Guanine in Extracts of *Girella nigricans*. The Specificity of Guanase." *Proceedings of the National Academy of Sciences of the United States of America* 30 (1944), 294–297.

G. H. Hitchings; E. A. Falco; M. B. Sherwood. "The Effect of Pyrimidine on the Growth of *Lactobacillus casei*." *Science* 102:2645 (1945), 251–252.

G. H. Hitchings; C. P. Rhoads. "6-Mercaptopurine." *Annals of the New York Academy of Sciences* 60 (1954), 183.

K. Imai; I. Nonaka; H. Takeuchi, editors. "Managing the Product Development Process: How Japanese Companies Learn and Unlearn." In *The Uneasy Alliance: Managing the Productivity-Technology Dilemma*. Boston: Harvard Business School Press, 1985.

D. A. Jabs; C. Newman; S. De Bustros; B. F. Polk. "Treatment of Cytomegalovirus Retinitis with Ganciclovir." *Ophthalmology* 94:7 (1987), 824–830.

I. D. Jenkins; J. P. Verheyden; J. G. Moffatt. "4'-Substituted Nucleosides. 2. Synthesis of the Nucleoside Antibiotic Nucleocidin." *Journal of the American Chemical Society* 98:11 (1976), 3346–3357.

H. E. Kaufman; C. Heidelberger. "Therapeutic Antiviral Action of 5-Trifluoromethyl-2'-Deoxyuridine in Herpes Simplex Keratitis." *Science* 145 (1964), 585.

H. E. Kaufman; A. B. Nesburn; E. D. Maloney. "IUdR Therapy for Herpes Simplex." *Archives of Ophthalmology* 67 (1962), 583.

W. N. Kelley; M. A. Randolph. "Executive Summary." In *Careers in Clinical Research: Obstacles and Opportunities*. Edited by W. N. Kelley and M. A. Randolph. Washington, D.C.: National Academy Press, 1994, 1–21.

A. J. Kostakis; D. J. White; R. Y. Calne. "Prolongation of Rat Heart Allograft Survival by Cyclosporin A." *IRC Medical Science: Surgery and Transplantation* 5 (1977), 280.

———. "Toxic Effects in the Use of Cyclosporin A in Alcoholic Solution as an Immunosuppressant of Rat Heart Allografts." *IRC Medical Science: Surgery and Transplantation* 5 (1977), 243.

S. Lee; M. E. Treacy. "Information Technology Impacts on Innovation." *R&D Management* 18:3 (1988), 257–271.

Y. Lis; S. R. Walker. "Novel Medicines Marketed in the UK (1960–87)." *Brittish Journal of Clinical Pharmacology* 28 (1989), 333–343.

D. C. Lorz; G. H. Hitchings. "Specificity of Xanthine Oxidase." *American Society of Biological Chemists* 9 (1950), 197.

D. E. Loveday. "Factors Affecting the Management of Interdisciplinary Research in the Pharmaceutical Industry." *R&D Management* 14:2 (1984), 93.

J. C. Martin; C. A. Dvorak; D. F. Smee; T. R. Matthews; J. P. Verheyden. 9-[(1,3-Dihydroxy-2-propoxy)methyl]guanine: A New Potent and Selective Antiherpes Agent." *Journal of Medicinal Chemistry* 26:5 (1983), 759–761.

P. McMaster; A. Procyshyn; R. Y. Calne; et al. "Prolongation of Canine Pancreas Allografts with Cyclosporin A." *Transplant Proceedings* 12:2 (1980), 275–277.

———. "Prolongation of Canine Pancreas Allograft Survival with Cyclosporin A: Preliminary Report." *BMJ* 280:6212 (1980), 444–445.

L. B. Mohr. "Determinants of Innovation in Organization." *American Political Science Review* 63:4 (1969), 111–126.

M. L. Murphy; R. R. Ellison; D. A. Karnofsky; J. H. Burchenal. "Clinical Effects of the Dichloro and Monochlorophenyl Analogue of Diamino Pyrimidine: Antagonists of Folic Acid." *Journal of Clinical Investigation* 33 (1954), 1388.

J. E. Murray; J. P. Merrill; J. H. Harrison; R. E. Wilson; G. J. Dammin. "Prolonged Survival of Human-Kidney Homografts by Immunosuppressive Drug Therapy." *New England Journal of Medicine* 268:24 (1963), 1315–1323.

R. A. Pearson. "Innovation Strategy." *Technovation* 10:3 (1990), 185–192.

F. S. Philips; S. S. Sternberg; L. Hamilton; D. A. Clarke. "The Toxic Effects of 6-Mercaptopurine and Related Compounds." *Annals of the New York Academy of Sciences* 60 (1954), 283–295.

F. S. Philips; J. B. Thiersch. "Actions of 2,6-Diaminopurine in Mice, Rats, and Dogs." *Proceedings of the Society for Experimental Biology and Medicine* 72 (1949), 401–408.

J. L. Pierce; M. Delbeq. "Organization Structure, Individual Attitudes, and Innovation." *Academy of Management Review* 2:1 (1977), 27–37.

R. A. Prentis; Y. Lis; S. R. Walker. "Pharmaceutical Innovation by the Seven UK-owned Pharmaceutical Companies (1964–1985)." *British Journal of Clinical Pharmacology* 25 (1988), 387–396.

N. Rosenberg. "Critical Issues in Science Policy Research." *Science and Public Policy* 18:6 (1991), 335–346.

N. Rosenberg; A. C. Gelijns; H. Dawkins, editors. *Sources of Medical Technology: Universities and Industry.* Medical Innovations at the Crossroads, vol. 5. Washington, D.C.: National Academy Press, 1995.

R. W. Rundles. "The Development of Allopurinol for the Treatment of Gout and Hyperuricemia." In *Design and Achievements in Chemotherapy: A Symposium in Honor of George H. Hitchings.* Research Triangle Park, N.C.: Science & Medicine Publishing, 1976, 27–34.

R. W. Rundles; E. N. Metz; H. R. Silberman. "Allopurinol in the Treatment of Gout." *Annals of Internal Medicine* 64:2 (1966), 229–257.

R. W. Rundles; J. B. Wyngaarden; G. H. Hitchings; H. R. Silberman. "Effects of a Xanthine Oxidase Inhibitor on Thiopurine Metabolism, Hyperuricemia, and Gout." *Transactions of the Association of American Physicians* 76 (1963), 126.

F. M. J. Schabel. "The Antiviral Activity of 9-Beta-D-arabinofuranosyladenine (ARA-A)." *Chemotherapy* 13:6 (1968), 321–338.

L. P. Schacter; C. Anderson; R. M. Canetta, et al. "Drug Discovery and Development in the Pharmaceutical Industry." *Seminars in Oncology* 19:6 (1992), 613–621.

H. J. Schaeffer; L. Beauchamp; P. de Miranda, et al. "9-(2-Hydroxyethoxymethyl) Guanine Activity against Viruses of the Herpes Group." *Nature* 272:5654 (1978), 583–585.

S. P. Schinke; M. A. Orlandi. "Technology Transfer." *NIDA Research Monograph* 107 (1991), 248–263.

R. S. Schwartz. "Perspectives on Immunosuppression." In *Design and Achievements in Chemotherapy : A Symposium in Honor of George H. Hitchings.* Research Triangle Park, N.C.: Science & Medicine Publishing, 1976, 39–41.

R. Schwartz; W. Dameshek. "Drug-induced Immunological Tolerance." *Nature* 183 (1959), 1682–1683.

R. Schwartz; J. Stack; W. Dameshek. "Effect of 6-Mercaptopurine on Antibody Production." *Proceedings of the Society for Experimental Biology and Medicine* 99 (1958), 164–167.

P. J. Selby; R. L. Powles; B. Janeson, et al. "Parenteral Acyclovir Therapy for Herpesevirus Infections in Man." *Lancet* 2:8155 (1979), 1267–1270.

R. K. Shaw; R. N. Shulman; J. D. Davidson; D. P. Rall; E. I. Frei. "Studies with the Experimental Antitumor Agent 4-Aminopyrazolo(3,4-d) Pyrimidine." *Cancer* 13 (1960), 482.

H. E. Skipper. "Some Thoughts on Cancer Chemotherapy." In *Design and Achievements in Chemotherapy : A Symposium in Honor of George H. Hitchings.* Research Triangle Park, N.C.: Science & Medicine Publishing, 1976, 12–17.

D. F. Smee; J. C. Martin; J. P. Verheyden; T. R. Matthews. "Anti-herpesvirus Activity of the Acyclic Nucleoside 9-(1,3-Dihydroxy-2-propoxymethyl)guanine." *Antimicrobial Agents and Chemotherapy* 23:5 (1983), 676–682.

K. O. Smith; K. S. Galloway; W. L. Kennell; K. K. Ogilvie; B. K. Radatus. "A New Nucleoside Analog, 9-[[2-Hydroxy-1-(hydroxymethyl)ethoxyl]methyl]guanine, Highly Active In Vitro against Herpes Simplex Virus Types 1 and 2." *Antimicrobial Agents and Chemotherapy* 22:1 (1982), 55–61.

W. F. Souder. "Effectiveness of Nominal and Interacting Group Decision Processes for Interacting R&D and Marketing." *Management Science* 23:6 (1977), 595–605.

R. Talley; V. K. Vaitkevicius. "Megaloblastosis Produced by Cytosine Antagonist, 1-ß-D-Arabinofuranocylcytosine." *Blood* 21 (1963), 352.

S. O. Thier. "The Social and Scientific Value of Biomedical Research." *Journal of the Royal College of Physicians of London* 28:6 (1994), 541–543.

R. L. Thompson; M. L. Price; S. A. Menton; G. B. Elion; G. H. Hitchings. "Effects of Purine Derivatives and Analogs on Multiplication of the Vaccinia Virus." *Journal of Immunology* 65 (1950), 529–534.

R. L. Thompson; M. L. Wilkin; G. B. Elion, et al. "The Effects of Antagonists on the Multiplication of Vaccinia Virus in Vitro." *Science* 110 (1949), 454.

J. P. Verheyden. "Ganciclovir." In *Chronicles of Drug Discovery.* Vol. 3. Edited by D. Lednicer. New York: Wiley, 1993, 2–27.

J. P. Verheyden. Personal communication. 1994.

J. P. Verheyden; I. D. Jenkins; G. R. Owen, et al. "Synthesis of Certain 4'-Substituted Nucleosides." *Annals of the New York Academy of Sciences* 255 (1975), 151–165.

J. P. Verheyden; J. G. Moffatt. "Halo Sugar Nucleosides. I. Iodination of the Primary Hydroxyl Groups of Nucleosides with Methyltriphenoxyphosphonium Iodide." *Journal of Organic Chemistry* 35:7 (1970), 2319–2326.

———. "Halo Sugar Nucleosides. II. Iodination of Secondary Hydroxyl Groups of Nucleosides with Methyltriphenoxyphosphonium Iodide." *Journal of Organic Chemistry* 35:9 (1970), 2868–2877.

———. "4'-Substituted Nucleosides. I. Synthesis of 4'-Methoxyuridine and Related Compounds." *Journal of the American Chemical Society* 97:15 (1975), 4386–4395.

———. "The Synthesis of a 4',5'-Unsaturated Nucleoside. *Journal of the American Chemical Society* 88:23 (1966), 5684–5685.

D. J. White. Letters to J. F. Borel. 27 April, 21 May, 3 November, and 24 November 1976; 27 January, 9 June, 5 September, 14 September, 27 September, and 23 December, 1977. David White's personal collection.

D. J. White; R. Y. Calne; A. Plumb. "Mode of Action of Cyclosporin A: A New Immunosuppressive Agent." *Transplantation Proceedings* 11:1 (1979), 855–859.

F. R. White. "4-Aminopyrazolo(3,4-d) Pyrimidine and Three Derivatives." *Cancer Chemotherapy Reports* 3 (1959), 26.

D. D. Woods. "The Relation of p-Aminobenzoic Acid to the Mechanism of the Action of Sulphanilamide. *British Journal of Experimental Pathology* 21 (1940), 74–90.

T. F. Yü; A. B. Gutman. "Effect of Allopurinol (4-Hydroxypyrazolo-(3,4-d)pyrimidine) on Serum and Urinary Uric Acid in Primary and Secondary Gout. *American Journal of Medicine* 37 (1964), 885–898.

Glossary

acetylcholine. A neurotransmitter in autonomic and central nervous systems.
acetylcholinesterase. An enzyme that catalyzes the formation of choline and acetic acid from acetylcholine.
achromatic microscope. Combines lenses made from different materials to prevent light from decomposing into its component colors.
acidosis. A pathologic condition resulting from accumulation of acid in the blood and body tissues.
adrenergic. Activated by or secreting epinephrine or norepinephrine, used usually in reference to nerve endings or receptors.
adrenoceptor. Adrenergic receptor.
adrenocortical. Pertaining to the adrenal cortex, the outer layer of adrenal gland.
affinity. Attraction to a specific element, organ, or structure.
agonist. A drug that has affinity for or mimics the physiologic activity of naturally occurring substances at the receptors.
agranulocytosis. A symptom complex characterized by a decrease in the number of nongranular leukocytes in the blood.
alkaloid. Nitrogen-containing basic substance usually found in plants.
anabolic. Pertaining to anabolism.
anabolism. Constructive metabolism, which forms body constituents.
anaerobic. Growing, living, or occurring in the absence of molecular oxygen.
analgesic. Drug that relieves or prevents pain.
anaphylaxis. A manifestation of immediate hypersensitivity to a specific antigen, characterized by skin eruption and angioedema and followed by vascular collapse and shock.
anastomosis. A connection between two vessels.
androgen. Any substance that promotes masculinization.
anemia. Condition characterized by a reduced number of erythrocytes or a reduced quantity of hemoglobin in each erythrocyte.

anemia, aplastic. A bone marrow disorder characterized by a reduction of stem cells in the bone marrow, which are precursors of blood cells.
anesthetic. A substance that produces loss of sensation.
angina. Spasmodic, choking, or suffocating pain.
angina pectoris. Paroxysmal thoracic pain, caused by a lack of oxygen supply to the heart muscle.
angioedema. Localized accumulation of fluid in the deeper layers of the skin.
antagonist. A substance that tends to nullify the action of another.
anthrax. An infectious bacterial disease, caused by *Bacillus anthracis* and transmitted mostly by contact with infected animals or their discharges.
antibiosis. An association of two organisms, which is detrimental to one of them.
antibody. An immunoglobulin that interacts only with the antigen that induced its synthesis.
anticoagulant. A substance that prevents blood from clotting.
antigen. Any substance that is capable of inducing a specific immune response and of reacting with the products of this response; antigens are usually foreign proteins.
antineoplastic. A substance that inhibits abnormal growth of tissue.
antiprotozoal. Drug that destroys protozoa or interferes with their growth or multiplication.
antipyretic. Drug that reduces fever.
antiseptic. A substance capable of preventing the growth and development of microorganisms without necessarily destroying them.
antitoxin. Antibody against a toxin or a purified serum from animals immunized with a toxin.
aplasia. Lack of development of an organ or tissue.
aplastic. Pertaining to aplasia.
arteriosclerosis. A group of diseases characterized by thickening and loss of elasticity of arterial wall.
aseptic. Free from infection or infectious material.
ataxia. Failure of muscular coordination.
atherosclerosis. A form of arteriosclerosis in which plaques containing cholesterol are formed in the walls of large and medium-sized arteries.
atrial fibrillation. An arrhythmia in which areas of atrial myocardium contract in a chaotic pattern.
attention-deficit disorder. Mental disorder of childhood characterized by disruptive behavior, inability to follow instructions, excessive talking, and difficulty remaining seated.
autonomic. Self-controlling, virtually independent; used in reference to the part of the nervous system that is not under the control of the individual (animal or human).

Bacillus. A genus of rod-shaped bacteria, consisting of forty-eight different species, three of which are known to cause disease.
bactericidal. Capable of destroying bacteria.
benign. Non-malignant, not recurrent, favorable for recovery.
beri-beri. A disease caused by thiamine (vitamin B_1) deficiency, characterized by polyneuritis, edema, and cardiac pathology.
botulism. Food poisoning caused by *Clostridium botulinum.*
brucellosis. A generalized infection of humans caused by bacteria of the *Brucella* genus, usually transmitted by domestic animals or their products; it is characterized by fever, sweating, and malaise.

carbohydrates. Group of neutral compounds composed of carbon, hydrogen, and oxygen; it includes sugars, dextran, and glycogen.
carbonic anhydrase. An enzyme present in blood and certain tissues that catalyzes the formation of carbonic acid from carbon dioxide and water.
catalyst. A substance capable of speeding chemical reactions.
cerebellum. A posterior (rear) portion of the brain concerned with muscle coordination and maintenance of equilibrium.
channel. A protein usually located in the cell membrane that allows passage of ions in or out of the cell.
Chlamydia. A genus of round, gram-negative bacteria that multiply only within a host cell, cause infections in domestic animals, and may affect humans.
cholera. Acute infectious disease caused by *Vibrio cholerae*, characterized by massive loss of fluid leading to acidosis and shock.
cholestasis. Suppression of bile flow.
cholinergic. Stimulated, activated, or transmitted by choline or acetylcholine, applied to nerve fibers that liberate acetylcholine, or to receptors activated by acetylcholine.
cinchona alkaloids. Derived from the bark of South American trees of the same name; known cinchona alkaloids include quinidine, quinine, cinchonine, and cinchonidine.
clot. A semisolidified mass of blood or lymph that forms in blood vessels or in a container outside the organism; it should be differentiated from a thrombus, which forms only in an organism and involves vessel wall elements.
clotting, or coagulation, factor. Substance in blood that is essential for the formation of blood clots.
coking. Producing coke (residue from carbonized coal used as a fuel).
colitis. Inflammation of the colon (part of the large intestines).
conjugate. A product of the joining together of two compounds.
conjunctivitis. Inflammation of the conjunctiva, a membrane that covers the eyelids and the exposed part of the sclera.
corticosteroids. Steroids produced by the adrenal cortex.

Digitalis. An herb genus; foxglove. The leaves of *Digitalis purpurea* or *D. lanata* contain glycosides used in the treatment of heart failure.

disaccharide. A substance belonging to the class of sugars; it consists of two monosaccharides, e.g., sucrose, lactose, or maltose.

distillation, fractional. Successive separation of volatile substances in accordance with their volatility.

DNA. Deoxyribonucleic acid; constitutes the primary genetic material of all cellular organisms as well as viruses.

dopa. 3,4-dihydroxyphenylalanine; an amino acid, precursor of epinephrine, norepinephrine, dopamine, and melanin.

dopa decarboxylase. An enzyme that catalyzes the removal of carbon dioxide from dopa.

dyskinesia. Distortional involuntary movements.

dysmenorrhea. Painful menstruation.

efficacy. Ability of a drug to produce maximal desirable therapeutic effect; it is independent from potency, which expresses the amount of the drug required to produce the desired effect.

embolism. Blockage of an artery by a clot formed elsewhere.

encephalitis. Inflammation of the brain.

encephalomyelitis. Inflammation of the brain and spinal cord.

endogenous. Originating within the organism.

endometriosis. A condition in which tissue containing uterine membrane lining occurs aberrantly in the pelvic cavity or other organs.

endorphin. Neuropeptide present in the brain, which binds to opioid receptors and has potent analgesic activity.

endothelium. The layer of cells that lines the blood vessels or cavities of the heart and other organs.

enkephalin. Pentapeptide present in the brain, which binds to opioid receptors, functions as a neuromodulator, and plays a role in pain perception.

fibrillation. Chaotic contractions of fibrils; the term used to describe specific cardiac arrhythmias.

fibrin. An insoluble protein, which forms the essential portion of a blood clot.

fibrinolysis. Dissolution of fibrin by enzymatic action.

generic. Term used for drug names not protected by patents or trademarks.

glaucoma. A group of eye diseases, characterized by elevation of intraocular pressure.

globulins. A class of plasma proteins that are insoluble in water.

glomerulonephritis. Inflammation of glomeruli in the kidneys, often caused by streptococci.

glomerulus. Cluster of blood vessels and nerves present in the kidneys.

glucocorticoids. Steroids from the adrenal cortex that regulate the metabolism of carbohydrates, lipids, and proteins.
glycosides. Compounds that contain a sugar moiety.
gout. A disorder of purine metabolism, characterized by deposition of uric acid crystals in the joints.
gram-negative. Bacteria that lose Gram's stain in the presence of alcohol.
gram-positive. Bacteria that retain Gram's stain in the presence of alcohol.
granulocytopenia. A condition characterized by a reduced number of granular leukocytes in the blood.
guanylate cyclase. An enzyme that catalyzes the formation of cyclic guanosine phosphates.

hemolysis. Disruption of the red cell membrane.
herpes zoster. Acute infectious disease caused by the chickenpox virus; it is characterized by inflammation of the affected nerve and vesicles on the skin over the affected nerve.
hexose. A monosaccharide containing six carbon atoms (e.g., glucose or fructose).
hirsutism. Abnormal growth of hair.
Hodgkin's disease. A form of malignant lymphoma characterized by progressive enlargement of lymph nodes and spleen.
hydrolyze. To split a compound into fragments by adding water.
hyperkalemia. Abnormally high potassium concentration in the blood.
hyperlipemia. Abnormally high concentrations of lipids in the blood.
hyperlipoproteinemia. An excess of lipoproteins in the blood.
hypokalemia. Abnormally low potassium concentration in the blood.
hypolipemia. An abnormally low amount of all lipids in the blood.

immunocompromised. Having attenuated immune response because of disease, malnutrition, drug use, or irradiation.
immunodeficiency. A disorder characterized by a deficient immune response.
immunogenicity. Capability of a substance to provoke an immune response.
infarction. An area of local tissue damage caused by obstruction of blood flow.

lactone. A cyclic organic compound in which the chain is closed by ester formation between a carboxyl and a hydroxyl group in the same molecule.
Legionnaires' disease. A highly fatal acute bacterial infection, first discovered in 1976 during an American Legion convention in Philadelphia; it is characterized by high fever, gastrointestinal pain, and pneumonia.
leishmaniasis. A disease caused by a protozoan of the genus *Leishmania*; it is subdivided into three types—cutaneous, mucocutaneous, or visceral—according to the location of the lesions.
leprosy. Chronic infectious disease caused by *Mycobacterium leprae*; it is

characterized by lesions in the skin, mucous membranes, bones, nerves, and internal organs.

leukemia. A progressive malignant disease of blood-forming organs, characterized by distorted proliferation of leukocytes and their precursors in blood and bone marrow.

ligand. A molecule that binds to another molecule, usually used to describe binding of an antigen to an antibody, of a hormone, or of a drug or a neurotransmitter to a receptor.

lipoprotein. Lipid-protein complex, in which lipids are transported in the blood.

lymphocytes. Mononuclear leukocytes that are immunologically competent.

lymphogranuloma. Nodular swelling of a lymph node; Hodgkin's disease is a malignant lymphogranuloma.

lymphoma. Malignant tumor of lymphoid tissue.

lyophilize. To create a stable preparation of a biological material by rapid freezing and dehydration under high vacuum.

madder. A plant with small yellowish flowers, or a dye obtained from the root of the madder plant.

meninges. Membranes that surround the brain and spinal cord.

meningitis. A bacterial or viral infection of meninges.

meningococci. Organisms of the species *Neisseria meningitidis.*

meningoencephalitis. Inflammation of brain and brain membranes.

methemoglobin. An oxidation product of hemoglobin.

microtubules. Tubular structures found in the cytoplasm of most cells; they are involved in the maintenance of cell shape, mitosis, and movements of various cellular components.

mitosis. Division of cell nuclei.

monoamine oxidase (MAO). An enzyme that catalyzes the destruction of monoamines (e.g., epinephrine, norepinephrine, serotonin).

mucous. Pertaining to, resembling, or relating to mucus, a product of gland secretion covering body membranes.

muscarinic. Denotes the effects of drugs on receptors that can be activated by muscarine, an alkaloid from *Amanita muscaria*, a poisonous mushroom.

Mycobacteria. A genus of bacteria; gram-positive, acid-fast rods that include pathogenic organisms that cause tuberculosis and leprosy.

Mycoplasma. A family of nonmotile microorganisms that are intermediate between bacteria and viruses.

myocarditis. Inflammation of the muscular walls of the heart.

nanogram. A unit of weight of the metric system; equals one billionth of a gram.

necrosis. The sum of cellular changes leading to cell death, caused by enzymes.

neuritis. Inflammation of a nerve.
neurotransmitter. Any substance released on excitation from the terminal of a neuron in the central (brain) or peripheral nervous system (e.g., norepinephrine, acetylcholine, dopamine, glutamic acid, serotonin).
neutropenia. A decrease in the number of neutrophils (white blood cells with phagocytic function).
nicotinic. Denotes the effects of drugs on receptors that can be activated at low and blocked at high concentrations of nicotine.

opioid. A synthetic narcotic that has opium-like activity but is not derived from opium.
orchitis. Inflammation of the testes.
osteoarthritis. Degenerative joint disease, characterized by degeneration of cartilage and hypertrophy of bone.

pancreatitis. Inflammation of the pancreas.
pandemic. Widely epidemic.
panencephalitis. Inflammation of the brain, which affects gray and white matter simultaneously.
papillary muscles. Small muscular columns in the ventricles of the heart that are attached through tendons to cardiac valves.
parasympathetic. A division of the autonomic nervous system that connects with the spinal cord in the cranial and sacral regions.
Parkinson's disease. Progressive neurodegenerative disease, usually occurring later in life; it is characterized by resting muscular tremor, mask-like face, and slowing of voluntary movements.
parotitis. Inflammation of the parotid gland.
pathogen. Any disease-producing microorganism.
penicillinase. An enzyme that catalyzes the cleavage of penicillin.
pheochromocytoma. A benign vascular tumor of the adrenal medulla; its cardinal symptom is hypertension caused by an increased formation of epinephrine.
plaque. An elevated lesion in the skin or blood vessel lumen.
plasma. The fluid portion of the blood; unlike serum, plasma contains fibrinogen.
plasmin. An enzyme that catalyzes the hydrolysis of peptide bonds; it solubilizes fibrin clots and degrades various proteins, including fibrinogen.
plasminogen. A precursor of plasmin, present in plasma.
poliomyelitis. An acute infectious disease caused by a virus.
polysaccharide. A carbohydrate containing a large number of saccharide groups (e.g., starch).
proteolysis. Splitting of proteins by hydrolysis of the peptide bond.
Protozoa. An animal subkingdom that comprises the simplest organisms.

receptor. A molecular structure on the cell surface or in the cell that selectively binds a specific substance.
recombinant. A cell or an organism with a new combination of genes.
retroviruses. A family of viruses with a unique replication system.
Reye's syndrome. An acute disease of childhood, characterized by fatty liver and encephalitis. The cause is unknown, but viruses or toxic agents, including salicylates, have been implicated.
rickets. A disorder caused by a deficiency or malabsorption of vitamin D and defective calcium and phosphorus homeostasis; it is manifested by growth retardation, bone disease, and convulsions.
rickettsia. Intracellular parasites the size of bacteria; they are usually maintained in nature by a cycle that involves an insect vector and an animal reservoir. Rickettsial diseases include typhus, Q fever, and Rocky Mountain spotted fever.
rubella. An acute viral disease, also known as German measles.

sclerosing. Causing sclerosis, a hardening of tissue (e.g., brain or arteries) from increased formation of connective tissue.
septicemia. A systemic disease associated with the persistent presence of microorganisms or their toxins in the blood.
serum. The clear liquid portion of blood that separates from blood after formation of a clot; it does not contain fibrinogen.
soma. The body as distinguished from the mind. Used also to describe the body of a cell.
somatic. Pertaining to the body.
stereoisomers. Compounds with the same composition but a different arrangement of atoms.
streptokinase. A protein produced by hemolytic streptococci that binds plasminogen and causes its cleavage to plasmin.
sulfonamides. Drugs containing the chemical group SO_2NH_2. Derivatives of sulfanilamide, also called sulfa drugs; some of them have antibacterial properties, while others are diuretics or antidiabetic agents.
sympathetic. A division of the autonomic nervous system that connects with the spinal cord in the thoracolumbar region.
sympathomimetic. A drug that produces effects similar to those produced by the sympathetic nervous system.

tachyphylaxis. Decreasing response to a drug after repeated administration.
tardive dyskinesia. Disorder caused by antipsychotic drugs, characterized by continuous chewing motion with intermittent movements of the tongue and extremities.
teratogenic. Causing anomalies in developing embryos.

tetanus. Acute and often fatal infection caused by the anaerobic bacillus *Clostridium tetani*; it is characterized by local or generalized muscle spasms.
thiamine. Vitamin B_1, present in meat, vegetables, and grain; it is composed of a substituted pyrimidine linked to a thiazole ring.
thrombin. An enzyme derived from prothrombin, which converts fibrinogen to fibrin.
thrombosis. Formation or presence of a thrombus.
thrombus. An aggregate of blood and vessel wall components that causes vascular obstruction.
thymus. A gland located at the base of the neck that plays a major role in the immunological response.
toxoid. A modified bacterial toxin that has lost toxicity but retained the ability to stimulate the formation of antitoxin.
trachoma. Chronic infectious eye disease caused by the bacterium *Chlamydia trachomatis.*
transdermal. Entering through dermis (skin); used to describe a drug application to the skin.
Trichomonas. A genus of parasitic protozoa found in the intestinal and genitourinary tracts of animals and humans.
triglyceride. A substance consisting of three molecules of fatty acid esterified to glycerol.
Trypanosoma. A genus of protozoa, some of which are pathogenic to humans.
tubulin. A constituent protein of microtubules, thought to be involved in the motility of phagocytes.
tularemia. An infectious disease found primarily in rodents and wildlife, which can also affect humans; it is caused by *Francisella tularensis* and can be transmitted by the bite of deer flies or ticks or by ingestion of contaminated food.
tumorigenic. Giving rise to malignant or benign tumors.
typhus. A group of acute diseases caused by *Ricksettsiae*. Epidemic typhus is transmitted from person to person by human body lice; it is characterized by high fever, headache, general malaise, and skin eruptions.

uric acid. The end product of purine metabolism.
uricosuric. Property of drugs that causes an increase in the excretion of uric acid in urine.

vaccine. A suspension of attenuated or killed microorganisms or antigenic proteins derived from these microorganisms used for prevention, amelioration, or treatment of a disease.
varicella. Chickenpox.
variolation. Inoculation with an unmodified smallpox virus to produce immunity.

ventricular fibrillation. Cardiac arrhythmia characterized by uncoordinated, minute contractions of the fibrils in the ventricle of the heart.
vermicide. A drug destructive to intestinal parasites.
Vibrio. A genus of anaerobic, gram-negative, rod-shaped bacteria.
virulence. The degree of pathogenicity of microorganisms.

xanthine oxidase. An enzyme that catalyzes the oxidation of hypoxanthine to xanthine and of xanthine to uric acid; its inhibition reduces the formation and excretion of uric acid.

yaws. A tropical disease caused by *Treponema pertenue*, a spirochete that enters through abrasion in the skin; it causes destructive lesions in the skin, bones, and joints.

Index

Note: The letter "f" after a page number refers to a figure; the letter "t" after a page number refers to a table. *See also* refers to related topics.

Contributors

Basil Achilladelis, PhD
Athens, Greece

Iain Cockburn, PhD
Professor
Boston University
Boston, Massachusetts
Research Associate
National Bureau of Economic Research
Cambridge, Massachusetts

Arthur R. DeSimone, MD
Instructor in Clinical Medicine
New York University Medical Center
New York, New York

Christopher R. Flowers, MD
Department of Internal Medicine
University of Washington School of Medicine
Seattle, Washington

Lewis E. Gasorek
President
Listowel, Inc.
New York, New York

Rebecca Henderson, PhD
Eastman Kodak Professor
Massachusetts Institute of Technology
Sloan School of Management
Cambridge, Massachusetts
Research Associate
National Bureau of Economic Research
Cambridge, Massachusetts

Ralph Landau, ScD
Senior Fellow
Stanford Institute for Economic Policy Research (SIEPR)
Consulting Professor of Economics
Stanford University
Stanford, California
Research Fellow
Kennedy School of Government
Harvard University
Cambridge, Massachusetts

Kenneth L. Melmon, MD
Associate Dean of Postgraduate Medical Education
Professor of Medicine and Molecular Pharmacology
Stanford University Medical School
Stanford, California

Alexander Scriabine, MD
Department of Pharmacology
Yale University
New Haven, Connecticut

Director of Publications: Frances Coulborn Kohler
Publications Manager: Shelley Wilks Geehr

Text designed by Patricia Wieland
Cover designed by Joel Katz design associates
Printed by Braun-Brumfield, Inc.